Texts in Computational Science and Engineering

8

Editors

Timothy J. Barth
Michael Griebel
David E. Keyes
Risto M. Nieminen
Dirk Roose
Tamar Schlick

For further volumes:
www.springer.com/series/5151

Bertil Gustafsson

Fundamentals
of Scientific
Computing

 Springer

Bertil Gustafsson
Information Technology
Uppsala University
PO Box 337
751 05 Uppsala
Sweden
bertil.gustafsson@it.uu.se

ISSN 1611-0994
ISBN 978-3-642-19494-8   e-ISBN 978-3-642-19495-5
DOI 10.1007/978-3-642-19495-5
Springer Heidelberg Dordrecht London New York

Library of Congress Control Number: 2011929632

Mathematics Subject Classification (2000): 65-01

Cover design: deblik, Berlin

Printed on acid-free paper

Springer is part of Springer Science+Business Media (www.springer.com)

Preface

Computers play an increasingly important role in our society. A breakdown of all computer systems would cause a breakdown of almost all activities of daily life. Furthermore, personal computers are available in almost every home in the industrialized world. But there is one sector where computers have a more strategic role, and that is in science and technology. A large number of physical and engineering problems are solved by the use of advanced computers. The first aircraft were designed by very clever individuals who understood the basic principles of aerodynamics, but today this is not enough. No manufacturer would start building a new aeroplane without extensive computer simulations of various models. Another example where computer simulation is a necessary tool is weather prediction. We know that these predictions are not completely accurate, but are still good enough to get a fairly good idea about the weather for the next few days. The question then is: how is it at all possible to predict the future of a physical system like the atmosphere around the globe? Or in the first example: how is it possible to predict the flight properties of an aircraft that has not yet been built, and where not even a model of the aircraft is available to put in a wind tunnel? No matter how powerful the computers are, we have to provide them with a program that tells them how to carry out the simulation. How is this program constructed?

The fundamental basis for these algorithms is a mathematical model of some kind that provides certain relations between the state variables. These relations lead to a set of equations, and in most cases these equations are *differential equations*. The problem is that these differential equations must be solved, and in most cases they are too difficult to be solved by any mathematician, no matter how sharp. Unfortunately, this is true even for the most powerful computer. This difficulty is overcome by constructing an approximation to the mathematical model, arriving at a numerical model that has a simpler structure based on simple operations like addition and multiplication. The problem usually requires an enormous number of such operations, but nowadays we have access to very fast computers. The state variables, like air pressure and velocity for the weather prediction, are computed by using the numerical model and, if the computer is faster than the weather proceeds in real time, a forecast can be presented for the general public.

This book is about the principles of mathematical and numerical models. We shall put most emphasis on the construction of numerical models, and how it leads to computational mathematics and scientific computing, which is now an indispensable tool in science and engineering. For many applications, the mathematical models were developed one or two centuries ago. There were also numerical models long ago, but the more powerful and robust methods didn't arise until there were electronic computers that could carry out the arithmetic fast enough. Mathematical modeling is usually called applied mathematics, and there are many new areas where this type of mathematics comes into use and is further developed. Numerical modeling is called numerical analysis, or numerical mathematics, and the development is very fast. The success of computer simulation should in fact be credited in equal parts to the development of fast computers and to the development of new numerical methods.

The famous mathematician, physicist and astronomer (and musician) Galileo Galilei (1564–1642) said that "the book of nature is written in the language of mathematics". This is certainly true but, in order to make practical use of it, we also need numerical analysis. By using simplified examples, it is our hope to catch the basics of mathematical/numerical modeling, and in that way explain how the more complicated problems can be solved. In this book we will explain the underlying mathematics in more detail than is usually done in textbooks. Anybody with senior high school mathematics should be able to understand most of the material, but it helps to have basic college mathematics. Scientists and engineers with no or little knowledge about computational mathematics is another group that hopefully will benefit from reading this book as an introduction to the topic. But they can skip the part about basic calculus in Chaps. 2–5. These chapters have a more tutorial style, and are written for those who have forgotten their calculus, or maybe never had much of it.

Uppsala, Sweden   Bertil Gustafsson

Acknowledgements

This book was written after my retirement from the chair at the Division of Scientific Computing at Uppsala University. However, the university has provided all the necessary infrastructure including help by the staff at the Department of Information Technology when trouble with certain software sometimes emerged.

Martin Peters and Ruth Allewelt at Springer have given much appreciated and effective support during the whole project. Furthermore, I would like to thank Michael Eastman who corrected language deficiencies and misprints. He also pointed out some missing details concerning the mathematical content.

Finally I would like to thank my wife Margareta for accepting my full engagement in still another book-writing project.

Contents

Part II   Fundamentals in Numerical Analysis

Part III   Numerical Methods for Differential Equations

Part IV   Numerical Methods for Algebraic Equations

Abbreviations

BVP	Boundary Value Problems
DFT	Discrete Fourier Transform
DG	Discontinuous Galerkin Method(s)
FDM	Finite Difference Method(s)
FEM	Finite Element Method(s)
FFT	Fast discrete Fourier Transform
FORTRAN	FORmula TRANslation (programming language)
IBVP	Initial–Boundary Value Problems
IVP	Initial Value Problems
MATLAB	MATrix LABoratory (programming language)
ODE	Ordinary Differential Equations
PDE	Partial Differential Equations

Part I
Models and Elementary Mathematics

Chapter 1
Introduction

How is it possible to use computer simulation to predict the behavior of some process that has not yet happened? How can the aerodynamic properties of an aircraft be computed before it has been built and even without wind tunnel experiments using a model. How can the computer predict the coming change in the weather when it has access solely to the weather data at the current point in time? The weather pattern for tomorrow has never occurred before.

The answer is that there are physical laws that are formulated as mathematical formulas that always hold, even in the future. This is a mathematical model. However, this model is certainly not a formula or graph where, in the case of weather prediction, we can plug in the coordinates of a certain place on earth to obtain the weather parameters such as air pressure, wind velocity etc. at some future point in time. It is rather a set of equations specifying certain relations between different variables. These equations are the fundamental physical laws expressed in mathematical terms. The problem is that these equations require a solution, and they are very difficult to solve. By using two examples, we shall indicate the basic ingredients in the computational process, without going into any details in this introductory chapter. In the last section we shall also give an overview of the content of the rest of the book.

1.1 Computational Example 1

When the Wright brothers constructed the first flying aircraft at the beginning of the 20th century, they were using basic aerodynamic knowledge, but not much quantitative computation. They were using the basic principle for flying: the wings are given a shape such that the air is flowing faster on the upper surface compared to the lower one, resulting in thinner air with lower pressure on top of the wing. If the difference between the pressure above and below is large enough, the resulting lifting force overcomes the gravitational pull downwards, such that the aircraft can lift off the ground and keep itself flying.

B. Gustafsson, *Fundamentals of Scientific Computing*,
Texts in Computational Science and Engineering 8,
DOI 10.1007/978-3-642-19495-5_1, © Springer-Verlag Berlin Heidelberg 2011

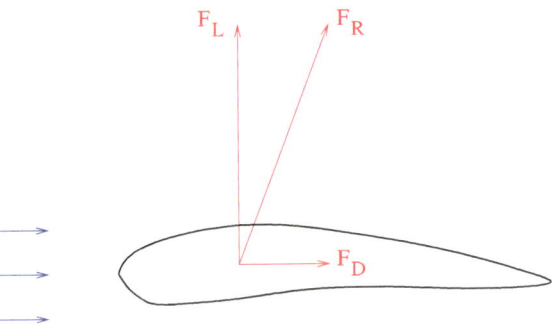

Fig. 1.1 Lift and drag forces on an airfoil

The situation is illustrated schematically in Fig. 1.1 showing a section through the wing called an *airfoil*. For practical reasons it is easier to consider the airfoil at rest with the air blowing from the left at the same speed as the aircraft is flying. The air is creating a force F_R acting upon it. This force can be split into one component F_D directed against the flight direction and one component F_L pointing upwards. The latter has to be larger than the gravitational pull that is trying to get the aircraft down. If the wing has the right geometry, and the engine is powerful enough, the lifting force becomes sufficiently large.

The question is now how to design the airfoil such that F_L becomes as large as possible, while at the same time F_D is kept as small as possible. Here is where the mathematical model is needed. The most convenient situation would be if the forces could be computed directly as explicit formulas, where the constant aircraft speed v_0 is a given parameter. This means that for a given geometry of the airfoil, the formulas could be written as

$$F_L = f_L(v_0),$$
$$F_D = f_D(v_0),$$

(1.1)

where $f_L(v_0)$ and $f_D(v_0)$ are certain mathematical expressions that can be evaluated for any given value of v_0. Complicated expressions don't create any problems, since we have computers that can do the computations easily and quickly. But the unfortunate fact is that no such formulas exist. No matter how complicated we allow them to be, as long as they consist of a series of elementary algebraic expressions and functions, they are not to be found.

It seems that we are stuck with an impossible situation. We have very large and fast computers at our disposal, and still we have no formulas to feed them for evaluation. But there is a way out. We can derive a mathematical model that doesn't have the simple explicit form that we would like, but is still the basis for further operations that finally lead to an approximate solution. In aerodynamics such models were derived long ago. They are based on the very early fundamental work by Isaac Newton (1643–1727) and Leonhard Euler (1707–1783). For an aircraft at a constant speed, a simplified model is a system of equations

$$\frac{\partial(\rho u)}{\partial x} + \frac{\partial(\rho v)}{\partial y} = 0,$$

$$\frac{\partial(\rho u^2 + p)}{\partial x} + \frac{\partial(\rho uv)}{\partial y} = 0,$$

$$\frac{\partial(\rho uv)}{\partial x} + \frac{\partial(\rho v^2 + p)}{\partial y} = 0,$$

$$\frac{\partial(u(E + p))}{\partial x} + \frac{\partial(v(E + p))}{\partial y} = 0.$$

These are the *Euler equations*, where it is assumed that the air has no viscosity. Furthermore we have assumed that there is no variation in the z-direction. These equations have a simple and, for a mathematician, nice-looking structure, but they are difficult to solve. Actually, it is impossible to solve them exactly. Let us first see what the various symbols stand for.

A standard Cartesian coordinate system has been introduced with x and y as the coordinates in space. The state variables representing physical quantities are

ρ　density,
u　velocity component in x-direction,
v　velocity component in y-direction,
E　total energy per unit volume,
p　pressure.

There are five state variables but only four equations. Therefore we need one more equation

$$p = p(\rho, u, v, E).$$

This is the *equation of state,* which is such that, for any given value of the variables within the parenthesis, the pressure p can be easily evaluated. The symbol ∂ represents the *differential operator*, and it stands for the *rate of change* of a variable. The expression $\partial(\rho u)/\partial x$ is called the *derivative* of ρu with respect to x, and it tells how the quantity ρu changes in the x-direction. Each equation states the connection between changes of certain variables in different directions. The first equation tells us that the changes of ρu in the x-direction and ρv in the y-direction cancel each other. This is called the *continuity equation*, and it is a way of expressing the basic physical fact that no material is created or destroyed.

This mathematical model is a *system of differential equations*, and this is probably the most common type of model in science and technology. Nature seems to be such that it is easier to find connections and relations between different variables and parameters if derivatives are involved. However, we have to pay a price for obtaining a simple looking model. In order to be of any use, we need to find a *solution* to the differential equation. The differential equations specify how different state variables are connected to each other, but their actual values are not known. We need the pressure $p = p(x, y)$ as a known function of the coordinates x and y. And here is where the difficulties show up. There is plenty of mathematical theory for differential equations, but there is certainly no theory available that makes it possible to write down the solution for our problem.

Fig. 1.2 Approximation of derivative

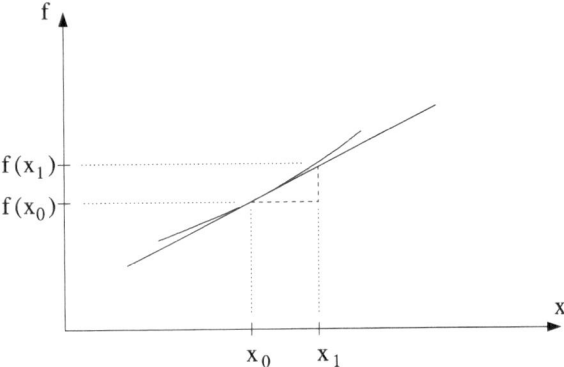

An analogy can be drawn with simpler algebraic equations. The equation

$$x^2 - 4 = 0 \qquad (1.2)$$

has the solutions

$$x_1 = 2, \qquad x_2 = -2. \qquad (1.3)$$

However, already for a slightly more complicated equation

$$x^5 - 3x + 1 = 0$$

we have a situation where no explicit formulas for the solutions exist. Mathematical theory tells us that there are 5 different solutions, but there are no explicit expressions for evaluation of them.

We have in principle the same situation for the differential equations above. We would like to find explicit computable expressions for the unknown variables ρ, u, v, E, p. However, if this is an impossible task already for such a simple algebraic equation as the one above, it is no surprise that there is no chance to solve the system of differential equations. So what do we do?

Apparently some approximation is required. The model must be given a form that leads to an algorithm consisting of elementary algebraic operations that the computer can carry out. Hopefully this approximation, also called a numerical model, is accurate enough for design purpose. One way of modifying the model is to substitute the derivatives with simpler expressions. In the next chapter we shall define the derivative of a function in more detail. Here we just note that the derivative of a function at a certain point is the slope of the curve at that point in the graph. The slope of the curve should be interpreted as the slope of the tangent to the curve as shown in Fig. 1.2 for a certain function $f(x)$. If the two points x_0 and x_1 are close together, the ratio

$$\frac{f(x_1) - f(x_0)}{x_1 - x_0} \qquad (1.4)$$

is a good approximation of the slope. By introducing such approximations all over the computational domain, we suddenly have a completely different situation. The

Fig. 1.3 Part of the
computational grid

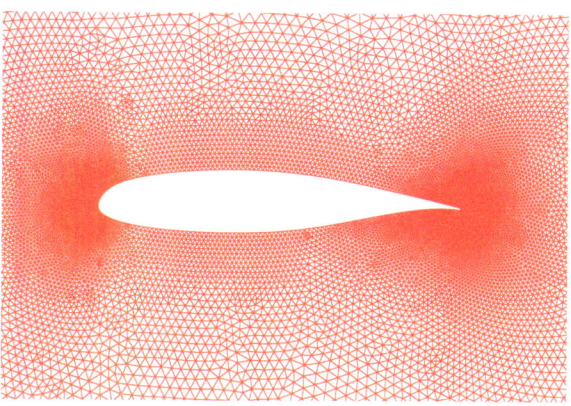

very difficult differential equations are transformed to a set of algebraic equations. But this requires that a computational grid is constructed with each state variable represented at the grid points. Figure 1.3 shows an example of such a grid in part of the computational domain. (This triangular pattern requires a special type of approximation, and we shall come back to this later in the book.)

The state variables are now numbered according to some rule such that we have the unknowns

$$\begin{array}{ccccc}
\rho_1, & \rho_2, & \rho_3, & \ldots, & \rho_N, \\
u_1, & u_2, & u_3, & \ldots, & u_N, \\
v_1, & v_2, & v_3, & \ldots, & v_N, \\
E_1, & E_2, & E_3, & \ldots, & E_N, \\
p_1, & p_2, & p_3, & \ldots, & p_N.
\end{array}$$

The approximation means that the variables are coupled to each other at neighboring grid points. Each differential equation is written down in its discretized form at each grid point, which results in a system of algebraic equations. This is called a *discretization* of the problem, since the variables are stored only at a finite set of discrete points. The grid shown in the figure is quite realistic for real cases, which means that N is a large number, typically of the order one million. The system of equations becomes huge, even in this simplified case where the third space dimension in the z-direction has been eliminated. Each equation couples only a few unknowns, but there are $4N$ equations. If each equation occupies one line, the system would require a book containing 80,000 pages!

Accordingly we have a new problem. How can one solve large systems of equations? Actually, the huge size of the system is not the real problem. As usual we can rely upon the high speed of the computer to overcome this difficulty. The problem is that the system is *nonlinear*. The equations

$$x + y = 2,$$
$$2x - y = -5$$

is an example of a linear system, since each unknown is multiplied by a given constant. It is very easy to solve. This is not the case for the nonlinear system

$$x^2 y + y = 2,$$
$$2x - y^2 = -5,$$

where different variables are multiplying each other or multiplied by itself. We saw earlier that even in the scalar case with only one unknown we quickly run into situations where there is no method for finding the exact solution if the equation is nonlinear. So here we go again: the tools of pure mathematics are not enough to find the solution, and we need to find a numerical method to compute an approximative, but sufficiently accurate, solution.

Later chapters in this book describe solution methods for this type of problem. Here we just mention the underlying basic technique, which is iteration. As an example we take the simple equation above:

$$x^5 - 3x + 1 = 0, \tag{1.5}$$

which can be rewritten in the form

$$x = \frac{x^5}{3} + \frac{1}{3}. \tag{1.6}$$

The idea with iteration is to start with an initial guess x_0, and then modifying it step by step according to a formula that takes the results $x_1, x_2, x_3, \ldots$ closer and closer to the true solution. These formulas can be chosen in many different ways, but it is not at all certain that they provide a sequence that approaches the true solution. And among those which do, we must make sure that they approach the solution sufficiently fast. For the rewritten version of our equation, it is natural to choose the iteration formula

$$x_{j+1} = \frac{x_j^5}{3} + \frac{1}{3}, \quad j = 0, 1, 2, \ldots. \tag{1.7}$$

If the sequence $\{x_j\}$ provides numbers that become almost equal for increasing j, i.e., $x_{j+1} \approx x_j$, then the equation is satisfied almost exactly. This iteration produces the numbers

$$x_0 = 1.00000$$
$$x_1 = 0.37723$$
$$x_2 = 0.33588$$
$$x_3 = 0.33476$$
$$x_4 = 0.33473$$
$$x_5 = 0.33473$$

It seems like we have obtained one of the solutions to the equation, and it is $x = 0.33473$. In fact one can prove that it is exact to five digits.

The full nonlinear system arising from the airfoil computation is huge, and it is important that the iterations converge fast, i.e., the error in the approximative solutions becomes small in a few iterations. This is a very active area of research in computational mathematics.

Fig. 1.4 Two different airfoil shapes

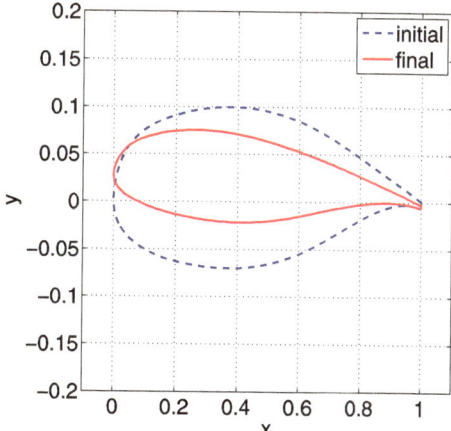

Fig. 1.5 Pressure distribution for the two airfoil shapes

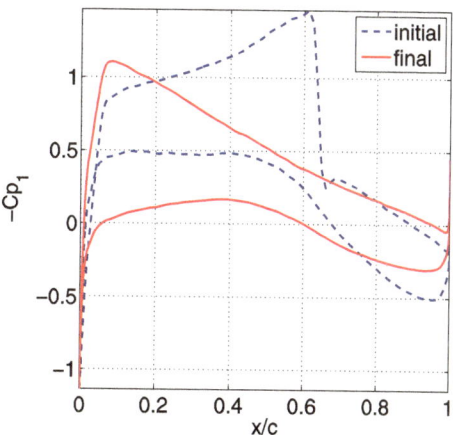

We shall come back with further details concerning the airfoil problem in Chap. 17. Here we shall present the result from a computation done by Olivier Amoignon at the Swedish Defense Research Institute, see [1]. Figure 1.4 shows two different airfoil geometries. Here the scale in the x-direction is modified by a factor 5 in order to better emphasize the difference in the shape.

The corresponding pressure distributions are shown in Fig. 1.5. The pressure is normalized around zero and, by tradition, $-p$ is presented in the graphs, i.e., moving upwards in the graph means decreasing pressure. The area between the upper and lower parts of the pressure curve represents the lifting force on the airfoil.

In the aircraft design problem described above we are interested in the pressure distribution at constant speed. This means that there are no changes in time of the state variables. We call this a *steady state problem*. In many other problems time variation is involved. Weather prediction is such an example. In this case we have a given state of the atmosphere which is known by measurements at a certain point in time. The problem is then to figure out how this state will change a few days forward in time. The differential equations are actually quite similar to the ones presented above for the aircraft design problem. After all, we are again dealing with aerodynamics, and the same laws apply. In addition we must keep track of the temperature, humidity and the presence of clouds. The continuity equation is the same, but now with the time derivatives involved. In three space dimensions it reads

$$\frac{\partial \rho}{\partial t} + \frac{\partial (\rho u)}{\partial x} + \frac{\partial (\rho v)}{\partial y} + \frac{\partial (\rho w)}{\partial z} = 0,$$

where w is the velocity component in the z-direction. The other equations contain time derivatives $\partial/\partial t$ as well. And the presence of time derivatives is the key to the process of advancing the variables forward in time. By discretizing the equations also in time, we get a connection between the variables at t_0 and $t_0 + \Delta t$, where Δt is a small time step, typically of the order a few minutes. If the state is completely known at $t = t_0$, for example 6 o'clock in the morning, the unknown variables at $t = t_1 = t_0 + \Delta t$ can be computed at all grid points in space, which represents a prediction of the new state. This operation is now repeated for a second time step, and so on. If the computer is fast enough, it produces the computed solution faster then the weather situation proceeds in real time, and we have a prediction that is useful.

The example in this introductory section serves as an illustration of the general situation. Mathematical models for many physical or other types of processes have long existed. But new models are being developed for new types of application all the time, for example in biology. The dominating type of model is a system of differential equations. It is a set of relations between derivatives of the variables, i.e., the rate of change in various directions in space and in time. The system has a solution, but there is no way of expressing it in a finite number of elementary algebraic expressions and functions, except in very simplified cases of little interest. In simple words: the differential equations cannot be solved by any human being or by any computer.

The way out of this situation is to introduce a *numerical model* that in some way is close to the mathematical model. There are essentially two different ways of representing the approximate solution u.

One way is to give up the principle of defining the solution everywhere, and let it be defined at a finite number of *grid points* $\mathbf{x}_j$, either by its point values $u(\mathbf{x}_j)$ or by some kind of average across a small neighborhood of the grid points. (For simplicity we let $\mathbf{x}$ denote all the independent variables.) A common technique for computing the discrete solution is to substitute the derivatives by finite difference approximations, as demonstrated in the example above.

The second approximation principle is to let the solution be represented by a finite sum

$$u(\mathbf{x}) = \sum_{j=1}^{N} a_j \phi_j(\mathbf{x}).$$

Here $\phi_j(\mathbf{x})$ are known functions that are simple to evaluate and at the same time have good approximation properties. The coefficients a_j are constants that are determined from some criterion that is closely connected to the differential equation.

Since we cannot expect a finite sum to satisfy the differential equation exactly, some discretization principle is involved here as well, and a computational grid has to be constructed.

Whatever principle we are choosing, the resulting numerical model leads to an algorithm where standard algebraic operations are used to compute the quantities that we need. But even if the operations are simple like additions and multiplications, there is such an enormous number of them that humans don't have a chance to carry them out within a lifetime. But computers can. This is where they have their strength. They don't have much of a brain, but they are extremely fast when it comes to carrying out simple algebraic operations.

Computers require programming before we get any results out of them. The following scheme describes the whole procedure.

Physical process

⇓

Mathematical model

⇓

Numerical model

⇓

Computer program

⇓

Computational results

It should be emphasized that this sketch represents the ideal situation. In practice, the process hardly ever follows such a straightforward pattern. At each stage there is a feedback to the previous stage resulting in modifications. For example, the results of the first version of the computer program may show that certain parts of the chosen numerical model must be modified. The procedure as a whole can be seen as a trial and error procedure, but based on solid theoretical analysis. Sometimes experimental results are available for a model problem, such as a wind tunnel experiment for a specific wing geometry in the example above. In such a case the mathematical/numerical model is considered to be satisfactory if the computational and experimental results agree within an acceptable error level. However, in most cases no measurements are available for comparison. A computation of the detailed bahaviour in the center of a black hole somewhere in the universe has to be judged by other criteria.

The programming of the computer is a major undertaking by itself. On the other hand, if the program can be made flexible enough, it can be used for many different applications with similar mathematical models. Many such systems are available either as commercial products, or as free software at universities and other research institutes. It is a smaller market than for example the computer game market, but supposedly it has a greater strategic significance.

Fig. 1.6 X-ray image of
human lungs

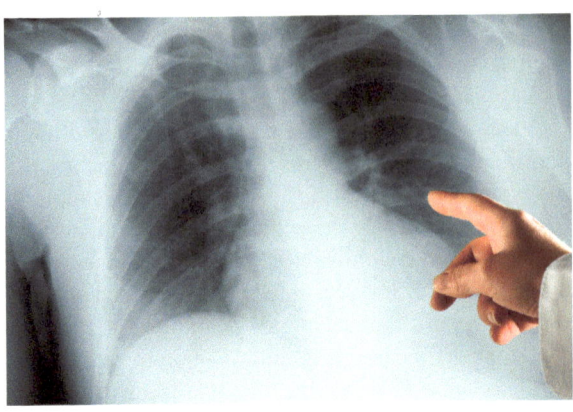

1.2 Computational Example 2

The problem of finding out what is hiding inside a living person is as old as the existence of man. The breakthrough came with the invention of the X-ray machine. Figure 1.6 shows an X-ray image of human lungs.

However, the image shows only a *projection* of the torso and the lungs. We know the location and geometry of the lungs quite well, but the image by itself doesn't show anything about the variation in the direction perpendicular to the image surface. The light parts represent variations in the tissue, but in principle these variations could be located anywhere between the front and back part of the torso.

More images taken from different angles give of course more information. By putting a series of X-ray projections together in some systematic way, it should be possible to get a more detailed apprehension of the inner structure. This is what tomography is about.

But how do we construct each such image? The technique is based on a mathematical/computational model that was developed by G.N. Hounsfield (1919–2004) and A.M. Cormack (1924–1998) independent of each other. The first computerized tomography system was patented in 1972, and they both received the Nobel Prize in medicine 1979. Today computer tomography is a standard tool with machines available at any fully equipped hospital. A typical machine is shown in Fig. 1.7

For each position along the "pipe", the machine revolves around the object resulting in a large number of projections, which in turn leads to a two-dimensional image of the layer corresponding to this position. The machine then moves to a slightly different position along the pipe, providing a new image, and so on. By putting all these images together, a full three-dimensional image of a certain part of the body is finally obtained.

The basis for X-ray diagnosis is that the energy of an X-ray beam is attenuated when it passes through matter, and the attenuation is stronger for higher density. The tissue has varying density, and a black and white X-ray image of the old type shows this variation on a grey scale. A tumor shows up as a lighter object in a darker neighborhood. But we don't know where it is located along the direction of

Fig. 1.7 Computer
tomography machine

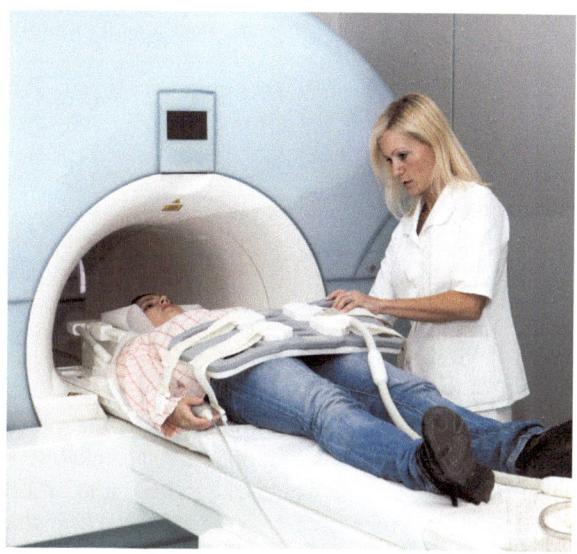

Fig. 1.8 An X-ray crossing
an object

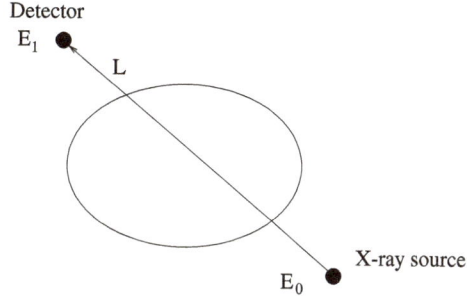

the X-ray beams. Let us now see what kind of mathematics is required for computer tomography leading to a complete image.

If we can find the density $u(x, y)$ at each point (x, y) within the object, the image is easily obtained by letting each grey level correspond to a certain value of u. An X-ray source emits the X-rays as shown in Fig. 1.8 and a detector on the other side of the object measures the energy. By arranging the sources and detectors as arrays and then turning them around, a complete set of data is obtained. By introducing a polar coordinate system as shown in Fig. 1.9, each beam is uniquely characterized by a pair of coordinates (r, θ).

The measurements of the X-ray energy (related to some reference energy) is obtained as a function $p(r, \theta)$. The formal mathematical equation is

$$p(r, \theta) = \int_{-\infty}^{\infty} \int_{-\infty}^{\infty} u(x, y)\, \delta(x \cos \theta + y \sin \theta - r)\, dx\, dy. \qquad (1.8)$$

Here we have a so called *integral equation* instead of a differential equation as in the previous example (integrals are discussed in Sect. 2.2.3). Some calculus is

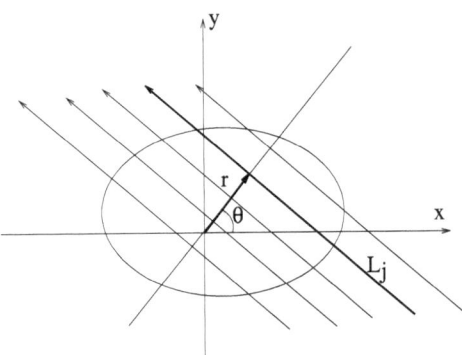

required to understand this formula, but at this point we omit this mathematical
discussion. The basic observation is that the unknown function $u(x, y)$ is part of the
right hand side, and we would like to have it as an explicit expression such that it
can be evaluated for each point (x, y). But we have the measured values for the left
hand side $p(r, \theta)$ at our disposal, not $u(x, y)$. The challenge is to solve the integral
equation for the unknown density u.

The right hand side of the equation is called the *Radon transform* of u after
the Austrian mathematician Johann Radon (1887–1956). He was able to find the
solution in a compact form:

$$u(x, y) = \frac{1}{2\pi^2} \int_0^\pi \int_{-\infty}^\infty \frac{\partial p / \partial r}{x \cos \theta + y \sin \theta - r} \, dr \, d\theta. \tag{1.9}$$

This is a beautiful compact formula but, for practical purposes, it has limited value.
Evaluation of it requires first of all that p is known everywhere in the domain covering the object, but there are only a finite number of X-ray detectors and a finite
number of angles θ. Some sort of discretization is required.

These days when almost everybody owns a digital camera and is watching digital
TV, the concept of pixel is well known. The image is composed of small squares,
where each one represents a constant color. For a black and white image with a
normalization such that $u = 0$ corresponds to white, and $u = 1$ corresponds to black,
each pixel has a value in the interval $0 \le u \le 1$. Mathematically, the image can then
be written as a finite sum

$$u(x, y) = \sum_{j=1}^N c_j \phi_j, \tag{1.10}$$

where ϕ_j is a basis function which is zero everywhere except in pixel j, where
it is one. A value of c_j near one means that pixel j is dark, while a small value
corresponds to a light pixel. The problem is to compute the coefficients c_j.

As we shall see in Sect. 2.2.3, integrals are closely connected to sums. Instead of
discretizing the Radon solution formula (1.9), we go back to the integral equation
(1.8) and approximate the double integral by a double sum with u substituted by the
representation (1.10). This results in a system of equations for the coefficients c_j.

Fig. 1.10 The Shepp–Logan phantom image

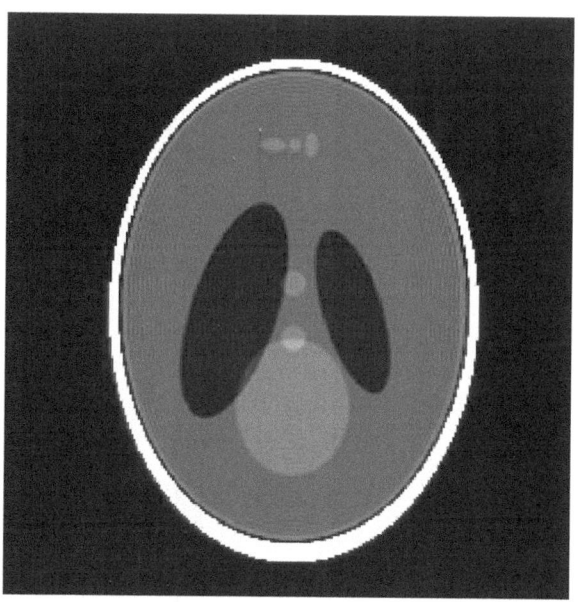

So here we go again: we need a fast solver of systems of equations. And this system is large for sure. Computer tomographs typically work with 1064×1064 pixels for each two-dimensional image, which means that we have a good million unknowns c_j. On the other hand, since each beam passes through only a small number of pixels compared to the total number, there is only a small number of unknowns in each equation. Just as in the aircraft design example, and in many other types of algorithm, we have a *sparse* system of equations.

In the original tomograph design, an iterative method was used for solving the systems. Later, many different solution methods have been developed, most of them using the fast Fourier transform (see Chap. 6). The latest tomographs have very effective algorithms that can keep up with the speed of the X-ray machine. The image is fully reconstructed and ready as soon as all the X-ray projections are finished, also in the full three-dimensional case.

Figure 1.10 shows the quality of a computed artificial image (the Shepp–Logan phantom image) that is often used as a test example.

1.3 Outline of the Book

In this book we shall introduce the basic principles for mathematical models and numerical solution methods with particular emphasis on differential equations, which is the dominating type of mathematical model. The book is divided into five parts, where Part III is the central one containing a presentation of the most common methods for the numerical solution of differential equations. The two parts preceding it contain material that is necessary in order to understand the derivation and character of the numerical methods. Part IV presents solution methods of certain subproblems that arise as a consequence of the methods in Part III, and the final part is about applications.

The book has the following structure in some more detail.

Part I: Models and Elementary Mathematics (Chaps. 1, 2, 3, 4, 5)

Scientific computing is based on mathematical models, and Chap. 2 is a basic introduction to such models, particularly differential equations. This in turn requires elementary differential and integral calculus, which is also included. When discussing methods for the numerical solution of differential equations, we also need some other mathematical concepts like linear algebra, infinite series and elementary functions. These concepts are discussed in Chaps. 3, 4 and 5.

Part I has a different character compared to the other parts of this book. Calculus and basic linear algebra are necessary tools for the development and understanding of the main material. To make it a little more convenient for those who have forgotten or had very little of this basic mathematics, Part I is included as a tutorial. It is quite elementary, and most of it can be ignored by those who have basic college mathematics in fresh memory. Each section in Part I is marked, either by * or **. One star means that the content is fairly elementary and should have been included either as advanced courses in senior high school or as basic college courses. Two stars mean that the content is a little tougher, and normally taught at university level, some of it even as advanced courses. (The exponential function has got the two star marking since it contains a subsection on complex arguments.)

Part II: Fundamentals in Numerical Analysis (Chaps. 6, 7, 8)

This part is about numerical fundamentals that must be understood before we go into methods for differential equations. Even if it is relatively easy to understand, it may not be very familiar to many with a more pure mathematical orientation. The Fourier transform in Chap. 6 is not only a very central tool for analysis of differential equations, but also an essential part of certain numerical algorithms.

Part III: Numerical Methods for Differential Equations (Chaps. 9, 10, 11, 12)

This is the central part of the book, and it is about the three fundamental classes of numerical methods for differential equations: finite difference methods, finite element methods and spectral methods.

Part IV: Numerical Methods for Systems of Equations (Chaps. 13, 14)

The numerical methods for differential equations lead to many subproblems, that in turn require numerical solution. For example, we must be able to solve huge systems of algebraic equations. In the linear case there are methods for finding the exact solution but, for most problems in applications, these are too expensive even with the fastest computers. Therefore we must construct numerical methods also for this purpose, and it is done in this part.

Part V: Applications (Chaps. 15, 16, 17, 18)

Throughout the book we are using simple examples in order to illustrate the basic principles for various models and methods. In particular, these examples are in most cases one-dimensional in space. In the later chapters we shall present some examples from various applications, where the models are more complicated, and have more space dimensions. Wave propagation, heat conduction and fluid dynamics have been chosen since they are all well established in computational mathematics, but also because they include all three classes of hyperbolic, parabolic and elliptic PDE.

We end the application part by a chapter about programming. The first part is about basic principles, and it can be ignored by those who have tried programming before. MATLAB is the ideal programming language when developing and analyzing numerical methods, and the second section is a brief introduction to its basic structure. In the final section we shall indicate some of the key programming problems that are caused by modern computer architectures containing many parallel processors.

At the end of the book there is an appendix containing some elementary formulas from mathematical calculus.

Many sections are ended with exercises, some of which can be solved by paper and pencil. However, some of them require a computer, and here we leave the choice of computer and programming language to the reader. The exercises in Part I are easy, and can be solved by a modern hand calculator. In the later parts, more general programming languages may be required, and MATLAB mentioned above is the recommended system.

Finally, a few comments on various concepts and labels related to computation and computer simulation. Unfortunately, there are no unique definitions of any central concept, and there is some confusion concerning the interpretation. We think of *Scientific Computing*, which is the topic of this book, as a wide field including mathematical modeling, construction and analysis of numerical methods as well as issues concerning programming systems and implementation on computers. However, others may think that mathematical modeling should not be included, but rather be considered solely as part of *Applied Mathematics*. *Computational Mathematics* is similar to Scientific Computing, but with less or no emphasis on programming issues.

There are a number of parallel fields related to specific application areas, for example: *Computational Physics*, *Computational Chemistry*, *Computational Biology* and *Computational Economics*. Here the emphasis is on the application of numerical methods to certain classes of problems within the field, and less on the development and theoretical analysis of these methods. Each field has many subfields, such as *Computational Fluid Dynamics* (*CFD*) within Computational Physics. Engineering can be thought of as another application area leading to *Computational Engineering*. All these areas form the field of *Computational Science*, which should not be confused with *Computer Science*. The latter area is the theory of computer systems including programming languages.

Chapter 2
Mathematical Models

In the previous chapter it was indicated how mathematical models can be used to compute the behavior of certain physical quantities of significance for the problem. Weather prediction is a typical example where the model can be used to simulate a process for future time. We can register all weather data at a certain time today, and then the mathematical model is the basis for computation of the weather tomorrow. In this chapter we shall give a very elementary description of mathematical models in general with the basic concepts illustrated by simple examples. The fundamental type of model is the differential equation, which specifies relations between *changes* in the variables and the variables themselves. An introduction to differential equations is the bulk of this chapter, but we begin with an introductory section containing a simple example for illustration of the most basic concepts.

The whole chapter, except possibly the sections about partial differential equations and well posed problems, can be skipped by the readers who have senior high school or basic college mathematics in fresh memory. These readers can skip Chaps. 3, 4, 5 as well, and go directly to Chap. 6.

2.1 *Basic Concepts

Let us start with a very simple example. If somebody walks with a constant velocity v meters per second, the distance u that he covers in t seconds is obtained by the formula

$$u = vt. \tag{2.1}$$

It can be used for evaluating the distance u when the velocity v and the time t are known. We can consider $u = u(t)$ as a *function* of time, and at any given time the value is obtained by a simple multiplication of two numbers. No computer is needed.

The function concept is fundamental in mathematics. With every value of the *variable* t in a certain interval, it associates a unique number $u(t)$. Almost all kinds of problem in science and technology lead to models where the solution is some

B. Gustafsson, *Fundamentals of Scientific Computing*,
Texts in Computational Science and Engineering 8,
DOI 10.1007/978-3-642-19495-5_2, © Springer-Verlag Berlin Heidelberg 2011

Fig. 2.1 Distance u as a function of time, $v = 1$ (—), $v = 1.5$ (– –)

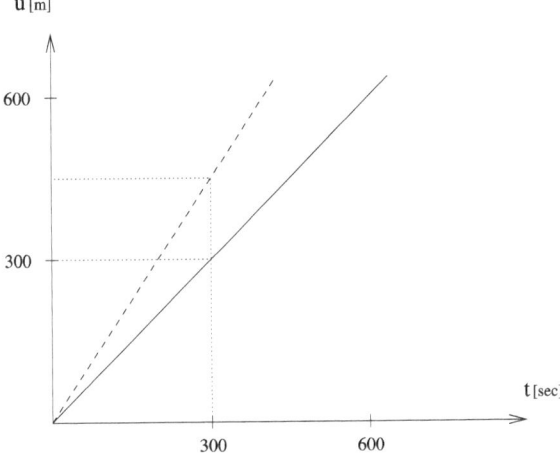

kind of function. This function is often defined in a domain containing several *independent variables*. In the aircraft example in the introductory chapter, the pressure $p = p(x, y)$ is one of the functions we are looking for, and it has two independent variables x and y. In the weather prediction example, there are four independent variables x, y, z, t. A typical domain may be Europe with z varying between the earth surface and the top of the atmosphere, and with t in a 24 hour span forward in time.

Functions are often represented as graphs. In our simple example above, we can make a graph of the distance $u(t)$ as a function of time. Figure 2.1 shows the distance for the two velocities 1 and 1.5 m/sec.

If the distance to the end point is known, and we have measured the time t it takes to walk there, the situation is in principle different. We want to figure out the velocity, but the relation (2.1) does not provide an explicit expression for v. Therefore it is called an *equation*, and it must be solved for the unknown quantity v, such that it appears by itself on one side of the equality sign. In this case it is easy, since we divide both sides of the equation by t, and obtain

$$v = \frac{u}{t}.$$

If the distance u and the time t are known, the number v can be computed. But for a given distance u we can also consider $v = v(t)$ as a function of time t for all values of t, and draw another graph as in Fig. 2.2, where $u = 300$ m. The function $v(t)$ is the *solution* of the problem, and it is found by solving the equation that is the mathematical model. The time t is the independent variable, and the velocity v is the dependent variable.

As another example we consider a tunnel with a cross section in the form of a semicircle as shown in Fig. 2.3, where the radius r is known. When moving x meters from the wall, we want to compute the height h between the floor and the ceiling. Pythagoras' theorem for the triangle gives the equation

$$(r - x)^2 + h^2 = r^2. \tag{2.2}$$

Fig. 2.2 Velocity v as a function of measured time t ($u = 300$ m)

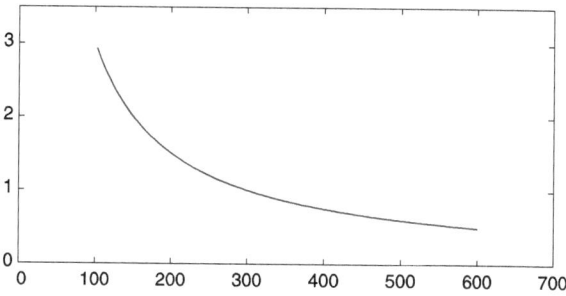

Fig. 2.3 Cross section of the tunnel

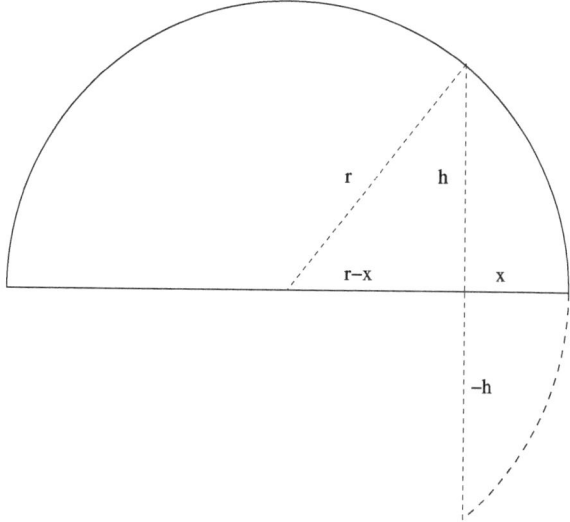

Again we have a mathematical model in the form of an equation which must be solved for the unknown h. We get

$$h = \pm\sqrt{2rx - x^2}.$$

Here we have a case with two solutions, but obviously we are interested only in the positive one. Is there something wrong with the mathematics that allows for another solution? Not at all. The negative solution is the distance downwards from the floor to an imaginative circular continuation of the tunnel wall. Equation (2.2) covers both cases, but the given physical problem accepts only one of them. This is a very common case in mathematical modeling, and we must be careful when analyzing a computed solution, making sure that we are dealing with the right one. Figure 2.4 shows the solution $h = h(x)$ as a function of x for $r = 5$ m.

In the two cases above, we were able to solve the equation without any computer. This is an unusual situation in real life, where the models are much more complicated. Let us first take a look at algebraic equations, which may be familiar from high school mathematics.

Fig. 2.4 The height h in the tunnel as a function of x

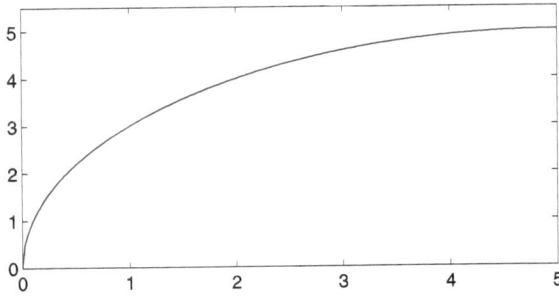

A quadratic equation

$$x^2 + ax + b = 0$$

has two solutions (also called roots)

$$x_1 = -\frac{a}{2} + \sqrt{\frac{a^2}{4} - b},$$

$$x_2 = -\frac{a}{2} - \sqrt{\frac{a^2}{4} - b}.$$

These are exact explicit formulas, which can be used for computing the two solutions, provided the constants a and b are known. We can also consider a and b as independent variables, and the solutions $x_1(a, b)$ and $x_2(a, b)$ as functions of two variables.

Such explicit (but more complicated) formulas for the solutions exist also for higher degree equations

$$x^3 + ax^2 + bx + c = 0$$

and

$$x^4 + ax^3 + bx^2 + cx + d = 0.$$

As usual in mathematics, one has to be careful before drawing any far-reaching conclusions from this. It is tempting to think that such explicit formulas exist for all equations of the form

$$x^n + ax^{n-1} + \cdots = 0 \tag{2.3}$$

for any integer n. But they don't, which was proven by the Norwegian mathematician Niels Henrik Abel (1802–1829). On the other hand it is known that n roots exist for any degree n, but there is no explicit form for them if $n > 4$. Here we have a typical example illustrating the difference between pure and applied mathematics. The pure mathematician proves the existence of solutions, but loses interest when it comes to computing them. The applied mathematician comes in here, and tries to construct methods for computing the solutions to any degree of accuracy by using a *numerical algorithm*. Usually this algorithm is a step-by-step procedure that uses simple algebraic operations like addition and multiplication, and from an initial guess it takes us closer and closer to the solution. We shall discuss such methods in Sect. 13.1.

Equation (2.3) is a mathematical model in the form of an algebraic equation, and we know that there is a solution to it (actually n solutions in this case). A numerical method is necessary for solving it and we need a computer. This one requires a program, which is a set of instructions that can be understood by the computer.

The mathematical models for many physical processes are well known for centuries. For example, Newton's law $F = ma$ in mechanics connecting force F, mass m and acceleration a was published 1687, and it is central in computations for a large number of different applications today. In fluid mechanics the basic models were found more than 150 years ago, and they serve as the central ingredient in the very challenging problems that researchers around the world are presently working on. However, new mathematical models are developed for new applications all the time. A typical area is biology, where there is now a very interesting and dynamic development of new models for the processes in the human body.

The most common type of mathematical model is the *differential equation* as described in the introductory chapter. If time is involved as an independent variable, one can use such differential equations to predict how certain state variables will behave at a later time given the initial state. This is the beauty of computational mathematics. Without physical experiments, one can still find out how a certain system behaves or is going to behave.

In this chapter we shall use a few very simple examples to give a brief introduction to differential equations and the numerical algorithms that are used for finding the solutions. In this way we hope to provide an understanding of the more complicated models discussed later in the book.

2.2 *Ordinary Differential Equations

In the above example relating velocity, time and distance, the velocity v is the rate of change of the distance. For many problems, this concept often plays an important role. The rate of change of a certain variable u is called the *derivative* of u. A mathematical model may be easily obtained by introducing derivatives and relating them to the variable itself, arriving at a differential equation. We shall begin this section by discussing derivatives.

2.2.1 Derivatives

In the above example connecting distance and velocity, the velocity v is a constant. Let us now assume that it varies with time such that $v = v(t)$ is a known function, for example $v(t) = 0.002\,t$. It is tempting to write down the standard formula "distance = velocity × time"

$$u = v(t)t = 0.002t \cdot t,$$

but of course this one is wrong. After $t = 300$ seconds, the formula gives the distance $u = 0.002 \cdot 300^2 = 180$ meters. But that is the distance we should have obtained with the maximal velocity $v = 0.002 \cdot 300$ *all the time*. Obviously we need another mathematical model, but this requires the introduction of derivatives and differential equations.

Assume that we are at a point $u = u(t)$ at a certain time t, and we want to know the position Δt seconds later when v is a constant. We have

$$u(t + \Delta t) = u(t) + v\Delta t,$$

which we write as

$$\frac{u(t + \Delta t) - u(t)}{\Delta t} = v.$$

This is a good starting point for generalization to more complicated problems. If the velocity changes with time, we must know exactly what the time variation is in the whole interval $[t, t + \Delta t]$. However, if Δt is small, the velocity doesn't change much, and we can write the approximate formula

$$\frac{u(t + \Delta t) - u(t)}{\Delta t} \approx v(t), \tag{2.4}$$

where the sign $\approx$ means that the formula is almost correct. Let us now make Δt smaller and actually approach zero. Figure 2.5 shows what happens near the point $t = t_0$. The vertical dashed lines represent the difference $u(t_0 + \Delta t) - u(t_0)$ for decreasing Δt, and the quotient on the left hand side of (2.4) is the ratio between the vertical and the horizontal sides of the corresponding "triangles" with a curved upper edge. This upper triangle edge looks more and more like a straight line, and coincides almost exactly with the tangent to the curve when Δt becomes very small. It is tempting to let Δt take the limit value zero, but we have a problem here, since we get a quotient $0/0$. As a constant, this value does not have any definition, but still we can give it a well defined value by using the limit process. We write the formal definition of the limit as

$$\frac{du}{dt} = \lim_{\Delta t \to 0} \frac{u(t + \Delta t) - u(t)}{\Delta t}. \tag{2.5}$$

(The notation t_0 has been substituted by t here, indicating that the definition holds for any point in time.) The "lim" notation means exactly the limit number as illustrated in the figure, and in geometrical terms it is the slope of the tangent to the curve. Note that du and dt do not mean anything by themselves, they have to occur together as a quotient. One often writes du/dt as u', and it is called the *derivative* of the function $u(t)$ as mentioned above. We say that u is *differentiated* with respect to t. This concept is one of the most fundamental parts of calculus.

The derivative of a function can of course be negative as well. A function is increasing with time when the derivative is positive, and it is decreasing when the derivative is negative.

It should be pointed out that limits of the type $0/0$ do not always exist. As an example we take a look at the function $u = (10^t - 1)/t^p$ near the point $t = 0$. Figure 2.6 shows this function for very small values of t. The lower curve is for $p = 1$,

Fig. 2.5 Difference in
function values for
decreasing Δt

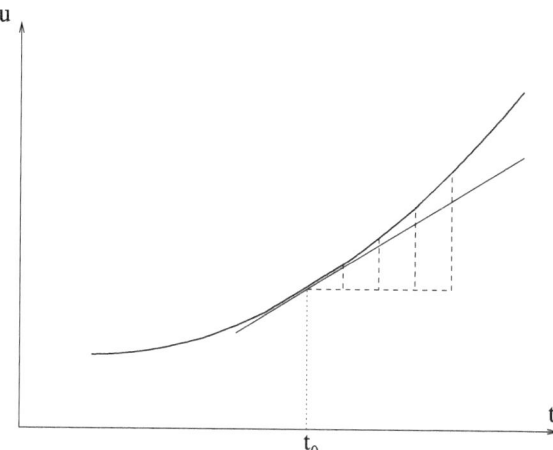

Fig. 2.6 $u = (10^t - 1)/t$ (—)
and $u = (10^t - 1)/t^{1.1}$ (- -)
for small t

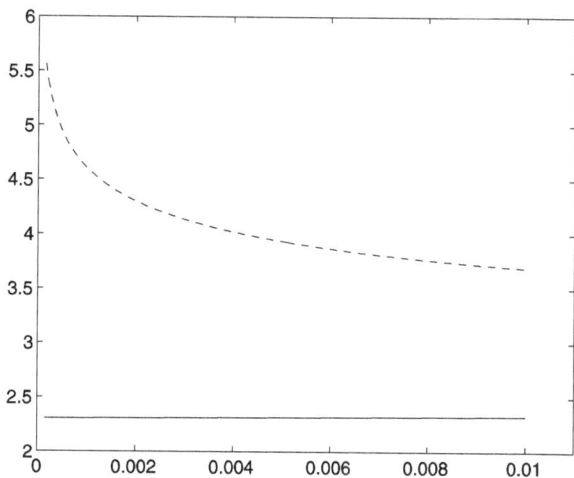

and there is a limit value $du/dt = 2.3026$. The upper curve shows that for the case $p = 1.1$, the function becomes larger and larger for decreasing t. The denominator tends to zero faster than the numerator. There is no limit value as $t \rightarrow 0$.

If the function itself does not exist at a certain point, we cannot expect the derivative to exist there either. But even if the function is well defined everywhere, it may well happen that the derivative is not. As an example we take the function $u = \sin(1/t) - \cos(1/t)/t$ shown in Fig. 2.7(a). (See Sect. 5.2 for the definition of the sine and cosine functions.) At the level of the plotter's resolution it seems like the function smoothes out when t approaches zero. But Fig. 2.7(b) shows the derivative, which does not exist at the limit point $t = 0$. Even if the magnitude of the function u tends to zero, it continues oscillating increasingly fast when t approaches zero.

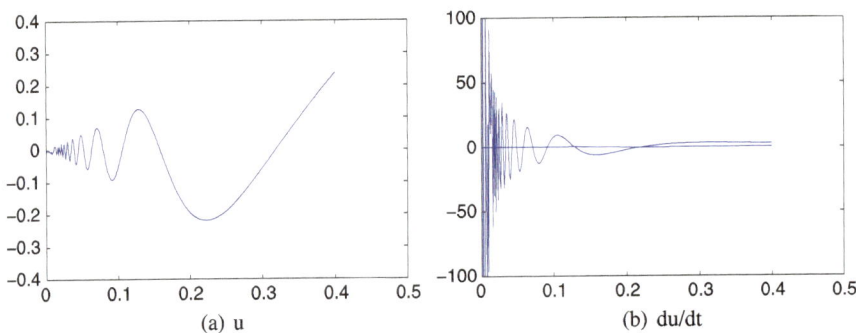

Fig. 2.7 The function $u = \sin(1/t) - \cos(1/t)/t$ and its derivative as functions of t

Now back to our original example. The approximate equation (2.4) becomes exact in the limit $\Delta t \to 0$, and we get the *differential equation*

$$\frac{du}{dt} = v(t). \tag{2.6}$$

The important point here is that, by introducing the derivative, the mathematical model becomes very simple to formulate, even if the velocity is a complicated function of time. (In this case the model is precisely the equation obtained from the definition of velocity.) But it remains to find the *solution* $u(t)$ of the equation, and this is where the main difficulty arises.

The derivative defined in (2.5) uses the limit defined from the right, which is also called a forward difference quotient. One can as well define it by a backward difference quotient:

$$\frac{du}{dt} = \lim_{\Delta t \to 0} \frac{u(t) - u(t - \Delta t)}{\Delta t}.$$

However, if at a certain point $t = t_0$ the derivative has a jump such that the two limits are different, the definition must be made more carefully.

Local maxima and minima of a function $u(t)$ are found by solving the equation $du/dt = 0$ for t, i.e., by finding those points where the tangent is horizontal. This problem may have several solutions, and if we want to find the global maximum or minimum, we should search among these points together with the end points of the interval of interest for t. Figure 2.8 shows a case with two local minima at $t = t_1$ and $t = t_3$, and one local maximum at $t = t_2$. The global maximum is at the end point $t = t_0$, and the global minimum is at $t = t_3$.

One can also define derivatives of higher order. The second derivative is defined in terms of the first derivative as

$$\frac{d^2u}{dt^2} = \frac{d}{dt}\left(\frac{du}{dt}\right) = \lim_{\Delta t \to 0} \frac{u'(t + \Delta t) - u'(t)}{\Delta t}.$$

If we use backward differences in the definition of u', we obtain

$$\frac{d^2u}{dt^2} = \lim_{\Delta t \to 0} \frac{u(t + \Delta t) - 2u(t) + u(t - \Delta t)}{\Delta t^2}. \tag{2.7}$$

Fig. 2.8 Local maxima and minima

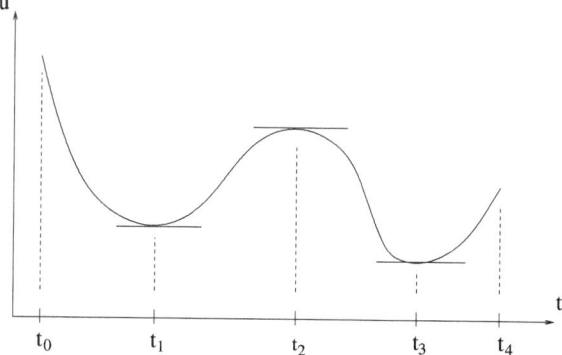

For all functions expressed in terms of explicit elementary functions, one can always find the derivative as a combination of elementary functions as well by using well known differentiation rules. A list of some of these rules is found in Appendix A.1.

Exercise 2.1 Local max- and min-points of $f(t)$ are characterized by $df/dt = 0$. Show that $d^2 f/dt^2 < 0$ at a max-point and $d^2 f/dt^2 > 0$ at a min-point.

Exercise 2.2 A *saddle point* of $f(t)$ is characterized by $df/dt = d^2 f/dt^2 = 0$. Draw a graph of a function with a saddle point.

2.2.2 Differential Equations and Initial Value Problems

A differential equation may be easy to derive but, in order for it to be of any practical value, we need to know how to solve it. For simple equations the solution can be found by methods taught in senior high school but, for more realistic ones, other methods must be used.

Let us now make a slight generalization of the first example above, and assume that the pedestrian walks the first 5 minutes with the speed 0.8 m/sec, and then increases the speed to 1.2 m/sec. The velocity $v = v(t)$ is defined by

$$v(t) = \begin{cases} 0.8, & 0 \le t \le 300, \\ 1.2, & 300 < t. \end{cases}$$

In order to solve the differential equation (2.6) we must define an *initial condition* for u, i.e., we must know where the pedestrian starts. The complete *initial value problem* is

$$\frac{du}{dt} = v(t), \quad t \ge 0,$$

$$u(0) = u_0,$$

(2.8)

Fig. 2.9 The distance u as a function of time with a speed increase at $t = 300$

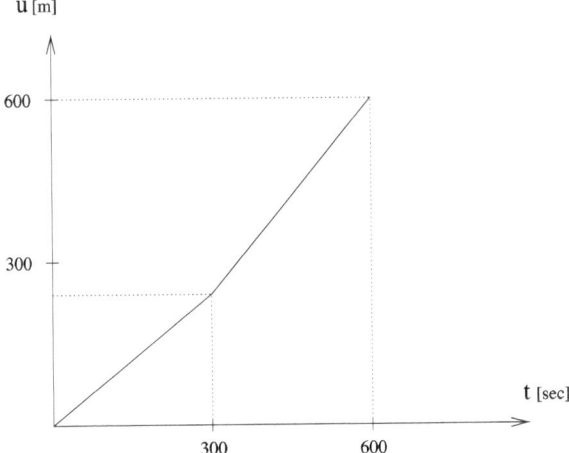

where in our case $u_0 = 0$. This problem can be solved in two steps. First we compute $u(t)$ for $0 \le t \le 300$, and with $u(300)$ known we can then compute $u(t)$ for $t > 300$. There is a slight formal complication here since there is a jump in the velocity at $t = 300$. Common sense tells us that the distance cannot take a sudden jump, but the mathematical definitions must be precise when switching from $v = 0.8$ to $v = 1.2$. We omit these details here. We have now solved the initial value problem (2.8), and the result is shown in Fig. 2.9.

The solution technique may be seen as a two step procedure, where the velocity is a constant during each step. We introduce the time step $\Delta t = 300$, and compute

$$u(300) = 0 + 0.8\Delta t = 240,$$
$$u(600) = 240 + 1.2\Delta t = 600.$$

This is a type of *discretization* of the differential equation, which makes it possible to solve the initial value problem by simple algebraic computations in a step-by-step procedure.

In the example above, there are only two steps in the computation and, furthermore, the computed solution is exact. Let us next generalize the problem a little further, and assume that the velocity is increased continuously such that $v(t) = 0.002t$ as in the beginning of Sect. 2.2.1. This is perhaps a somewhat unusual model for a pedestrian, but we can think of a certain accelerating vehicle. A solution can still be obtained by analytical means by introducing the first formula for differentiation. If a function $f(t)$ is defined by $f(t) = ct^p$, where c and p are constants, then $df/dt = cpt^{p-1}$. This rule may look like magic, but is not very complicated to derive. Take the example $p = 2$. By going back to the definition of derivatives, we consider a small time step Δt and compute f at the new time $t + \Delta t$:

$$f(t + \Delta t) = c(t + \Delta t)^2 = c(t^2 + 2t\Delta t + \Delta t^2) = f(t) + c\Delta t(2t + \Delta t).$$

We get

$$\frac{f(t + \Delta t) - f(t)}{\Delta t} = c2t + c\Delta t$$

and, by letting Δt approach zero, we get the final result

$$\frac{df}{dt} = c2t,$$

which agrees with the formula above. This is of course no proof that the general rule holds. If p is any other positive integer, the same type of technique can be used to prove the result, but for noninteger and negative p the proof is a little more complicated.

By using this rule for our differential equation (2.8), it follows that $u(t) = 0.001t^2$ is a solution, and obviously it satisfies the initial condition $u(0) = 0$ as well. Referring back to the introduction of this section, we note that the true solution is exactly one half of the one obtained by simply using the erroneous formula $u(t) = v(t)t$.

In this case the analytical solution is easily found but, in a case where it is not, could we compute the solution by a discretization procedure? A computer has electronic circuits that can do simple algebraic operations like addition and multiplication, but there are no circuits for differentiation of arbitrary functions. The velocity is not constant anywhere, so the computation with piecewise constant velocity does not give the true solution. However, if Δt is sufficiently small, then $v(t)$ does not change much in an interval $[t, t + \Delta t]$, and therefore we approximate it by a constant. We introduce *grid points*

$$t_n = n\Delta t, \quad n = 0, 1, \ldots,$$

and denote by u_n the approximation of the solution $u(t_n)$. The simplest approximation is to use the velocity $v_n = v(t_n)$ in the whole interval $[t_n, t_{n+1}]$. The discrete procedure becomes

$$u_{n+1} = u_n + v_n\Delta t, \quad n = 0, 1, \ldots,$$
$$u_0 = 0. \tag{2.9}$$

In this case we cannot expect the numerical solution to be exact, since the velocity v is approximated by a constant in each interval. However, if the time step is small, we expect the solution to be close to the correct one. For a computation over the whole interval $0 \le t \le 600$, the step size is defined as $\Delta t = 600/N$, i.e., there are N subintervals. Figure 2.10 shows the result for $N = 10$ and $N = 40$. As expected we get a better result for smaller step size.

The solution procedure based on discretization is closely connected to the definition of the derivative by means of the limit process $\Delta t \to 0$. We have simply switched direction of the process by going back from the derivative to the finite difference quotients that were used in the limit procedure.

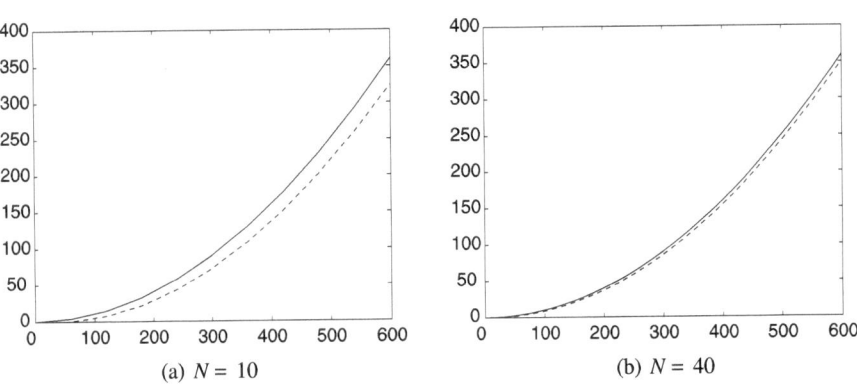

(a) $N = 10$ (b) $N = 40$

Fig. 2.10 Exact solution of (2.8) (—), approximate solution (2.9) (−−)

2.2.3 Integrals

Let us next assume that T is a fixed point in time, and that N steps are required to get there, i.e., $T = N\Delta t$. When summing up the formula (2.9), the approximation u_N is defined by

$$u_N = u_0 + v_0\Delta t + v_1\Delta t + \cdots + v_{N-1}\Delta t = u_0 + \sum_{n=0}^{N-1} v_n\Delta t.$$

The experiment above (where $u_0 = 0$) indicates that the sum becomes closer and closer to the correct value $u(t)$ for any t in the interval $0 \le t \le T$ as Δt becomes smaller, i.e., N becomes larger. For the purpose of analysis, it is convenient to introduce the infinity concept ∞. This is not a number, and it does not make sense to write $N = \infty$ for a certain variable N. Instead we write $N \to \infty$, i.e., N takes on values that become larger and larger without any bound.

In the limit, the number of terms in the sum becomes infinite and, just as for derivatives, we must be careful when considering the limit process. (We shall comment more upon this in Chap. 4.) In our case it seems like the sum converges to a certain value, and we call this process *convergence* as N tends to infinity. Actually, the sum is a so-called *Riemann sum*. The different terms can be interpreted as the area of the rectangles with sides Δt and v_j as illustrated in Fig. 2.11 for a general function $v(t)$.

The Riemann sum can be used to define an *integral* as the limit

$$\int_0^T v(t)\, dt = \lim_{N \to \infty} \sum_{n=0}^{N-1} v_n\Delta t.$$

The procedure above shows that the numerical solution approaches the true solution when N becomes larger and, by taking the limit, the solution of (2.8) is defined as

$$u(T) = u(0) + \int_0^T v(t)\, dt. \tag{2.10}$$

Fig. 2.11 The Riemann sum

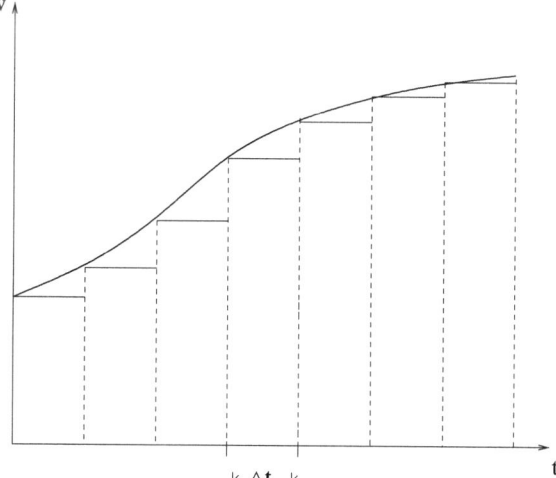

Note that dt is not a variable carrying any value, it is just a symbol to indicate that the integration is carried out along the t-axis. Apparently, the integral of a function $v(t)$ is the area under the curve $y = v(t)$. If $v(t)$ is negative in part of the domain, the corresponding contribution to the integral is negative.

We have now introduced the two basic concepts in calculus, namely derivatives and integrals. The formula above is actually an application of the *fundamental theorem of integral calculus* relating derivatives and integrals to each other

$$\int_{t_0}^{t_1} \frac{du}{dt}\, dt = u(t_1) - u(t_0), \tag{2.11}$$

for the special values $t_0 = 0$ and $t_1 = T$. The integral here with given end points t_0 and t_1 is called a *definite integral*. There is also the *indefinite integral*

$$\int u(t)\, dt$$

for a general function $u(t)$. This one is defined as any function $U(t)$ such that $dU/dt = u$. We call $U(t)$ a *primitive function* of $u(t)$. For example, if $u = t$, then we can take $U = t^2/2$, but also for example $U = t^2/2 + 1$ (the derivative of any constant is zero). In both cases we have

$$\int_{t_0}^{t_1} u\, dt = \int_{t_0}^{t_1} \frac{d(t^2/2)}{dt}\, dt = \frac{t_1^2}{2} - \frac{t_0^2}{2},$$

$$\int_{t_0}^{t_1} u\, dt = \int_{t_0}^{t_1} \frac{d(t^2/2 + 1)}{dt}\, dt = \int_{t_0}^{t_1} \frac{d(t^2/2)}{dt}\, dt = \frac{t_1^2}{2} - \frac{t_0^2}{2}.$$

The general rule is that, if $U(t)$ is *any* primitive function of $u(t)$, then

$$\int_{t_0}^{t_1} u\, dt = U(t_1) - U(t_0), \tag{2.12}$$

The key question is therefore how to find a primitive function $U(t)$ when it comes to computation of an integral. For certain classes of functions $u(t)$ there are a number of rules and tricks for finding $U(t)$, but in general integration is a difficult problem. And here we have another case where computers have difficulties as well, since there are no electronic circuits that can do general integration. There are program systems that can do differentiation as well as integration for certain classes of functions, but for general problems we have to find the answer by constructing numerical methods based on the principle of discretization. We saw how integrals can be approximated by finite sums corresponding to a sum of rectangle areas, but we can of course use more accurate approximations of the area corresponding to the integral.

Another rule that will be used quite frequently later in this book is *integration by parts*. Assume that $U(t)$ is a primitive function of $u(t)$. Then for any function $v(t)$

$$\int_{t_0}^{t_1} u(t)v(t)\,dt = U(t_1)v(t_1) - U(t_0)v(t_0) - \int_{t_0}^{t_1} U(t)\frac{dv}{dt}(t)\,dt. \qquad (2.13)$$

A special application is the relation

$$\int_{t_0}^{t_1} \frac{du}{dt}(t)v(t)\,dt = u(t_1)v(t_1) - u(t_0)v(t_0) - \int_{t_0}^{t_1} u(t)\frac{dv}{dt}(t)\,dt. \qquad (2.14)$$

This is a very useful formula, since it allows for shifting the derivative from one function to the other one in the product.

Exercise 2.3 The graph of the function $f(t)$ is a straight line cutting through the t-axis at $t = t_0$. Show that $\int_{t_0-a}^{t_0+a} f(t)\,dt = 0$ for all $a > 0$.

2.2.4 More on Differential Equations and Discretization

Going back to differential equations, we note that the accuracy of the numerical method (2.9) can be improved significantly by choosing a more accurate method for approximating $v(t)$ in each interval. In this way we can get away with fewer points t_n, i.e., larger Δt. This is an important consideration when it comes to real and more demanding problems, and we shall discuss it further in the next chapter.

The differential equation in our example is trivial in the sense that it can be solved by analytical means, and there is no need for discrete or *numerical methods*. More general differential equations have the form

$$\frac{du}{dt} = g(u), \qquad (2.15)$$

where $g(u)$ is a general function that depends on the solution u that we are looking for. The form of the equation may not be very different from the one we treated above, but the level of complexity when it comes to solving it has suddenly gone up dramatically. Actually, there may be cases when a solution does not even exist. Or it may exist for certain values of t, but not for others. For example, the equation

$$\frac{du}{dt} = u^2,$$

Fig. 2.12 Solution to a linear (—) and a nonlinear (− −) differential equation

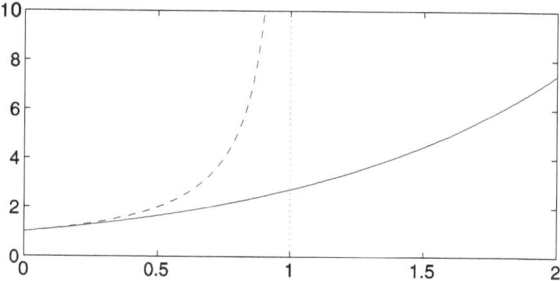

has the solution $u(t) = 1/(1 - t)$ (Exercise 2.4) if the initial value is $u(0) = 1$. But at $t = 1$ the solution does not exist, and we cannot continue further in time.

Outside the mathematical world one sometimes hears the expression "nonlinear growth" used for something that grows without any control. The equation above may be an example representing the basis for this expression.

When it comes to analytical solution techniques, the real distinction between easy and difficult differential equations is linear and nonlinear equations. A linear function has the form $g(u) = au + b$, where $a = a(t)$ and $b(t)$ are functions that may depend on t but not on u. Even if these functions are very large, there is always a solution to the differential equation for all t. Let us take a closer look at the example above. The differential equation is written as $du/dt = uu$, and we substitute the first factor on the right hand side by the midpoint value $u(t) = 1/(1 - 1/2) = 2$ of the solution in the interval $0 \leq t \leq 1$. It seems like a reasonable approximation, and it results in the new initial value problem

$$\frac{dv}{dt} = 2v,$$
$$v(0) = 1,$$

which has a well defined solution for all t. Figure 2.12 shows the two solutions. In the nonlinear case, the growth of the derivative is so strong that the solution tends to infinity very quickly.

On the other hand, many nonlinear differential equations have unique solutions (for a given initial value), at least for some finite time (as the one above). In fact, most physical phenomena are nonlinear, and we have to deal with them. But it is important to do the analysis correctly, and to draw the right conclusions about the mathematical properties. From now on we assume that $g(u)$ is such that a solution to the differential equation (2.15) exists, but we don't expect that there is any analytical solution in explicit form.

Differential equations may contain higher-order derivatives. Going back to the example with u representing distance, we note that the concept of acceleration a is related to velocity by the differential equation $dv/dt = a$, and therefore we have

$$\frac{d^2 u}{dt^2} = a.$$

We know that if a is a constant, then $v = at$ is the velocity for an object that is at rest at $t = 0$. When using the rule for differentiation, we find the solution

$u = 0.5at^2$ after one more integration. As an example we consider the acceleration $a = 9.81$ m/sec^2 caused by the Earth's gravitation. If we disregard the air resistance we can use this formula to figure out the speed when a falling body hits the ground. For example, if an apple is hanging two meters above the ground and comes loose, we get the equation $2 = 0.5 \cdot 9.81t^2$, which has the solution $t = 0.639$ seconds. The apple has the speed $v = 9.81 \cdot 0.639 = 6.27$ m/sec when it hits the ground.

This is a special case of the initial value problem

$$\frac{d^2u}{dt^2} = a(t), \quad t \geq 0,$$
$$u(0) = u_0,$$
$$u'(0) = v_0,$$

where $a(t)$ depends on time and u_0 and v_0 are the initial position and initial velocity respectively. A numerical difference scheme is obtained by taking a finite Δt in the formal definition (2.7) of the second derivative:

$$u_{n+1} = 2u_n - u_{n-1} + \Delta t^2 a(t_n).$$

Since three levels in time are involved, we need not only the given value u_0 but also u_1 in order to get it started. It is of course no coincidence that we have another initial condition available that can be used to provide this value: it simply reflects the connection between the differential equation and its discretization. The simplest way to obtain u_1 is to use the difference approximation $u_1 - u_0 = \Delta t v_0$ of the second initial condition.

If we observe the falling body and measure the position at two different points T_1 and T_2 in time, we get the *boundary value problem*

$$\frac{d^2u}{dt^2} = a(t), \quad T_1 \leq t \leq T_2,$$
$$u(T_1) = p_1,$$
$$u(T_2) = p_2.$$

With $N + 1$ grid points including the end points of the interval $[T_1, T_2]$, the corresponding difference scheme is

$$u_{n+1} - 2u_n + u_{n-1} = \Delta t^2 a(t_n), \quad n = 1, 2, \ldots, N - 1,$$
$$u_0 = p_1,$$
$$u_N = p_2.$$

Here we encounter a new problem. The values u_n cannot be computed explicitly starting with u_2, since u_1 is not known. We have a system of equations for the unknowns $u_1, u_2, \ldots, u_{N-1}$ and, for small time steps Δt, the system is a large one. Efficient techniques are known for such systems, and we come back to this topic later in this book.

So far we have been discussing scalar ODE, i.e., there is only one unknown dependent function u. In applications there are usually more than one function, and we have more than one differential equation. The system

$$\frac{du_1}{dt} = u_1 + cu_2,$$

$$\frac{du_2}{dt} = cu_1 + u_2,$$

couples the two dependent variables u_1 and u_2 to each other through the two differential equations. (The subscript notation here should not be mixed up with the discretization subscript above.) In order to define a unique solution, we must specify both variables in an initial condition. A more compact form is obtained if we use vector notation (see Chap. 3). In the general case we define a vector $\mathbf{u} = \mathbf{u}(t)$ containing the m elements $u_j = u_j(t)$. The derivative of a vector is defined as the vector of derivatives, i.e.,

$$\mathbf{u} = \begin{bmatrix} u_1 \\ u_2 \\ \vdots \\ u_m \end{bmatrix}, \qquad \frac{d\mathbf{u}}{dt} = \begin{bmatrix} du_1/dt \\ du_2/dt \\ \vdots \\ du_m/dt \end{bmatrix}.$$

A general initial value problem for a system of ODE can now be written in the form

$$\frac{d\mathbf{u}}{dt} = \mathbf{g}(t, \mathbf{u}), \quad t \geq 0,$$

$$\mathbf{u}(0) = \mathbf{f},$$

(2.16)

where $\mathbf{g}$ and $\mathbf{f}$ are vectors defined by

$$\mathbf{g}(t, \mathbf{u}) = \begin{bmatrix} g_1(t, u_1, u_2, \ldots, u_m) \\ g_2(t, u_1, u_2, \ldots, u_m) \\ \vdots \\ g_m(t, u_1, u_2, \ldots, u_m) \end{bmatrix}, \qquad \mathbf{f} = \begin{bmatrix} f_1 \\ f_2 \\ \vdots \\ f_m \end{bmatrix}.$$

For numerical methods the same compact notation can be used. If $\mathbf{u}_n$ denotes an approximation of $\mathbf{u}(t_n)$, the simple difference scheme (2.9) can be written in the form

$$\mathbf{u}_{n+1} = \mathbf{u}_n + \Delta t \mathbf{g}(t_n, \mathbf{u}_n), \quad n = 0, 1, \ldots,$$

$$\mathbf{u}_0 = \mathbf{f}.$$

Exercise 2.4 In Appendix A.1 we find that at^{a-1} is the derivative of t^a for any constant a. Use the chain rule to show that $u(t) = 1/(c-t)$ satisfies the differential equation $du/dt = u^2$.

Exercise 2.5 Show that $u(t)$ is a decreasing function of t for increasing $t > 0$ if

$$\frac{du}{dt} = -\frac{1}{1 - u^3},$$

$$u(0) = -1.2.$$

Exercise 2.6 Define the vectors $\mathbf{u}$ and $\mathbf{g}(\mathbf{u})$ such that the system

$$\frac{du}{dt} = au + bv,$$
$$\frac{dv}{dt} = cu^2$$

takes the form (2.16).

2.3 **Partial Differential Equations

So far we have discussed ordinary differential equations, where there is only one independent variable t. In most applications there are two or more independent variables, and then we must deal with *partial derivatives*.

2.3.1 Partial Derivatives

For a function $u(x,t)$ we define *partial derivatives* as limits of finite difference quotients as

$$\frac{\partial u}{\partial x} = \lim_{\Delta x \to 0} \frac{u(x + \Delta x, t) - u(x, t)}{\Delta x},$$
$$\frac{\partial u}{\partial t} = \lim_{\Delta t \to 0} \frac{u(x, t + \Delta t) - u(x, t)}{\Delta t}.$$

For a constant value of t, we can consider u as a function of the single variable x, and the derivative $\partial u / \partial x$ corresponds to the slope of the tangent of the function curve, just as for the one-dimensional case. In Fig. 2.13 the directions for measuring the slope are indicated by the short arrows marked Δx and Δt.

Partial differentiation of a function $u(x_1, x_2, \ldots, x_d)$ is in principle no more complicated than differentiation of a function of one variable. Differentiation with respect to x_1 is simply done by considering all the other variables $x_2, x_3, \ldots, x_d$ as constants, and then applying the ordinary differentiation rules to u as a function of x_1. As an example, for the function

$$u(x, y, z) = x^2 y^3 + 2yz^2 + xyz \tag{2.17}$$

we have

$$\frac{\partial u}{\partial x} = 2xy^3 + yz,$$
$$\frac{\partial u}{\partial y} = 3x^2 y^2 + 2z^2 + xz,$$
$$\frac{\partial u}{\partial z} = 4yz + xy.$$

Fig. 2.13 Partial derivatives
in (x, t)-space

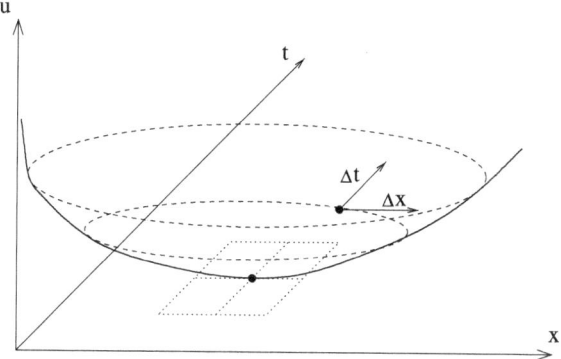

Higher order derivatives are defined in analogy with the 1D-case, for example

$$\frac{\partial^2 u}{\partial x^2} = \frac{\partial}{\partial x}\left(\frac{\partial u}{\partial x}\right) = 2y^3.$$

However, here we get the additional complication of getting *mixed derivatives*, for example

$$\frac{\partial^2 u}{\partial x \partial y} = \frac{\partial}{\partial x}\left(\frac{\partial u}{\partial y}\right) = \frac{\partial}{\partial y}\left(\frac{\partial u}{\partial x}\right) = 6xy^2 + z.$$

When looking for local minima or maxima of a function $u(x_1, x_2, \ldots, x_d)$, we try to find the points where all partial derivatives are zero:

$$\frac{\partial u}{\partial x_1} = \frac{\partial u}{\partial x_2} = \cdots = \frac{\partial u}{\partial x_d} = 0.$$

The tangent plane is horizontal at the minimum point for a function of two variables as shown in Fig. 2.13.

Figure 2.14 shows the function $u(x, y) = x^2 y^3$ and its partial derivative $\partial u/\partial x = 2xy^3$. For each constant $y = y_0$, the derivative $\partial u/\partial x$ is a straight line as shown in Fig. 2.14(c).

The partial derivatives of a function can be ordered as a vector, and we call it the *gradient*. It is denoted by $\operatorname{grad} u$ or ∇u, and is defined by

$$\nabla u = \begin{bmatrix} \partial u/\partial x_1 \\ \partial u/\partial x_2 \\ \vdots \\ \partial u/\partial x_d \end{bmatrix}.$$

We are also going to use the notation

$$\Delta = \nabla \cdot \nabla = \frac{\partial^2}{\partial x^2} + \frac{\partial^2}{\partial y^2} + \frac{\partial^2}{\partial z^2}$$

for the *Laplacian* operator named after the French mathematician Pierre Simon de Laplace (1749–1827). (Here we have used the dot-notation for the scalar product between vectors which will be discussed in Chap. 3.)

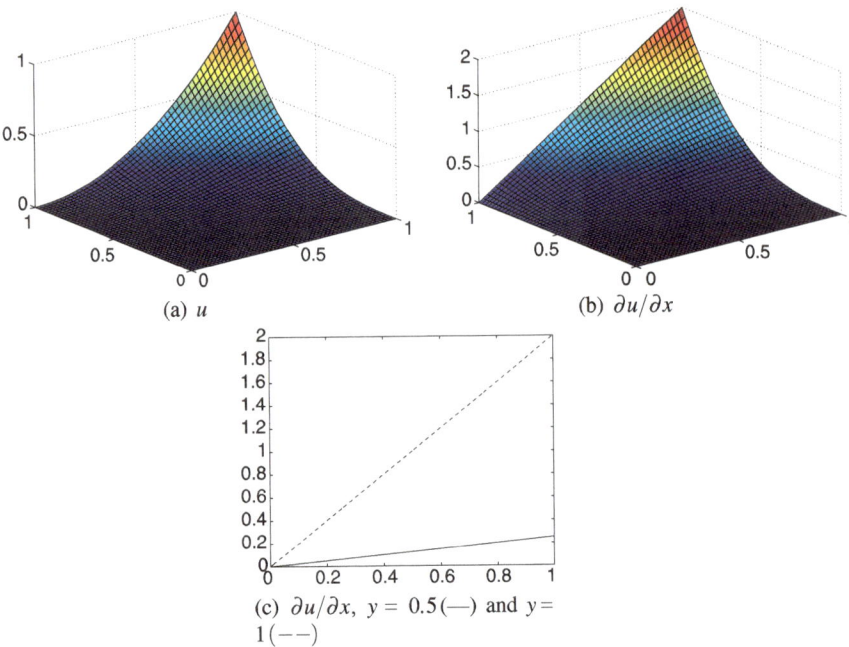

(a) u

(b) $\partial u / \partial x$

(c) $\partial u / \partial x$, $y = 0.5 (\text{—})$ and $y = 1 (--)$

Fig. 2.14 $u = x^2 y^3$ and $\partial u / \partial x$

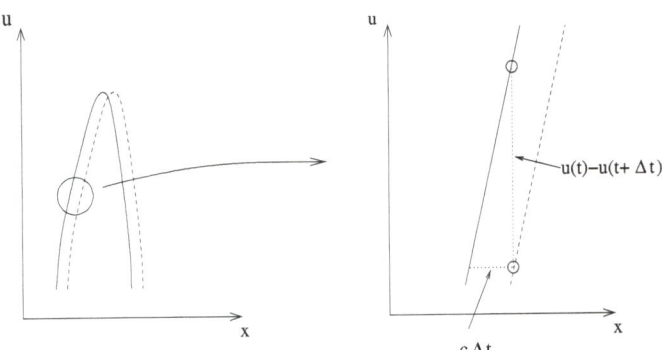

$u(t) - u(t + \Delta t)$

$c \Delta t$

Fig. 2.15 A moving pulse

It can be shown that the vector ∇u points in the direction where the function u has the steepest slope (positive or negative). The condition for a local maximum or minimum can be written as $\nabla u = \mathbf{0}$.

A partial differential equation contains relations between different partial derivatives. For illustration, we consider also here a very simple example, and we choose wave propagation. Consider the left part of Fig. 2.15 which could be a pressure pulse. A certain feature in the pressure distribution is moving to the right with speed c m/sec, and the figure shows the pulse at two different but close points in time. Part of the graph is magnified giving the right part of the figure, where a sufficiently small part of the curve can be considered as an almost straight line. The function u at a certain fixed point x has changed from $u(t)$ to the smaller value $u(t + \Delta t)$. Any feature has traveled a distance $c\Delta t$, and the figure shows that $(u(t) - u(t + \Delta t))/(c\Delta t)$ is the slope of the line. But this slope is by definition the derivative in the x-direction, i.e., $\partial u/\partial x$. For the true function $u(x, t)$ we must take the limit, and we have

$$\lim_{\Delta t \to 0} \frac{u(t + \Delta t) - u(t)}{c\Delta t} = -\frac{\partial u}{\partial x}.$$

The partial differential equation becomes

$$\frac{\partial u}{\partial t} + c\frac{\partial u}{\partial x} = 0. \tag{2.18}$$

It shows how changes of the function in the x-direction are related to changes in the t-direction. We note that any function $f(x - ct)$ is a solution since by the chain rule

$$\frac{\partial u}{\partial x} = f'(x - ct),$$

$$\frac{\partial u}{\partial t} = -cf'(x - ct),$$

and this is what we expect. The assumption was that any feature in the solution is moving with the velocity c, i.e., the solution is constant along any line $x = ct$ in the (x, t)-plane. These lines are called *characteristics* of the PDE.

For our example, there is no need to use the differential equation for describing the process, since we assumed from the beginning that the pulse is moving with the velocity c. In other words, we derived the differential equation from the knowledge of the true solution. But the nice thing is that (2.18) can easily be generalized to a more complicated situation. The differential equation holds locally at every point (x, t) for any c that may change with x and t. We simply change the equation to

$$\frac{\partial u}{\partial t} + c(x, t)\frac{\partial u}{\partial x} = 0,$$

where $c = c(x, t)$ is a given function. The generalized model is easy to understand, but the solution is much more difficult to obtain. Just as for ordinary differential equations, one numerical method presents itself by going back to the finite difference formulation that was the basis for the definition of the derivatives. We shall discuss such methods in Chap. 10.

Integrals can be defined for functions of several variables. If $u(x, y)$ is defined in a rectangle $a \le x \le b$, $c \le y \le d$, the *double integral* over this rectangle is

$$I = \int_c^d \int_a^b u(x, y)\,dx\,dy.$$

It is not necessary to define it via a discrete sum, since it can as well be defined by single integrals. The function

$$v(y) = \int_a^b u(x, y)\,dx$$

is well defined as a function of y, and we have

$$I = \int_c^d v(y)\,dy,$$

i.e.,

$$I = \int_c^d \left(\int_a^b u(x, y)\,dx \right) dy.$$

The integration can be expressed in the opposite order as well:

$$I = \int_a^b \left(\int_c^d u(x, y)\,dy \right) dx.$$

The definition of multidimensional integrals for any number of variables should be clear from this.

If the domain is no longer a rectangle (or hyper-rectangle in higher dimensions), it is not that easy to define the integrals as a sequence of one-dimensional integrals. However, integrals can be defined along curves as well as along straight lines. Assume that a certain curve Γ in the x/y-plane is defined by

$$\Gamma = \{(x, y) : x = x(s), \ y = y(s), \ 0 \le s \le S\},$$

where s is the *arc length*. This means that the coordinates of the curve Γ are well defined for all s in the interval $[0, S]$, and for each point $s = s^*$ the length of the curve from the starting point is s^*. The total length of the curve is S. If a function $u(x(s), y(s))$ is defined on Γ, the *line integral* (or *curve integral*) along Γ is

$$\int_\Gamma u\big(x(s), y(s)\big)\,ds = \int_0^S u\big(x(s), y(s)\big)\,ds,$$

which is well defined as a one-dimensional integral. Likewise, we can define integrals on a surface in the three-dimensional space, and so on.

For 1D-integrals we had the fundamental theorem (2.11) relating derivatives to integrals. In several dimensions there is no such simple formula, but there are many relations that are similar, one of them being the *divergence theorem*. It says that, for any two functions $u(x, y)$ and $v(x, y)$ defined in the domain Ω with boundary Γ, we have

$$\int\int_\Omega \left(\frac{\partial u}{\partial x} + \frac{\partial v}{\partial y} \right) dx\,dy = \int_\Gamma (un_x + vn_y)\,ds. \tag{2.19}$$

Here n_x and n_y are the components of the outward pointing normal $\mathbf{n}$ to the curve Γ (with $n_x^2 + n_y^2 = 1$), see Fig. 2.16.

The theorem brings down the dimension of the double integral one step ending up with a line integral. This is analogous to the 1D-integral which was brought down to two boundary points. The divergence theorem can be generalized to any number of dimensions d, such that the result is an integral in $d - 1$ dimensions.

In Chap. 11 we shall use the divergence theorem when discussing finite element methods.

Fig. 2.16 The outward
normal

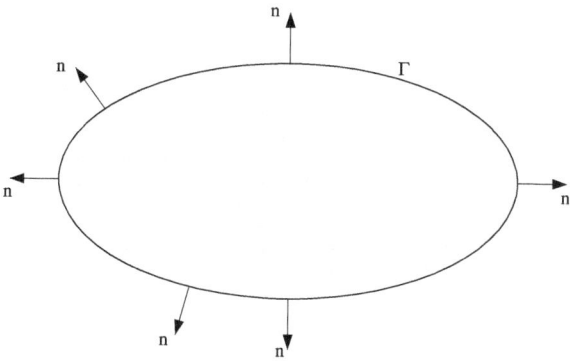

Exercise 2.7 Calculate the six different second derivatives for u defined in (2.17).

Exercise 2.8 Find the local minimum of the function $(x-a)^4 + (y-b)^4$, and prove that it is also the global minimum.

2.3.2 Boundary and Initial-Boundary Value Problems

In order to illustrate initial-boundary value problems, we shall use the heat conduction problem as an example (it will be discussed in more detail in Chap. 16). It leads to a partial differential equation of higher order. Figure 2.17 shows a wall, where the red layers on the sides symbolize some device that keeps the temperature $u(x, t)$ constant. If the wall is very long and high, we can consider the temperature as a constant in the y- and z-directions, and the problem becomes one-dimensional. This is a very common simplification in mathematical modeling. The computation becomes so much simpler, and the results may still be accurate enough. And even if the 1D-simplification is not very accurate, it serves as a suitable problem for analysis and understanding of basic properties.

The partial differential equation

$$\frac{\partial u}{\partial t} = a \frac{\partial^2 u}{\partial x^2}$$

is a simple version of heat conduction corresponding to the case where the heat conduction coefficient a is constant. The material in the wall is identical everywhere so that the heat conduction properties are the same everywhere. Since the temperature at the surface on both sides is kept constant, we have the *boundary conditions*

$$u(0, t) = u_0,$$
$$u(1, t) = u_1,$$

where it is assumed that the boundaries are located at $x = 0$ and $x = 1$. If we also include the known heat distribution as an initial condition $u(x, 0) = f(x)$, we have

Fig. 2.17 Heat conduction in
a wall

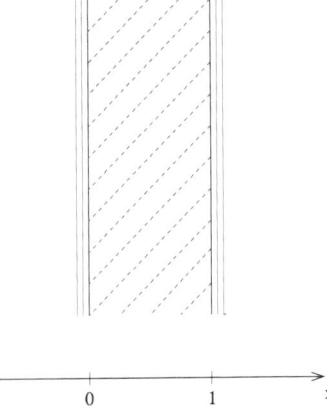

a complete *initial-boundary value problem*. If the temperature is known on three sides, the solution can be computed in the whole (x, t)-domain for any time $t \geq 0$. We shall discuss numerical methods for this and other PDE problems in Chap. 9.

It is reasonable to assume that after some time the temperature distribution no longer changes, and we reach a *steady state*. This means that the time derivative is zero, and there is no initial condition. The boundary value problem is

$$\frac{d^2u}{dx^2} = 0, \quad 0 \leq x \leq 1,$$
$$u(0) = u_0,$$
$$u(1) = u_1.$$

Note that there are no partial derivatives here, and therefore we use the standard notation d^2/dx^2 for the second derivative.

Indeed this boundary value problem, is very easy to solve. If the second derivative is zero, the solution must be a straight line $u(x) = b + cx$, where the constants b and c are determined by the boundary conditions. By letting $x = 0$ we get $b = u_0$, and then it follows that $c = u_1 - u_0$ by letting $x = 1$.

The simple form of the problem is a result of the assumption that the heat conduction coefficient is constant, and that the variation of the solution u is limited to the x-direction. By considering a wall that is finite in the y-direction with given temperature at $y = y_0$ and $y = y_1$, the steady state problem for $u(x, y)$ is

$$\frac{\partial^2 u}{\partial x^2} + \frac{\partial^2 u}{\partial y^2} = 0, \quad x_0 \leq x \leq x_1, \; y_0 \leq y \leq y_1,$$
$$u(x, y_0) = g_0(x), \quad x_0 \leq x \leq x_1,$$
$$u(x_0, y) = h_0(y), \quad y_0 \leq y \leq y_1,$$
$$u(x, y_1) = g_1(x), \quad x_0 \leq x \leq x_1,$$
$$u(x_1, y) = h_1(y), \quad y_0 \leq y \leq y_1,$$

see Fig. 2.18.

Fig. 2.18 Heat equation in a pillar

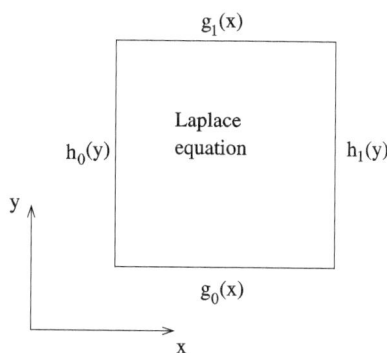

By measuring the temperature at all the edges, it is possible to figure out the temperature in the whole wall. The differential equation is the well known *Laplace equation* in 2D. It occurs in many other applications, and is one of the most studied partial differential equations ever. Even if there are many ways of deriving analytical solutions for special cases, numerical methods are almost exclusively used for its solution. Such algorithms are called *Laplace solvers*.

2.4 **Well Posed Mathematical Models

In most computations we have to deal with perturbations of various kinds. For example, an initial value problem that is used for predicting a certain quantity requires initial data. These are seldom exact, and therefore it is important that a small error, or perturbation, does not cause large errors in the solution at later time. We call such a problem *well posed*, and we shall now discuss this concept with the heat conduction problem as an illustration:

$$\frac{\partial u}{\partial t} = \frac{\partial^2 u}{\partial x^2}, \quad 0 \le x \le 1, \ t_1 \le t,$$
$$u(0, t) = 1,$$
$$u(1, t) = 1,$$
$$u(x, t_1) = f_1(x).$$

(2.20)

We can also formulate the inverse problem. If the temperature is known at $t = t_2$, we can ask ourselves what the heat distribution was at the earlier point in time $t = t_1$. In other words, we want to solve *the heat equation backwards*

$$\frac{\partial u}{\partial t} = \frac{\partial^2 u}{\partial x^2}, \quad 0 \le x \le 1, \ t_1 \le t \le t_2,$$
$$u(0, t) = 1,$$
$$u(1, t) = 1,$$
$$u(x, t_2) = f_2(x).$$

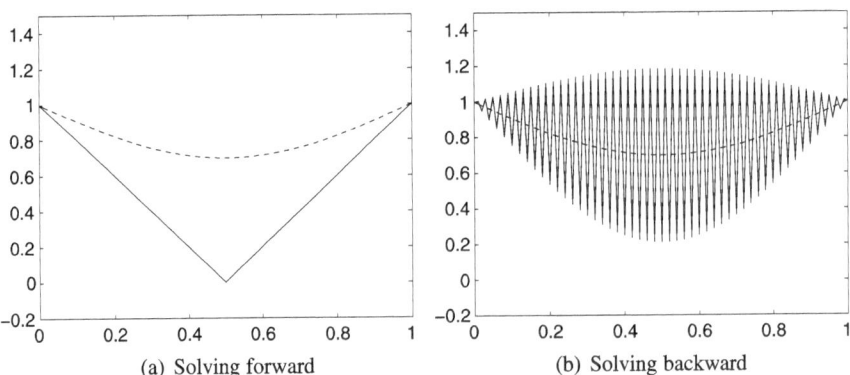

Fig. 2.19 The heat equation

Numerical methods for PDE will be discussed later in this book. Here we use a simple difference scheme for both problems. For the forward problem (2.20) we use the initial function

$$f_1(x) = |1 - 2x|$$

at $t_1 = 0$ and solve for $0 \le t \le 0.1$. The result in Fig. 2.19(a) shows the correct solution at $t = 0.1$ as the upper dashed curve. Next we use this solution as the function $f_2(x)$ for the backward problem, hoping to get the initial function $f_1(x)$ back as the solution at $t_1 = 0$. However, already after a few small steps, the solution is the oscillating one shown in Fig. 2.19(b). It is not even near the correct solution (which is smooth for all $t > 0$), and obviously something has gone wrong here. This will happen with any other numerical method applied directly to the backward problem. The solution may have a different form, which will be completely wrong after a short time. We are seeing an example of an *ill posed problem* in contrast to the forward heat equation, which is a *well posed problem*. The meaning of the latter concept is that if we perturb the initial data slightly, the solution at later time is also perturbed slightly. More precisely, if the initial perturbation is $\varepsilon(x)$ with $|\varepsilon(x)| \le \varepsilon_0$, where ε_0 is a small number, then the deviation $v(x, t)$ from the original solution $u(x, t)$ at later time satisfies an inequality $|v(x, t)| \le K\varepsilon_0$ where K is a constant of reasonable size. For an ill posed problem such a constant does not exist.

An ill posed problem is an almost hopeless case for a numerical approximation. Since the approximation by definition introduces perturbations of the solution, the result will be bad after a short time. However, by introducing certain filters that keep the solution within strict bounds, it is sometimes still possible to get reasonable solutions.

The concept of well-posedness is very important in applications. Many political decisions are based on mathematical modeling and computer simulations. As a drastic and simplified example, think of a country that is on the verge of introducing nuclear power as a new energy source, and that a new type of reactor is to be used. One political party is against the introduction of nuclear power, and they refer to the results from a mathematical model that is used for simulation of the cooling procedure. This model shows that the temperature goes well above the security limit if a certain pipe is broken, and this shows that the reactor is not safe. The government asks the national energy agency for another simulation with slightly perturbed data. This shows that the temperature stays well below the security limit, and the opposition party is accused of presenting false information. Who is right?

After intense analysis, it is found that the model is ill posed, and consequently neither party is right. There is simply not enough information for a reliable conclusion. It may be difficult to explain to the general public what an ill posed problem is, but clearly the decision about going ahead with the project has to be postponed.

An ill posed problem should not be confused with a model with poor accuracy. The backwards heat equation is a correct model when it comes to stating the true relation between the rate of change in time and in space. The difficulty arises when we want to solve it for computing the temperature at an earlier time starting from measurements at a later time. Then the model overreacts to small perturbations, and has no practical use. We need accurate *and* well posed models.

Chapter 3
Basic Linear Algebra

Linear algebra deals with vector spaces and linear transformations represented by *matrices*. It may seem to be a field of little interest when it comes to the solution of differential equations. However, the fact is that it is a central and indispensable field that is necessary for the analysis and understanding of the problems and algorithms. One reason is that most numerical methods lead to the solution of linear systems of algebraic equations, which is an immediate application of linear algebra. Another reason is that systems of differential equations require linear algebra for analysis even before discretization.

In this chapter we shall give a brief survey of linear algebra. It turns out that it is difficult to deal with matrices when limited to the use of real numbers as we know them from daily life. Therefore we shall introduce so-called complex numbers in the section preceding the matrix section.

3.1 *Vector Spaces

The Euclidean three-dimensional space is used to describe the three-dimensional world we are living in. A coordinate system with 3 parameters is used to describe any point in space. With the position of the origin $(0, 0, 0)$ known, the triplet (x, y, z) describes how far away in the three perpendicular directions the point is located. The numbers x, y, z are called (Cartesian) *coordinates* in the three-dimensional Euclidean space. In mathematics it is convenient to generalize this concept to more than three dimensions. In modern physics one actually talks about a universe that has more than three dimensions, but this is not what we should have in mind here. We just think of any phenomenon that can be characterized by a set of parameters that we order in a vector

$$\mathbf{x} = \begin{bmatrix} x_1 \\ x_2 \\ \vdots \\ x_N \end{bmatrix}.$$

B. Gustafsson, *Fundamentals of Scientific Computing*,
Texts in Computational Science and Engineering 8,
DOI 10.1007/978-3-642-19495-5_3, © Springer-Verlag Berlin Heidelberg 2011

An example could be a certain sound source consisting of N different tones with frequencies $f_1, f_2, \ldots, f_N$ that are known. The vector $\mathbf{x}$ could then be used to describe the strength x_j of each frequency.

The normal form of a vector is a *column vector*, i.e., the elements are ordered in a vertical pattern. Sometimes we shall work with *row vectors*, where the elements are ordered in a horizontal pattern. For the relation between the two, we use the transpose label T, such that

$$\begin{bmatrix} x_1 & x_2 & \cdots & x_N \end{bmatrix}^T = \begin{bmatrix} x_1 \\ x_2 \\ \vdots \\ x_N \end{bmatrix}.$$

Two vectors with the same number of elements can be added to each other:

$$\mathbf{x} + \mathbf{y} = \begin{bmatrix} x_1 + y_1 \\ x_2 + y_2 \\ \vdots \\ x_N + y_N \end{bmatrix}.$$

Two vectors cannot be multiplied with each other. However, if c is a scalar, we have

$$c\mathbf{x} = \begin{bmatrix} cx_1 \\ cx_2 \\ \vdots \\ cx_N \end{bmatrix}.$$

A general set of M vectors $\mathbf{x}_j$ are called *linearly dependent* if there are coefficients c_j such that

$$c_1\mathbf{x}_1 + c_2\mathbf{x}_2 + \cdots + c_M\mathbf{x}_M = \mathbf{0},$$

where the c_j are not all zero. The vectors are *linearly independent* if such a relation is possible only if all coefficients are zero. Figure 3.1 shows two linearly dependent vectors $\mathbf{x}_1$ and $\mathbf{x}_2$. The two pairs of vectors $\mathbf{x}_1, \mathbf{x}_3$ and $\mathbf{x}_2, \mathbf{x}_3$ are both linearly independent.

Linear independence plays a fundamental role in mathematical and numerical analysis as we shall see. The analysis often leads to equalities of the type

$$\sum_{j=1}^{M} a_j\mathbf{x}_j = \sum_{j=1}^{M} b_j\mathbf{x}_j,$$

where the vectors $\mathbf{x}_j$ are linearly independent. This leads to the very simple set of relations

$$a_j = b_j, \quad j = 1, 2, \ldots, M.$$

The set of all vectors with N elements is called a *vector space* of dimension N. There can never be more than N linearly independent vectors in an N-dimensional

Fig. 3.1 Linearly dependent
and linearly independent
vectors

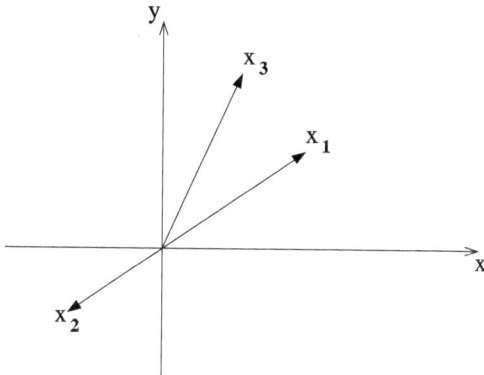

space. Any set of vectors $\{\mathbf{v}_j\}_{j=1}^{N}$ can be used as *basis vectors* provided that they are linearly independent. Any vector can be expressed in terms of a set of basis vectors as

$$\mathbf{x} = c_1\mathbf{v}_1 + c_2\mathbf{v}_2 + \cdots + c_N\mathbf{v}_N,$$

where the coefficients c_j are the coordinates associated with the given basis. The most convenient basis is given by the unit vectors

$$\mathbf{e}_1 = \begin{bmatrix} 1 \\ 0 \\ 0 \\ \vdots \\ 0 \end{bmatrix}, \quad \mathbf{e}_2 = \begin{bmatrix} 0 \\ 1 \\ 0 \\ \vdots \\ 0 \end{bmatrix}, \quad \mathbf{e}_3 = \begin{bmatrix} 0 \\ 0 \\ 1 \\ \vdots \\ 0 \end{bmatrix}, \quad \ldots, \quad \mathbf{e}_N = \begin{bmatrix} 0 \\ 0 \\ 0 \\ \vdots \\ 1 \end{bmatrix},$$

leading to the representation of $\mathbf{x}$

$$\mathbf{x} = x_1\mathbf{e}_1 + x_2\mathbf{e}_2 + \cdots + x_N\mathbf{e}_N.$$

The scalar product of two vectors $\mathbf{x}$ and $\mathbf{y}$ is defined as

$$(\mathbf{x}, \mathbf{y}) = x_1 y_1 + x_2 y_2 + \cdots + x_N y_N.$$

The notation $\mathbf{x} \cdot \mathbf{y}$ is also used. The scalar product is *bilinear*, i.e., for any vectors $\mathbf{x}, \mathbf{y}, \mathbf{z}$ and any constant c

$$(\mathbf{x}, \mathbf{y} + c\mathbf{z}) = (\mathbf{x}, \mathbf{y}) + c(\mathbf{x}, \mathbf{z}),$$
$$(\mathbf{x} + c\mathbf{y}, \mathbf{z}) = (\mathbf{x}, \mathbf{z}) + c(\mathbf{y}, \mathbf{z}).$$

Two vectors are *orthogonal* if $(\mathbf{x}, \mathbf{y}) = 0$. We note that the unit vectors $\mathbf{e}_j$ defined above are mutually orthogonal, i.e.,

$$(\mathbf{e}_j, \mathbf{e}_k) = 0 \quad \text{if } j \neq k.$$

The length (or *norm*) of a vector is

$$\|\mathbf{x}\| = \sqrt{(\mathbf{x}, \mathbf{x})} = \sqrt{x_1^2 + x_2^2 + \cdots + x_N^2},$$

and we note that $\|\mathbf{e}_j\| = 1$ for all j. When a set of vectors are not only orthogonal, but also has length one like the unit vectors, the vectors are called *orthonormal*. The set of unit vectors $\mathbf{e}_j$ is a special case of mutually orthonormal vectors. By introducing the *Kronecker* δ defined by

$$\delta_{jk} = \begin{cases} 0 & \text{if } j \neq k, \\ 1 & \text{if } j = k, \end{cases}$$

we have

$$(\mathbf{e}_j, \mathbf{e}_k) = \delta_{jk}.$$

Since the norm is a generalization of the vector length in the Euclidean 3D-space, we call it the *Euclidean norm* sometimes denoted by $\|\mathbf{x}\|_2$.

For a sum of real vectors we have

$$\|\mathbf{x} + \mathbf{y}\|^2 = (\mathbf{x} + \mathbf{y}, \mathbf{x} + \mathbf{y})$$
$$= \|\mathbf{x}\|^2 + (\mathbf{x}, \mathbf{y}) + (\mathbf{y}, \mathbf{x}) + \|\mathbf{y}\|^2 = \|\mathbf{x}\|^2 + 2(\mathbf{x}, \mathbf{y}) + \|\mathbf{y}\|^2.$$

(For complex vectors this relation takes a different form, see below.)

One can define other type of norms, for example $\max_j |x_j|$, but in this book we mainly stick to the Euclidean norm. All norms must be such that $\|\mathbf{x}\| = 0$ if and only if $\mathbf{x} = 0$. Whatever norm we are using, the *triangle inequality*

$$\|\mathbf{x} + \mathbf{y}\| \leq \|\mathbf{x}\| + \|\mathbf{y}\| \tag{3.1}$$

always holds for any vectors $\mathbf{x}$ and $\mathbf{y}$.

The basis vectors must be linearly independent, but from a numerical point of view this condition is not sufficient. We take an example in the 2D-space. The natural choice of basis vectors are the usual orthogonal unit vectors along the x- and y-axes

$$\mathbf{e}_1 = \begin{bmatrix} 1 \\ 0 \end{bmatrix}, \qquad \mathbf{e}_2 = \begin{bmatrix} 0 \\ 1 \end{bmatrix}.$$

This choice guarantees that if a certain vector is perturbed slightly, the perturbation of the coefficients for the basis vectors is small. The vector $[a \ b]^T$ has the representation

$$\begin{bmatrix} a \\ b \end{bmatrix} = a \begin{bmatrix} 1 \\ 0 \end{bmatrix} + b \begin{bmatrix} 0 \\ 1 \end{bmatrix},$$

while the perturbed vector $[a \ b + \delta]^T$ has the representation

$$\begin{bmatrix} a \\ b + \delta \end{bmatrix} = a \begin{bmatrix} 1 \\ 0 \end{bmatrix} + (b + \delta) \begin{bmatrix} 0 \\ 1 \end{bmatrix}.$$

The difference is in the second coefficient, and it has the same size δ as the perturbation in the given vector. This is the perfect situation from a numerical point of view, and we say that the problem of finding the coefficients is *well conditioned*.

Let us now change the basis to the new vectors

$$\mathbf{v}_1 = \begin{bmatrix} 1 \\ 0 \end{bmatrix}, \qquad \mathbf{v}_2 = \frac{1}{\sqrt{1 + \varepsilon^2}} \begin{bmatrix} 1 \\ \varepsilon \end{bmatrix},$$

where $|\varepsilon|$ is small. Here it is obvious that the two vectors are almost linearly de-
pendent, but in the general high dimensional case we must use some mathematical
criterion. We know that orthogonal vectors $\mathbf{u}, \mathbf{v}$ satisfy $(\mathbf{u}, \mathbf{v}) = 0$, while the scalar
product of two identical vectors of unit length is $(\mathbf{u}, \mathbf{u}) = 1$. Therefore the magnitude of the scalar product between two vectors is a measure of the degree of linear
independence, with decreasing value when two vectors tend towards the strongest
degree of linear independence, which is orthogonality. In our case

$$(\mathbf{v}_1, \mathbf{v}_2) = \frac{1}{\sqrt{1+\varepsilon^2}},$$

which is very close to one. (Actually the scalar product is $\cos\theta$, where θ is the angle
between the vectors, see Sect. 5.2.) The vector $[a\ b]^T$ has the representation

$$\begin{bmatrix} a \\ b \end{bmatrix} = c_1\mathbf{v}_1 + c_2\mathbf{v}_2,$$

where

$$c_1 = a - \frac{b}{\varepsilon}, \qquad c_2 = \frac{b\sqrt{1+\varepsilon^2}}{\varepsilon}.$$

With the same perturbation of the vector as above we get the new coefficients

$$\tilde{c}_1 = a - \frac{b+\delta}{\varepsilon}, \qquad \tilde{c}_2 = \frac{(b+\delta)\sqrt{1+\varepsilon^2}}{\varepsilon},$$

i.e., the perturbation is

$$\tilde{c}_1 - c_1 = -\frac{\delta}{\varepsilon}, \qquad \tilde{c}_2 - c_2 = \frac{\delta\sqrt{1+\varepsilon^2}}{\varepsilon},$$

which is large if ε is small. The problem of solving for the coefficients for the
basis vectors is *ill conditioned*. And it is so even before we have decided upon any
particular numerical method for computing the coefficients.

It is not a coincidence that the orthogonal unit vectors are the basis vectors in
the Cartesian coordinate system. We are used to see this system as natural from
a graphical point of view. But we should also keep in mind that the orthogonality
guarantees that this system is the best one from a numerical/computational point of
view as shown by our example.

Later we shall generalize the concept of vector basis to function basis, with a
given function represented as a linear combination of certain basis functions. We
shall demonstrate that the choice of bases is a central issue for computational mathematics. One such choice is orthogonal functions.

Exercise 3.1 The vectors

$$\mathbf{x} = \begin{bmatrix} 1 \\ 1 \end{bmatrix}, \qquad \mathbf{y} = \begin{bmatrix} 2 \\ a \end{bmatrix}$$

are given, where a is a constant.

(a) For what values of a are the vectors linearly independent?
(b) For what values of a are the vectors orthogonal?

(c) What is the norm of the vectors?

(d) Scale the orthogonal vectors in (b) such that they become orthonormal.

Exercise 3.2 Prove the triangle inequality (3.1) for the Euclidean vector norm.

3.2 *Complex Numbers

Sometimes it is not enough to use real numbers, and we shall briefly introduce *complex* numbers. Consider the equation $x^2 + 1 = 0$. Formally we can write one solution as $x = \sqrt{-1}$, but what does the square-root of a negative number mean? We are looking for a number whose square is -1. If we take any number, either positive or negative, the square of it is positive. The way out of this difficulty is to simply define a new type of number whose square is -1. When first invented, this number was thought of as an imaginary number that actually didn't exist, and therefore the notation i was introduced. It is a special case of a *complex number*, which has the general form $z = x + iy$, where x and y are real numbers as we know them. The numbers x and y are the real and imaginary parts denoted by

$$\operatorname{Re} z = x,$$
$$\operatorname{Im} z = y.$$

Algebraic rules apply as before:

$$z_1 + z_2 = x_1 + iy_1 + x_2 + iy_2 = (x_1 + x_2) + i(y_1 + y_2),$$

$$z_1 z_2 = (x_1 + iy_1)(x_2 + iy_2) = (x_1 x_2 - y_1 y_2) + i(x_1 y_2 + x_2 y_1),$$

$$\frac{z_1}{z_2} = \frac{x_1 + iy_1}{x_2 + iy_2} = \frac{(x_1 + iy_1)(x_2 - iy_2)}{x_2^2 + y_2^2} = \frac{x_1 x_2 + y_1 y_2}{x_2^2 + y_2^2} + i \frac{x_2 y_1 - x_1 y_2}{x_2^2 + y_2^2}.$$

When changing the sign of the imaginary part, we call it the *complex conjugate* of z:

$$\bar{z} = \overline{x + iy} = x - iy.$$

We also have

$$|z|^2 = z\bar{z} = (x + iy)(x - iy) = x^2 + y^2,$$
$$|z| = \sqrt{x^2 + y^2},$$
$$|z_1 z_2| = |z_1||z_2|,$$
$$\overline{z_1 z_2} = \bar{z_1}\, \bar{z_2}.$$

Complex numbers have a geometric interpretation. Any number z corresponds to a point (or a vector) in the x, y plane, where x and y are the real and imaginary parts respectively. The absolute value of $|z| = \sqrt{x^2 + y^2}$ is the distance to the origin $x = y = 0$. A circle with radius r and centered at the origin is characterized by the equation $x^2 + y^2 = r^2$. This can be written in the simple form $|z| = r$. The

Fig. 3.2 Complex numbers

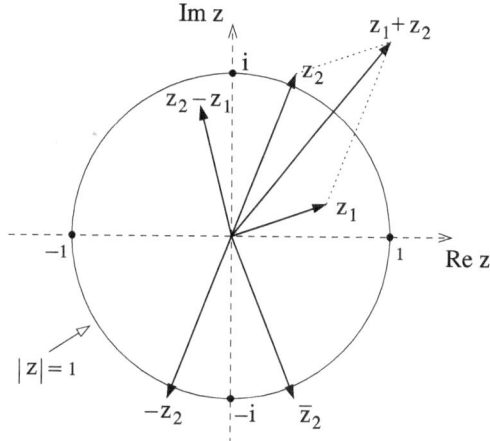

special case $|z| = 1$ is the *unit circle*. Figure 3.2 shows the geometric interpretation of complex numbers.

When it comes to complex vectors, the basic concepts described above for real vectors do not change, except for the definition of the scalar product and norm. For two complex vectors $\mathbf{u}$ and $\mathbf{v}$ we have

$$(\mathbf{u}, \mathbf{v}) = \bar{u}_1 v_1 + \bar{u}_2 v_2 + \cdots + \bar{u}_N v_N,$$
$$\|\mathbf{u}\|^2 = (\mathbf{u}, \mathbf{u}),$$

with the rules

$$\overline{(\mathbf{u}, \mathbf{v})} = (\mathbf{v}, \mathbf{u}),$$
$$\|\mathbf{u} + \mathbf{v}\|^2 = \|\mathbf{u}\|^2 + 2\,\mathrm{Re}(\mathbf{u}, \mathbf{v}) + \|\mathbf{v}\|^2.$$

(The definition $(\mathbf{u}, \mathbf{v}) = u_1 \bar{v}_1 + u_2 \bar{v}_2 + \cdots + u_N \bar{v}_N$ is used in some literature.)

We note that $\|\mathbf{u}\|$ is always real, since

$$\|\mathbf{u}\|^2 = \sum_{j=1}^{N} \bar{u}_j u_j = \sum_{j=1}^{N} |u_j|^2.$$

Complex numbers are very useful in just about every application area, when it comes to analysis. We shall come back to them in Sect. 5.3 when discussing the exponential function.

Exercise 3.3 Let z_1, z_2 be two complex numbers.

(a) Use geometrical arguments to prove that $|z_1|$ is the length of the vector z_1 in Fig. 3.2.
(b) Use the definition of $|z|$ to prove $|z_1 z_2| = |z_1|\,|z_2|$.
(c) Use the definition of $\bar{z}$ to prove $\overline{z_1 z_2} = \bar{z}_1\, \bar{z}_2$.

Exercise 3.4 If x_1 and x_2 are two positive real numbers with $|x_1| > |x_2|$, then $|x_1 - x_2| = |x_1| - |x_2|$. If z_1 and z_2 are two complex numbers with positive real parts and $|z_1| > |z_2|$, is it true that $|z_1 - z_2| = |z_1| - |z_2|$?

3.3 *Matrix Algebra

In linear algebra we are dealing with linear transformations between vectors. A given vector $\mathbf{x}$ is transformed to another one by an operator A such that $\mathbf{y} = A\mathbf{x}$. Each element in the new vector $\mathbf{y}$ is specified as a combination of all the elements in $\mathbf{x}$. The operator A is called a *matrix*, and is represented by a set of numbers that uniquely determine the new vector. We shall concentrate on square matrices, which is by far the most important class in the type of application that we have in mind. In this case the vectors $\mathbf{x}$ and $\mathbf{y}$ have the same number of elements.

For illustration we first assume a 4-dimensional vector space with vectors

$$\mathbf{x} = \begin{bmatrix} x_1 \\ x_2 \\ x_3 \\ x_4 \end{bmatrix}.$$

The numbers a_{jk}, $1 \le j \le 4$, $1 \le k \le 4$ form a matrix A in a square pattern

$$A = \begin{bmatrix} a_{11} & a_{12} & a_{13} & a_{14} \\ a_{21} & a_{22} & a_{23} & a_{24} \\ a_{31} & a_{32} & a_{33} & a_{34} \\ a_{41} & a_{42} & a_{43} & a_{44} \end{bmatrix},$$

often denoted by (a_{jk}). The transformation represented by A is now defined by the new vector

$$A\mathbf{x} = \begin{bmatrix} a_{11} & a_{12} & a_{13} & a_{14} \\ a_{21} & a_{22} & a_{23} & a_{24} \\ a_{31} & a_{32} & a_{33} & a_{34} \\ a_{41} & a_{42} & a_{43} & a_{44} \end{bmatrix} \begin{bmatrix} x_1 \\ x_2 \\ x_3 \\ x_4 \end{bmatrix} = \begin{bmatrix} a_{11}x_1 + a_{12}x_2 + a_{13}x_3 + a_{14}x_4 \\ a_{21}x_1 + a_{22}x_2 + a_{23}x_3 + a_{24}x_4 \\ a_{31}x_1 + a_{32}x_2 + a_{33}x_3 + a_{34}x_4 \\ a_{41}x_1 + a_{42}x_2 + a_{43}x_3 + a_{44}x_4 \end{bmatrix}.$$

There are several rules for calculations with the matrices involved. Two matrices can be added by adding the corresponding elements:

$$A + B = \begin{bmatrix} a_{11} + b_{11} & a_{12} + b_{12} & a_{13} + b_{13} & a_{14} + b_{14} \\ a_{21} + b_{21} & a_{22} + b_{22} & a_{23} + b_{23} & a_{24} + b_{24} \\ a_{31} + b_{31} & a_{32} + b_{32} & a_{33} + b_{33} & a_{34} + b_{34} \\ a_{41} + b_{41} & a_{42} + b_{42} & a_{43} + b_{43} & a_{44} + b_{44} \end{bmatrix}.$$

In order to define multiplication, we first define the row vectors of A as

$$\mathbf{a}_j = \begin{bmatrix} a_{j1} \, a_{j2} \, a_{j3} \, a_{j4} \end{bmatrix}, \quad j = 1, 2, 3, 4,$$

and the column vectors of B as

$$\mathbf{b}_k = \begin{bmatrix} b_{1k} \\ b_{2k} \\ b_{3k} \\ b_{4k} \end{bmatrix}, \quad k = 1, 2, 3, 4.$$

With this notation we have

$$A = \begin{bmatrix} \mathbf{a}_1 \\ \mathbf{a}_2 \\ \mathbf{a}_3 \\ \mathbf{a}_4 \end{bmatrix}, \quad B = \begin{bmatrix} \mathbf{b}_1 & \mathbf{b}_2 & \mathbf{b}_3 & \mathbf{b}_4 \end{bmatrix}.$$

Multiplication of two matrices can now be defined by using scalar products of vectors as

$$AB = \begin{bmatrix} (\mathbf{a}_1, \mathbf{b}_1) & (\mathbf{a}_1, \mathbf{b}_2) & (\mathbf{a}_1, \mathbf{b}_3) & (\mathbf{a}_1, \mathbf{b}_4) \\ (\mathbf{a}_2, \mathbf{b}_1) & (\mathbf{a}_2, \mathbf{b}_2) & (\mathbf{a}_2, \mathbf{b}_3) & (\mathbf{a}_2, \mathbf{b}_4) \\ (\mathbf{a}_3, \mathbf{b}_1) & (\mathbf{a}_3, \mathbf{b}_2) & (\mathbf{a}_3, \mathbf{b}_3) & (\mathbf{a}_3, \mathbf{b}_4) \\ (\mathbf{a}_4, \mathbf{b}_1) & (\mathbf{a}_4, \mathbf{b}_2) & (\mathbf{a}_4, \mathbf{b}_3) & (\mathbf{a}_4, \mathbf{b}_4) \end{bmatrix}.$$

(Here we have used row vectors as the left member in the scalar product.) For the 2×2 case we have

$$AB = \begin{bmatrix} a_{11} & a_{12} \\ a_{21} & a_{22} \end{bmatrix} \begin{bmatrix} b_{11} & b_{12} \\ b_{21} & b_{22} \end{bmatrix} = \begin{bmatrix} a_{11}b_{11} + a_{12}b_{21} & a_{11}b_{12} + a_{12}b_{22} \\ a_{21}b_{11} + a_{22}b_{21} & a_{21}b_{12} + a_{22}b_{22} \end{bmatrix}.$$

For addition the commutative rule $A + B = B + A$ applies, but for multiplication it does not. For general matrices we have $AB \neq BA$. However, for some very special matrices we may have equality, in which case we say that the matrices commute.

Addition and multiplication can be defined for rectangular matrices as well. Assume that A has m rows and n columns, and that B has r rows and s columns. (A is an $m \times n$ matrix and B is an $r \times s$ matrix.) Then $A + B$ is well defined if $m = r$ and $n = s$. Multiplication AB is well defined if $n = r$, and the result is an $m \times s$ matrix, see Fig. 3.3.

A vector $\mathbf{x}$ is a special case of rectangular matrix with only one column. A matrix/vector multiplication as defined above is therefore consistent with the general multiplication rule. Note that the multiplication is defined only when the matrix is on the left hand side of the vector. The expression $\mathbf{x}A$ has no meaning.

There is one case where multiplication is well defined even when the number of columns in the left matrix does not equal to the number of rows in the right matrix. That is multiplication of a matrix by a scalar c, which is defined by

$$cA = Ac = (ca_{jk}),$$

i.e., each element is multiplied by the scalar. With this rule added to the rules above, it follows that A is a linear operator:

$$A(c\mathbf{x} + \mathbf{y}) = cA\mathbf{x} + A\mathbf{y}$$

for all vectors $\mathbf{x}$ and $\mathbf{y}$.

Fig. 3.3 Matrix
multiplication

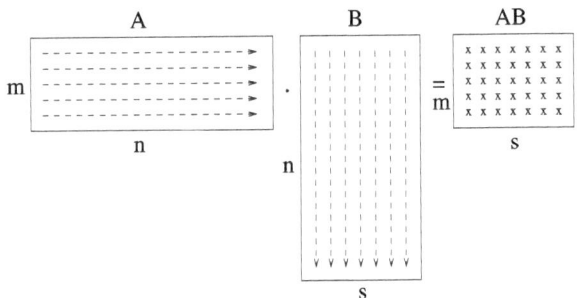

In Sect. 3.1 we defined the scalar product $(\mathbf{x}, \mathbf{y})$ of two vectors. Obviously it is a special case of the matrix product:

$$(\mathbf{x}, \mathbf{y}) = \mathbf{x}^T \mathbf{y} = \mathbf{y}^T \mathbf{x},$$
$$\|\mathbf{x}\|^2 = \mathbf{x}^T \mathbf{x}.$$

For vectors the definition of the norm is quite natural as a generalization of the length in a 3D-Euclidean space. For matrices it is difficult to find any natural concept corresponding to length. Instead we think of a matrix as a transformation of a vector $\mathbf{x}$ to another vector $\mathbf{y} = A\mathbf{x}$. One question is then: what is the length of $\mathbf{y}$ compared to $\mathbf{x}$? Furthermore, what is the maximal ratio $\|\mathbf{y}\|/\|\mathbf{x}\|$ when allowing all possible vectors $\mathbf{x}$? This maximal ratio is called the norm of the matrix, and is formally defined by

$$\|A\| = \max_{\mathbf{x} \neq 0} \frac{\|A\mathbf{x}\|}{\|\mathbf{x}\|}.$$

Since $\|c\mathbf{x}\| = c\|\mathbf{x}\|$ and $\|Ac\mathbf{x}\| = c\|A\mathbf{x}\|$ for any positive constant c, we get

$$\|A\| = \max_{\mathbf{x} \neq 0} \frac{\|Ac\mathbf{x}\|}{\|c\mathbf{x}\|}.$$

By choosing $c = 1/\|\mathbf{x}\|$ and $\mathbf{y} = c\mathbf{x}$ we have $\|\mathbf{y}\| = 1$, and the norm can be defined as

$$\|A\| = \max_{\|\mathbf{y}\|=1} \frac{\|A\mathbf{y}\|}{\|\mathbf{y}\|}.$$

Obviously, the inequality

$$\|A\mathbf{x}\| \leq \|A\| \, \|\mathbf{x}\|$$

holds for all matrices and vectors. One can also prove the relations

$$\|cA\| = |c| \, \|A\|, \quad c \text{ scalar constant,}$$

$$\|A + B\| \leq \|A\| + \|B\|,$$

$$\|AB\| \leq \|A\| \, \|B\|.$$

For other vector norms, the corresponding related matrix norm can be defined in exactly the same way. Once the vector norm is chosen, the matrix norm is called the *subordinate norm*.

There are several special types of matrices. The matrix

$$A = \begin{bmatrix} a_{11} & a_{12} & a_{13} & a_{14} \\ 0 & a_{22} & a_{23} & a_{24} \\ 0 & 0 & a_{33} & a_{34} \\ 0 & 0 & 0 & a_{44} \end{bmatrix}$$

is *upper triangular*, and

$$A = \begin{bmatrix} a_{11} & 0 & 0 & 0 \\ a_{21} & a_{22} & 0 & 0 \\ a_{31} & a_{32} & a_{33} & 0 \\ a_{41} & a_{42} & a_{43} & a_{44} \end{bmatrix}$$

is *lower triangular*. The matrix

$$A = \begin{bmatrix} a_{11} & 0 & 0 & 0 \\ 0 & a_{22} & 0 & 0 \\ 0 & 0 & a_{33} & 0 \\ 0 & 0 & 0 & a_{44} \end{bmatrix} \equiv \mathrm{diag}(a_{11}, a_{22}, a_{33}, a_{44})$$

is *diagonal*. The *identity matrix*

$$I = \begin{bmatrix} 1 & 0 & 0 & 0 \\ 0 & 1 & 0 & 0 \\ 0 & 0 & 1 & 0 \\ 0 & 0 & 0 & 1 \end{bmatrix}$$

is a special diagonal matrix. A *tridiagonal* matrix has the form

$$A = \begin{bmatrix} a_{11} & a_{12} & 0 & 0 \\ a_{21} & a_{22} & a_{23} & 0 \\ 0 & a_{32} & a_{33} & a_{34} \\ 0 & 0 & a_{43} & a_{44} \end{bmatrix}.$$

This is a special case of a *sparse* matrix, which is an important class of matrices, when the order N of the matrix is large. A sparse matrix has few nonzero elements compared to the total number of elements. For example, a tridiagonal matrix with N^2 elements has at most $3N - 2$ nonzero elements. The tridiagonal matrix is a special case of a *band matrix*, where all the elements outside a band around the diagonal are zero:

$$a_{jk} = 0 \quad \text{for } k < j - k_1 \text{ and } k > j + k_2.$$

The *bandwidth* is $k_1 + k_2 + 1$.

The transpose of a matrix $A = (a_{jk})$ is defined by $A^T = (a_{kj})$, i.e., the column vectors have been turned 90 degrees to become row vectors in the same order. If $A = A^T$, the matrix is *symmetric* and, if $A = -A^T$, it is *skew-symmetric*.

In applications *positive definite* matrices occur quite frequently. They are symmetric, and satisfy $(\mathbf{x}, A\mathbf{x}) > 0$ for all vectors $\mathbf{x} \neq \mathbf{0}$. We shall discuss them further in Chap. 14.

Let us next consider systems of equations, where we first use the case $N = 4$ for illustration. The system is

$$a_{11}x_1 + a_{12}x_2 + a_{13}x_3 + a_{14}x_4 = b_1,$$
$$a_{21}x_1 + a_{22}x_2 + a_{23}x_3 + a_{24}x_4 = b_2,$$
$$a_{31}x_1 + a_{32}x_2 + a_{33}x_3 + a_{34}x_4 = b_3,$$
$$a_{41}x_1 + a_{42}x_2 + a_{43}x_3 + a_{44}x_4 = b_4,$$

which is written in matrix/vector form as

$$A\mathbf{x} = \mathbf{b},$$

where the elements of $\mathbf{b}$ are known values. The solution is written as

$$\mathbf{x} = A^{-1}\mathbf{b}.$$

The matrix A^{-1} is called the *inverse* of A, and is a generalization of division by numbers. The inverse is defined by the requirement $A^{-1}A = I$, and here the commuting property $A^{-1}A = AA^{-1} = I$ holds.

The existence of a unique solution $\mathbf{x}$ is equivalent to the existence of the inverse A^{-1}. But not all systems of equations have a unique solution. For example, consider the system

$$x + 2y = c,$$
$$2x + 4y = 6.$$

After dividing the second equation by 2 and subtracting it from the first one, we are left with the equation $0 = c - 3$. This means that if $c \neq 3$, there is no solution at all. On the other hand, if $c = 3$, then we can choose y arbitrarily, and get the solution $x = 3 - 2y$. Consequently, the system has either no solution at all, or it has infinitely many solutions. What is the problem here?

If the second equation is divided by 2, we get the equation $x + 2y = 3$. But the first equation says that $x + 2y = c$, and we can see right away that c must equal 3. Another point of view is obtained by ordering the coefficients on the left hand side as two vectors

$$\mathbf{a}_1 = \begin{bmatrix} 1 \\ 2 \end{bmatrix}, \qquad \mathbf{a}_2 = \begin{bmatrix} 2 \\ 4 \end{bmatrix}.$$

Obviously the relation

$$\mathbf{a}_1 - 0.5\mathbf{a}_2 = \mathbf{0}$$

holds. The two vectors are linearly dependent. When that happens, the system is called *singular*. Let now A be a general matrix with the vectors $\mathbf{a}_j$ as columns

$$A = [\mathbf{a}_1 \ \mathbf{a}_2 \ \dots \ \mathbf{a}_N],$$

and consider the system

$$A\mathbf{x} = \mathbf{b},$$

Fig. 3.4 Vectors rotated by
an orthogonal matrix A

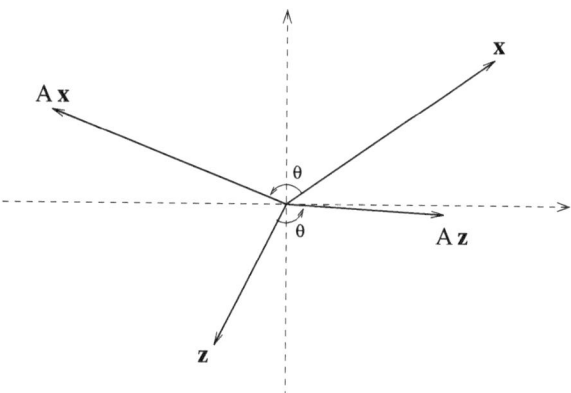

where $\mathbf{x}$ is the vector of unknowns

$$\mathbf{x} = \begin{bmatrix} x_1 \\ x_2 \\ \vdots \\ x_N \end{bmatrix}.$$

One can prove that the system has a unique solution if and only if the coefficient
matrix A is *nonsingular*, i.e., the vectors $\mathbf{a}_j$ are linearly independent. One can also
prove the generalization of the 2×2 example above: if A is singular, then there is
either no solution at all, or there are infinitely many solutions.

One can also define the rows in the matrix A as row vectors, and define the
concept linear dependence in analogy with the definition above for column vectors.
A singular matrix can be characterized by linearly dependent row vectors as well.

The solution of linear systems of equations is a central problem, and it takes
up most of the time and storage consuming part in many algorithms arising from
applications. One way of obtaining the solution $A^{-1}\mathbf{b}$ would be to compute the
inverse A^{-1} and then obtain $\mathbf{x}$ as a simple matrix/vector multiplication. However, in
Chap. 14 we shall present much more effective methods.

A special class of matrices is obtained when $A^{-1} = A^T$, or equivalently,
$A^T A = I$. According to the rules for multiplication, this means that $(\mathbf{a}_j, \mathbf{a}_k) = \delta_{jk}$,
where $\mathbf{a}_j$ are the column vectors of A. Clearly, these vectors are orthogonal (actually
orthonormal), and we call the matrix A orthogonal. When such a matrix is applied
to a vector, the norm of the vector doesn't change, just the direction. In the 2×2
case, every orthogonal matrix can be written in the form

$$A = \begin{bmatrix} \cos\theta & -\sin\theta \\ \sin\theta & \cos\theta \end{bmatrix},$$

where θ is the rotation angle as shown in Fig. 3.4.

Many numerical methods are constructed such that orthogonal matrices arise,
leading to much simplified computations.

For complex matrices all the rules above still apply except the orthogonality condition. We saw that the scalar product for complex vectors is generalized such that

the elements of the first vector are entering in its complex conjugated form. For a general matrix it is therefore convenient to introduce a generalization of the transpose. If $A = (a_{jk})$, then we define $A^* = (\bar{a}_{kj})$. This means that, for the special case of a vector $\mathbf{u}$, we have

$$\mathbf{u}^* = [\bar{u}_1 \, \bar{u}_2 \, \ldots \, \bar{u}_N],$$

and we can write the scalar product as

$$(\mathbf{u}, \mathbf{v}) = \mathbf{u}^* \mathbf{v}.$$

Orthogonality is defined as before by $(\mathbf{u}, \mathbf{v}) = 0$. For a matrix with orthonormal column vectors, we get $A^* A = I$, and such a matrix is often called orthogonal also in the complex case. It also goes under the name *unitary*. The conjugate transpose form is also called the *adjoint* matrix and, if $A = A^*$, then A is called *selfadjoint*. Another name for the same property is *Hermitian*. If $A = -A^*$, then A is called *skew-Hermitian* or *anti-Hermitian*. (In some literature one is using "symmetric" and "skew-symmetric" as notation for Hermitian and skew-Hermitian matrices.)

A *normal* matrix commutes with its adjoint: $AA^* = A^* A$. (In the real case $AA^T = A^T A$.)

We shall now introduce the determinant of a matrix, and we consider first a 2×2 matrix A. If the column vectors $\mathbf{a}_1$ and $\mathbf{a}_2$ of the matrix are linearly dependent, we have

$$\begin{bmatrix} a_{11} \\ a_{21} \end{bmatrix} + c \begin{bmatrix} a_{12} \\ a_{22} \end{bmatrix} = \begin{bmatrix} 0 \\ 0 \end{bmatrix}$$

for some nonzero constant c. We multiply the first element of the vector by a_{22}, and the second by a_{12} to obtain

$$a_{22} a_{11} + c a_{12} a_{22} = 0,$$
$$a_{12} a_{21} + c a_{12} a_{22} = 0.$$

When subtracting the second equation from the first one, we get

$$\text{Det}(A) \equiv a_{11} a_{22} - a_{12} a_{21} = 0.$$

Here $\text{Det}(A)$ stands for the *determinant* of A, and in this case it is zero. Furthermore we note that it is zero precisely when the two vectors are linearly dependent, i.e., when A is singular. The matrix is nonsingular if and only if $\text{Det}(A) \neq 0$.

The determinant can be defined for any matrix A. For the 3×3 case we have

$$\text{Det}(A) = \text{Det} \left(\begin{bmatrix} a_{11} & a_{12} & a_{13} \\ a_{21} & a_{22} & a_{23} \\ a_{31} & a_{32} & a_{33} \end{bmatrix} \right)$$
$$= a_{11} a_{22} a_{33} + a_{12} a_{23} a_{31} + a_{13} a_{21} a_{32}$$
$$- a_{13} a_{22} a_{31} - a_{12} a_{21} a_{33} - a_{11} a_{23} a_{32}.$$

The rule can be seen as follows.

- Move the lower left triangle (excluding the diagonal) to the right of the matrix. Three full diagonals are obtained.

Fig. 3.5 Computation of the determinant

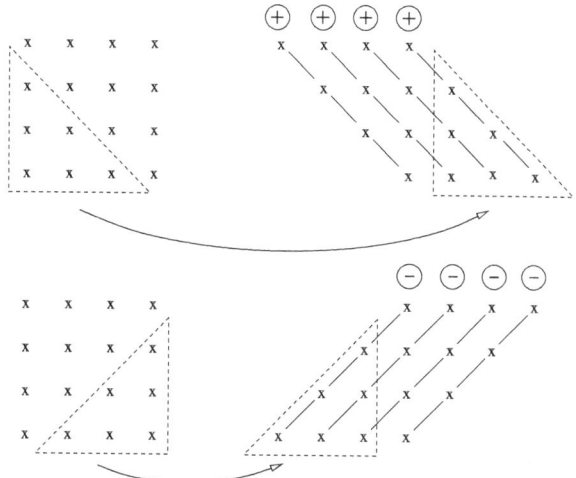

- Multiply together the three elements in each diagonal and add the three products so obtained, giving the number S.
- Move the lower right triangle (excluding the "counter" diagonal) to the left of the matrix. Three full diagonals are obtained.
- Do the same as before, giving the number T. The determinant is computed as $S - T$.

The rule can be applied to any $N \times N$ matrix, just by substituting "three" in the description above by "N". In Fig. 3.5 the procedure is illustrated for the case $N = 4$.

It is harder to see the connection to the concept of linearly dependent column vectors here, but it can be shown that the matrix is singular if and only if $\mathrm{Det}(A) = 0$.

The determinant $\mathrm{Det}(A + B)$ of a sum is in general different from the sum $\mathrm{Det}(A) + \mathrm{Det}(B)$ of the determinants. On the other hand, perhaps somewhat surprisingly, the determinant of a product is the product of the determinants: $\mathrm{Det}(AB) = \mathrm{Det}(A)\,\mathrm{Det}(B)$.

The general rule defining the determinant is usually no good for computation. There are many other and better ways of computing it, but here we leave those out.

Exercise 3.5 Compute AB, where

$$A = \begin{bmatrix} 1 & 2 \\ -1 & 1 \end{bmatrix}, \qquad B = \begin{bmatrix} -1 & 1 \\ 2 & -1 \end{bmatrix}.$$

Show that $\mathrm{Det}(AB) = \mathrm{Det}(A)\,\mathrm{Det}(B)$ by direct computation of the determinants.

Exercise 3.6 Let the vectors $\mathbf{x}$ and $\mathbf{y}$ defined in Exercise 3.1 be the column vectors in the matrix A.

(a) Find the value of a such that $\text{Det}(A) = 0$.
(b) Show that this is precisely the value of a that makes the two vectors linearly dependent.
(c) Try to solve the linear system

$$A\mathbf{z} = \begin{bmatrix} 0 \\ 1 \end{bmatrix}$$

for this value of a by first eliminating z_2. Verify that there is no solution z_1, and consequently no solution $\mathbf{z}$.

Exercise 3.7 Assume that $\mathbf{x}$ is a real vector and that A and B are real orthogonal matrices. Prove the relations

(a) $\|A\mathbf{x}\| = \|\mathbf{x}\|$.
(b) A^T is orthogonal.
(c) A^{-1} is orthogonal.
(d) AB is orthogonal. (Use the rules in Appendix A.4.)

Exercise 3.8 Assume that $\mathbf{x}$ is a real vector and that A is a normal matrix. Prove that $\|A\mathbf{x}\| = \|A^T\mathbf{x}\|$.

Exercise 3.9 Prove that an Hermitian matrix is normal.

3.4 **Eigenvalues

Eigenvalues of a matrix play an important role for the analysis of many numerical methods. We ask ourselves: given a matrix A, are there certain non-zero vectors $\mathbf{x}$ such that multiplication by A retains the vector $\mathbf{x}$, multiplied by a constant? This is an eigenvalue problem

$$A\mathbf{x} = \lambda\mathbf{x},$$

where the constant λ is an *eigenvalue*, and the vector $\mathbf{x}$ is an *eigenvector*. If there are such eigenvectors, they are perfect as basis vectors, since multiplication by A becomes such a simple procedure. Figure 3.6 shows an example of eigenvectors geometrically in 2D.

The equation is rewritten as

$$(A - \lambda I)\mathbf{x} = 0,$$

and we note that there is always one solution, namely $\mathbf{x} = \mathbf{0}$. However, this one is not of much use, and we are looking for nontrivial vectors $\mathbf{x} \neq \mathbf{0}$. But since we already know one solution, the coefficient matrix clearly must be singular for other solutions to exist, i.e., the determinant must be zero:

$$\text{Det}(A - \lambda I) = 0.$$

Fig. 3.6 Eigenvectors x_1 and x_2 and a general vector y

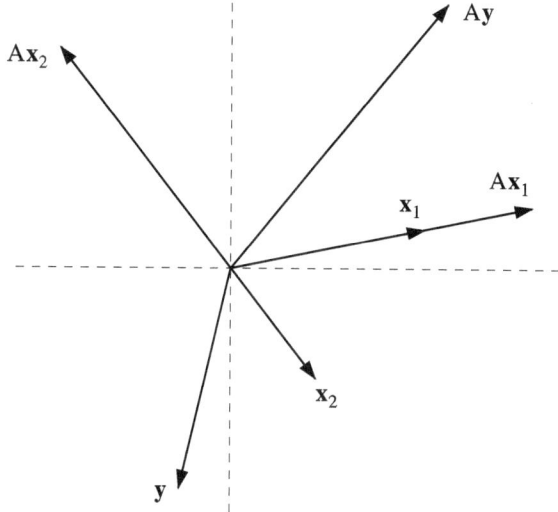

This equation is called the *characteristic equation*. In the definition of the determinant, each term contains N factors, and the highest power of λ that occurs is λ^N. Accordingly, we have a polynomial of degree N, and we know that there are N roots λ_j. We note that these may be complex even if the matrix A is real. This is one of the reasons for introducing complex numbers above.

Assume now that λ_j is an eigenvalue. Then there is an eigenvector $\mathbf{x}_j$ such that

$$(A - \lambda_j I)\mathbf{x}_j = 0.$$

However, any vector $c\mathbf{x}$, where $c \neq 0$ is a constant, is also an eigenvector. Since all of these are linearly dependent, we keep only one, and normalize it such that $\|\mathbf{x}_j\| = 1$.

The next question is whether we can find N linearly independent eigenvectors, which we can use as a basis for the N-dimensional space. The answer is that for certain matrices we can, but for others we cannot. Consider the matrix

$$A = \begin{bmatrix} 1 & 2 \\ 0 & 1 \end{bmatrix}.$$

The eigenvalues are obtained as the solutions to

$$\text{Det}(A - \lambda I) = \text{Det} \begin{bmatrix} 1-\lambda & 2 \\ 0 & 1-\lambda \end{bmatrix} = (1-\lambda)^2 - 0 = 0,$$

which has the double root $\lambda_{1,2} = 1$. The equation for the eigenvectors is

$$(A - I)\mathbf{x} = \begin{bmatrix} 0 + 2x_2 \\ 0 + 0 \end{bmatrix} = \begin{bmatrix} 0 \\ 0 \end{bmatrix}.$$

The first equation shows that $x_2 = 0$, but then there is no further condition on x_1. Accordingly, there is only one linearly independent eigenvector $\mathbf{x}_1$, which we can normalize such that $\mathbf{x}_1 = [1\ 0]^T$.

One can show that, if the characteristic equation has N distinct roots, then there are N linearly independent eigenvectors. But this condition on the roots is not necessary. A trivial counter-example is the identity matrix $A = I$. The characteristic equation has the only root $\lambda = 1$, which has multiplicity N. Still, *any* vector $\mathbf{x}$ is an eigenvector, since $I\mathbf{x} = \mathbf{x}$.

If T is a nonsingular matrix, a *similarity transformation* of A is defined by $T^{-1}AT = B$. If $A\mathbf{x} = \lambda\mathbf{x}$, then

$$B(T^{-1}\mathbf{x}) = T^{-1}ATT^{-1}\mathbf{x} = T^{-1}A\mathbf{x} = \lambda T^{-1}\mathbf{x},$$

i.e., λ is an eigenvalue of the new matrix as well (but the eigenvector is different). A special case is obtained if we choose the column vectors of T as the eigenvectors $\mathbf{x}_j$ of A such that

$$AT = A[\mathbf{x}_1\ \mathbf{x}_2\ \ldots\ \mathbf{x}_N] = [\lambda_1\mathbf{x}_1\ \lambda_2\mathbf{x}_2\ \ldots\ \lambda_N\mathbf{x}_N]. \qquad (3.2)$$

Referring back to the multiplication rule for matrices, we note that the last matrix can be written as

$$T\Lambda = T \begin{bmatrix} \lambda_1 & 0 & \ldots & 0 \\ 0 & \lambda_2 & \ldots & 0 \\ \vdots & & \ddots & \vdots \\ 0 & \ldots & \ldots & \lambda_N \end{bmatrix},$$

i.e., Λ is the diagonal matrix containing the eigenvalues in the diagonal. Accordingly, when multiplying (3.2) from the left by T^{-1}, we obtain

$$T^{-1}AT = \Lambda.$$

We call this transformation a *diagonalization* of A.

We have seen that even if the matrix A is real, the eigenvalues may be complex. It is therefore natural to introduce complex matrices when dealing with eigenvalue problems. The whole discussion about eigenvalues above holds also for complex matrices.

There are special cases where it is known a priori that the eigenvalues are real. For example, it can be shown that a symmetric (or Hermitian) matrix has real eigenvalues. Furthermore, such matrices have orthogonal eigenvectors $\mathbf{v}_j$. If these are normalized such that $\|\mathbf{v}_j\| = 1$, and ordered as column vectors in the matrix T, we have $\|T\| = \|T^{-1}\| = \|T^*\| = 1$. We also have

$$\|A\| = \max_j |\lambda_j|,$$

$$\|A^{-1}\| = \frac{1}{\min_j |\lambda_j|}.$$

There is a special notation for the largest eigenvalue magnitude. It is called the *spectral radius* and is denoted by

$$\rho(A) = \max_j |\lambda_j|.$$

Many fundamental concepts and problems can be described in terms of the eigenvalues. For example, it is quite easy to show that a matrix is singular if and only if there is at least one zero eigenvalue.

Eigenvalue problems arise in many different applications, and there are many efficient numerical methods for finding them. Here we shall not go into these methods, since we shall only use the theoretical aspects of eigenvalues in the analysis of certain problems connected to differential equations.

Exercise 3.10 Find the eigenvalues of the matrices

$$\begin{bmatrix} 1 & 0 & 0 \\ 0 & 2 & 0 \\ 0 & 0 & 3 \end{bmatrix}, \quad \begin{bmatrix} 1 & 2 \\ 1 & -1 \end{bmatrix}, \quad \begin{bmatrix} 1 & 2 & 0 \\ 1 & -1 & 0 \\ 0 & 0 & 2 \end{bmatrix}.$$

Exercise 3.11 Let A and B be defined by

$$A = \begin{bmatrix} 0 & 1 \\ 1 & 0 \end{bmatrix}, \quad B = \begin{bmatrix} 0 & 1 \\ -1 & 0 \end{bmatrix}.$$

(a) Find the eigenvalues and eigenvectors of A and B.
(b) Find the norm of A and B by using the definition of the norm.
(c) Verify that $\rho(A) = \|A\|$ and $\rho(B) = \|B\|$ for these matrices.

Exercise 3.12 Prove that a matrix is singular if and only if it has at least one zero eigenvalue.

Exercise 3.13 Assume that A is an Hermitian matrix. Prove that

(a) A has real eigenvalues.
(b) The eigenvectors are orthogonal.
(c) $\|A\| = \rho(A)$.
(d) $\|A^{-1}\| = 1/\min_j |\lambda_j|$.

Chapter 4
Analysis Tools

The development of numerical methods is based on fundamental mathematical analysis, and this chapter includes some of the basic concepts that are essential for later chapters. Even if numerical methods lead to a finite number of algebraic operations, it turns out that the theoretical analysis benefits from the introduction of the infinity concept, which will be discussed in the first section.

As we have seen in Chap. 2, we are dealing with small perturbations and small steps Δx when dealing with differential equations and their numerical solution. This makes it natural to discuss Taylor expansions, which we do in the second section.

The content of this chapter corresponds roughly to part of a first college course in mathematical analysis, and can be skipped by the reader who has this subject fresh in memory.

4.1 *Infinite Processes

Many numerical methods for solving differential equations give rise to algebraic equations or systems of equations that must be solved efficiently. These solution methods are almost exclusively based on iteration, i.e., starting from an initial guess, an iterative algorithm produces approximative solutions $\mathbf{u}_n$ that become closer to the true solution for each step. In practice, the process has to be stopped somewhere, and the question is how we can make sure that the error is small at that point. A necessary condition is that the solution $\mathbf{u}_n$ *converges* to a *limit solution*

$$\lim_{n \to \infty} \mathbf{u}_n = \mathbf{u}_\infty,$$

that is a good approximation of the true solution $\mathbf{u}$. Even if we cannot iterate for ever, it is a significant help to understand the theoretical infinite process, and we shall devote the next two subsections to discussing some simple processes and their basic properties.

B. Gustafsson, *Fundamentals of Scientific Computing*,
Texts in Computational Science and Engineering 8,
DOI 10.1007/978-3-642-19495-5_4, © Springer-Verlag Berlin Heidelberg 2011

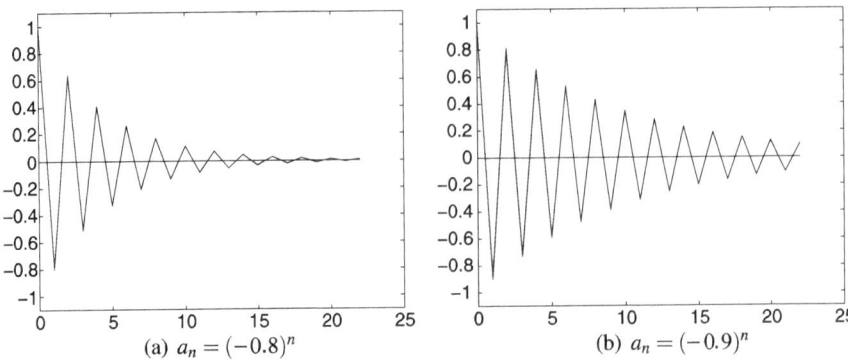

Fig. 4.1 Converging number sequences

4.1.1 Infinite Number Sequences

We begin by considering a sequence of numbers

$$\{a_n\}_{n=0}^{N} = a_0, a_1, a_2, \ldots, a_N.$$

Here it is assumed that a_n is an expression that is well defined for any given integer n. Let us look at an example.

We consider first the case $a_n = (-0.8)^n$ which is depicted in Fig. 4.1(a) with n on the horizontal axis (the points are connected by straight lines). Obviously the numbers approach zero for increasing n, and this is the true limit value. We write it as

$$\lim_{n \to \infty} a_n = \lim_{n \to \infty} (-0.8)^n = 0.$$

Strict mathematically, this condition means that no matter how small a deviation ε we prescribe, there is alway a number N such that $|a_n| < \varepsilon$ for all n with $n \geq N$. If we choose $\varepsilon = 0.01$ in our example, we can choose $N = 21$, while $\varepsilon = 0.001$ requires $N = 31$. We say that the number sequence in such a case is *convergent*.

Let us next choose $a_n = (-0.9)^n$ shown in Fig. 4.1(b). The numbers still approach zero, but now at a slower rate. Assuming that we compute the powers by using the formula

$$a_{n+1} = -0.9a_n, \quad n = 0, 1, \ldots,$$
$$a_0 = 1,$$

we have to work harder to obtain a good approximation of the limit value $a_\infty = 0$. The first sequence converges faster than the second sequence.

For the case $a_n = (-1.1)^n$ there is no convergence at all. The numbers become larger and larger for increasing n as shown in Fig. 4.2(a), and finally they will exceed the largest number that can be represented in the computer. The sequence is *diverging*.

For the border case $a_n = (-1)^n$ shown in Fig. 4.2(b), there is no divergence, but neither is there any convergence. The condition for convergence of a sequence $\{c^n\}$ is $|c| < 1$. (The case $c = 1$ is special with convergence to one.)

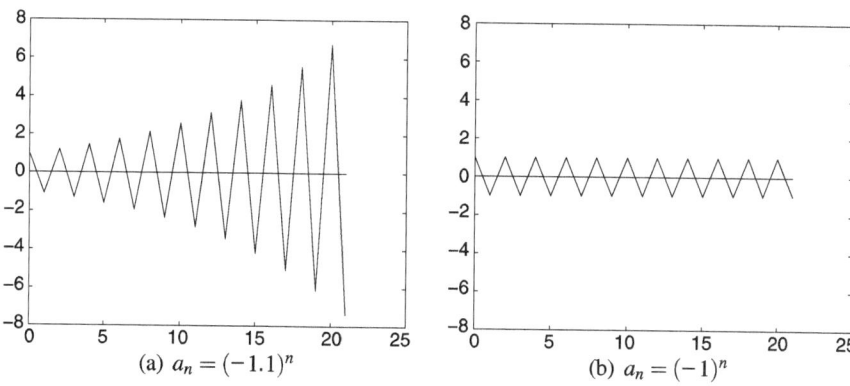

(a) $a_n = (-1.1)^n$ (b) $a_n = (-1)^n$

Fig. 4.2 Nonconverging number sequences

In general, the true limit value of the sequence is obtained from the iterative formula. For our example we have

$$a_{n+1} = ca_n, \quad n = 0, 1, \dots.$$

We simply take the limit of both sides and obtain with the notation $a_\infty = \lim_{n \to \infty} a_n$

$$a_\infty = ca_\infty, \tag{4.1}$$

or equivalently

$$(1 - c)a_\infty = 0.$$

If $|c| < 1$, the correct solution of this equation is $a_\infty = 0$. However, also for the case $|c| > 1$ the solution is $a_\infty = 0$. Is something wrong with the mathematics here, since we have shown above that there is no limit value in this case? Certainly not. When writing down the formula (4.1), we *assume* that there is a limit a_∞. But for the case $|c| > 1$ this assumption is wrong. The limit $\lim_{n \to \infty} a_n$ doesn't exist, and a_∞ doesn't mean anything.

This example shows that mathematics, as usual, is very strict. In particular, one has to be very careful when dealing with infinite sequences.

4.1.2 Infinite Series

The analysis of infinite number sequences is fairly simple. The situation becomes more complicated when dealing with infinite series defined as a sum of infinitely many terms. Given a number sequence $\{a_n\}_{n=0}^N$ we define the sum

$$\sum_{n=0}^N a_n = a_0 + a_1 + a_2 + \dots + a_N,$$

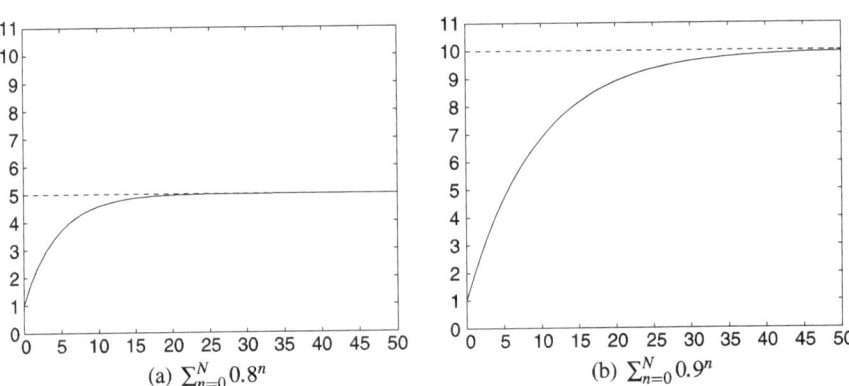

Fig. 4.3 Converging sums

and ask ourselves what happens if we keep adding more and more terms without stopping. Is the sum approaching a certain finite value? If this is true, we write the limit value as

$$\lim_{N \to \infty} \sum_{n=0}^{N} a_n = \sum_{n=0}^{\infty} a_n.$$

If all numbers a_n are positive, it seems that the sum will never settle down, since the sum increases for every new term. If we just keep adding new terms, doesn't the sum exceed any given value, no matter how large?

Indeed it does not. Consider the sequence in the previous section, but with all terms positive: $a_n = c^n$. Figure 4.3(a) shows the sum $\sum_{n=0}^{N} c^n$ as a function of the number of terms for $c = 0.8$. The dashed horizontal line represents the value 5, and it seems that the sum approaches this value when N increases. Figure 4.3(b) shows the same sum, but now for $c = 0.9$. Even in this case there seems to be a limit value, and now it is 10. Indeed, one can easily figure out the theoretical limit value by analytical means. It is a *geometric series*, and it occurs frequently in applications. For the finite sum $S_N = \sum_{n=0}^{N} c^n$ we have

$$c S_N = \sum_{n=1}^{N+1} c^n = \sum_{n=0}^{N} c^n - 1 + c^{N+1} = S_N - 1 + c^{N+1},$$

i.e.,

$$\lim_{N \to \infty} S_N = \lim_{N \to \infty} \frac{1 - c^{N+1}}{1 - c} = \frac{1}{1 - c}, \tag{4.2}$$

which agrees with the graphs. The sum converges as $N \to \infty$. It is no surprise that the convergence is faster when c is smaller, since each added term a_n is smaller.

There is of course no need to try to compute the sum if $c > 1$, since a necessary condition for convergence is that a_n approaches zero as n increases. However, this is not a sufficient condition for convergence. The numbers in the sequence must decrease sufficiently fast. For example, the sum converges for $a_n = 1/n^2$, but diverges for $a_n = 1/n$.

Another example of an infinite sum, is the well known paradox with a running rabbit that never passes the much slower turtle. We assume that the rabbit is one hundred times faster, but it starts 100 meters behind the turtle. When it reaches the starting point for the turtle, it sees the turtle one meter ahead. It runs this meter, and finds the turtle one centimeter ahead. When this centimeter is passed, the turtle is still one tenth of a millimeter ahead, etc. Even if we quickly arrive at extremely small distances, they are still positive, and the conclusion seems to be that the rabbit never passes the turtle. Obviously this is not true, but what is wrong with the argument? One way of explaining the paradox is to take time into account. Assume that the rabbit speed is 10 m/sec, and the turtle speed 1 dm/sec. The distance covered by the rabbit is 100 m for the first part of the race, 1 m for the second part, 1 cm for the third etc. The corresponding time intervals for each part are 10 seconds, 0.1 seconds, 0.001 seconds, and so on. This number sequence t_n is defined by

$$t_n = 10(0.01)^n, \quad n = 0, 1, 2, \ldots, \tag{4.3}$$

and according to the formula (4.2) we have

$$\sum_{n=0}^{\infty} t_n = \sum_{n=0}^{\infty} 10(0.01)^n = 10\frac{1}{0.99} = 10.101010\ldots. \tag{4.4}$$

Even if this number has an infinite number of digits, it never reaches 10.2, say. The way we have chosen to describe the process cannot be used for the whole race, since it doesn't allow time to continue after the rabbit has reached the turtle. The time computed above is simply the exact time for this passage (computed in a more complicated way than would be necessary). This is another example showing that one has to be very careful when dealing with the infinity concept.

Iterative algorithms that occur frequently in computational mathematics are of course more complicated than the examples we have discussed here. The typical situation is that each iteration produces a vector or a function that is characterized by many numbers instead of a single value as in our examples. However, the concept of convergence is fundamental also in this case. The vectors or functions must settle down and change very little if the process is carried out sufficiently far. In the limit they must reach a limit vector or a limit function.

The conclusion of this section is the following: In order to get an accurate solution when the infinite process is interrupted by the computer, we have to make sure that we are dealing with a converging process.

When introducing integrals in Sect. 2.2.3, we used Riemann sums in the derivation. They can be generalized such that one or both of the interval ends of the integral becomes infinite forming a *generalized integral*:

$$\int_a^\infty f(x)\,dx = \lim_{b\to\infty} \int_a^b f(x)\,dx.$$

Evidently, there are restrictions on the function $f(x)$ such that it tends to zero sufficiently fast when $x \to \infty$.

We also note that an integral may exist even if the function becomes infinite at one or more points. For example, it can be shown that the integral

$$\int_0^1 \frac{1}{\sqrt{1-x}} \, dx$$

is well defined despite the fact that $1/\sqrt{1-x}$ becomes infinite at $x = 1$. The function values are growing sufficiently slowly, such that the Riemann sum still exists in the limit.

Exercise 4.1 Use a computer program to study the convergence properties of the series

$$\sum_{n=0}^{N} \frac{1}{n^p}$$

for $p = 2, 3, 4$. For each p, find the number of terms $N + 1$ such that the first five digits don't change if more terms are added. What happens for $p = 1$?

4.2 *Taylor Series

Taylor series (or Taylor expansions) are a very useful tool for the analysis of numerical methods. The basic idea is to gain knowledge about functions that are known together with their derivatives at a certain point $x = a$. By using the *factorial* notation

$$n! = 1 \cdot 2 \cdot 3 \cdot \ldots \cdot n, \qquad 0! = 1,$$

the formal expansion is an infinite series

$$u(x) = \sum_{n=0}^{\infty} \frac{(x-a)^n}{n!} \frac{d^n u}{dx^n}(a).$$

Even if all the derivatives exist, the series does not necessarily converge to a finite value for any x. There are actually several possibilities:

• The series converges everywhere and equals $u(x)$. Then $u(x)$ is called *entire*.
• The series converges and equals $u(x)$ in a neighbourhood of $x = a$. Then $u(x)$ is called *analytic* in that neighbourhood.
• The series converges but is not equal to $u(x)$ at $x = a$.
• The series does not converge for any x.

The special case $a = 0$ is called a *Maclaurin series*.
Difference methods are associated with a small step size Δx, and we are interested in the approximation properties in the neighbourhood of $\Delta x = 0$:

$$u(x + \Delta x) = u(x) + \Delta x \frac{du}{dx}(x) + \frac{\Delta x^2}{2!} \frac{d^2 u}{dx^2}(x) + \frac{\Delta x^3}{3!} \frac{d^3 u}{dx^3}(x) + \cdots. \qquad (4.5)$$

Before going further, we introduce the $\mathcal{O}(\Delta x)$ concept. It is defined for positive Δx as a function $f(\Delta x)$ such that

$$\frac{|f(\Delta x)|}{\Delta x} \le c \quad \text{as } \Delta x \to 0, \tag{4.6}$$

where c is a positive constant. We say that a function $f(\Delta x) = \mathcal{O}(\Delta x)$ is of *the order* Δx. It follows that an error $\mathcal{O}(\Delta x^p)$ indicates a better approximation than an error $\mathcal{O}(\Delta x^q)$ if $p > q$.

If the Taylor expansion is truncated after the term of order Δx^{p-1}, we are left with the "tail" of the series, which is the error. With sufficiently strong regularity assumptions on the function, the error is of the same order as the leading term, which is of order Δx^p. If not stated otherwise we shall assume that this is the case, and we write

$$u(x + \Delta x) = \sum_{j=0}^{p-1} \frac{\Delta x^j}{j!} \frac{d^j u}{dx^j}(x) + \mathcal{O}(\Delta x^p).$$

Since we are interested in approximations of derivatives here, we note from the above that

$$\frac{u(x + \Delta x) - u(x)}{\Delta x} = \frac{du}{dx}(x) + \mathcal{O}(\Delta x). \tag{4.7}$$

The difference approximation on the left hand side has an error of order Δx, and we say that it is a first-order accurate approximation of the derivative. A better approximation is obtained if we also include a point to the left of x, and we have

$$u(x + \Delta x) = u(x) + \Delta x \frac{du}{dx}(x) + \frac{\Delta x^2}{2} \frac{d^2 u}{dx^2} + \frac{\Delta x^3}{6} \frac{d^3 u}{dx^3} + \mathcal{O}(\Delta x^4),$$

$$u(x - \Delta x) = u(x) - \Delta x \frac{du}{dx}(x) + \frac{\Delta x^2}{2} \frac{d^2 u}{dx^2} - \frac{\Delta x^3}{6} \frac{d^3 u}{dx^3} + \mathcal{O}(\Delta x^4).$$

If the second equation is subtracted from the first one, we get the *centered approximation*

$$\frac{u(x + \Delta x) - u(x - \Delta x)}{2\Delta x} = \frac{du}{dx}(x) + \mathcal{O}(\Delta x^2). \tag{4.8}$$

For the second derivative we derived an approximation in Sect. 2.2.4, and from the Taylor expansions above we get

$$\frac{u(x + \Delta x) - 2u(x) + u(x - \Delta x)}{\Delta x^2} = \frac{d^2 u}{dx^2}(x) + \mathcal{O}(\Delta x^2). \tag{4.9}$$

Higher-order approximations are obtained by introducing also the grid points $x \pm 2\Delta x$, and we can continue further, making the computing stencil wider and wider. Figure 4.4 shows the points involved for 2nd, 4th and 6nd order approximations. For d/dx the center point is not involved at all. Note however, that the indicated order of accuracy holds only at the center point. It is an effect of the symmetric character of the stencil with proper centering.

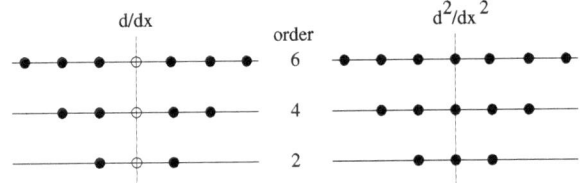

Fig. 4.4 Difference stencils for d/dx and d^2/dx^2, accuracy order $2, 4, 6$

By "high order difference methods" one usually means a higher order of accuracy than 2. Such methods have been used for a long time for the solution of ODE, and in Sect. 10.1 we shall discuss them further. Nowadays, high order methods are used quite frequently also for PDE.

We finally note that the "big $\mathcal{O}$" concept can be used also for large numbers N, where N typically is the number of grid points. As we shall see, the necessary number of arithmetic operations for a certain algorithm is often of the form $a_0 + a_1 N + \cdots + a_p N^p$ and, if N is large enough, the last term will dominate. Therefore we talk about an algorithm of order N^p. For a large parameter N we define the $\mathcal{O}(N)$ concept as a function $f(N)$ such that

$$\frac{|f(N)|}{N} \leq c \quad \text{as } N \to \infty, \tag{4.10}$$

where c is a constant. An $\mathcal{O}(N^p)$ algorithm is faster than an $\mathcal{O}(N^q)$ algorithm if $p < q$.

The Taylor expansion can be generalized to functions $u(x_1, x_2, \ldots, x_d)$ of several variables. The first terms are

$$u(x_1 + \Delta x_1, x_2 + \Delta x_2, \ldots, x_d + \Delta x_d)$$
$$= u(x_1, x_2, \ldots, x_d) + \Delta x_1 \frac{\partial u}{\partial x_1}(x_1, x_2, \ldots, x_d)$$
$$+ \Delta x_2 \frac{\partial u}{\partial x_2}(x_1, x_2, \ldots, x_d) + \cdots + \Delta x_d \frac{\partial u}{\partial x_d}(x_1, x_2, \ldots, x_d) + \mathcal{O}(\Delta x^2),$$

where it is assumed that all step sizes are of the same order:

$$\Delta x_j = \mathcal{O}(\Delta x), \quad j = 1, 2, \ldots, d.$$

A more compact notation is

$$u(\mathbf{x} + \Delta \mathbf{x}) = u(\mathbf{x}) + \big(\nabla u(\mathbf{x})\big)^T \Delta \mathbf{x} + \mathcal{O}(\Delta x^2),$$

where

$$\mathbf{x} = \begin{bmatrix} x_1 \\ x_2 \\ \vdots \\ x_d \end{bmatrix}, \qquad \Delta \mathbf{x} = \begin{bmatrix} \Delta x_1 \\ \Delta x_2 \\ \vdots \\ \Delta x_d \end{bmatrix}, \qquad \nabla u = \begin{bmatrix} \partial u / \partial x_1 \\ \partial u / \partial x_2 \\ \vdots \\ \partial u / \partial x_d \end{bmatrix}.$$

Exercise 4.2 Prove that $(u(x_0 + \Delta x) - u(x_0))/\Delta x$ is a second order approximation of du/dx at $x = x_0 + \Delta x/2$.

Exercise 4.3 Find the constants a_j such that $\sum_{j=-2}^{2} a_j u(x_j)/\Delta x$ is a fourth order approximation of du/dx at $x = x_0$.

Exercise 4.4 Find the constants b_j such that $\sum_{j=-2}^{2} b_j u(x_j)/\Delta x^2$ is a fourth order approximation of d^2u/dx^2 at $x = x_0$.

Exercise 4.5 The function $u(x) = x^8$ has the derivative $u'(x) = 8x^7$. Use a computer program to compute the derivative by a first order approximation (4.7) and a second order approximation (4.8) with $\Delta x = 0.1$ for $0 \le x \le 1$. Plot the two curves together with $u'(x)$ and compare the accuracy.

Chapter 5
Elementary Functions

There are many elementary functions that play important roles in mathematics, and whose properties have long been well known. Some of these are important also in the construction of numerical methods, and they occur in two different roles. One role is as part of the numerical algorithms where certain solutions are represented in terms of these functions for easy evaluation. The other role is as a tool for *analysis* of the algorithms. In this case the functions are not used as part of the computational algorithm, but they are used in order to understand the fundamental properties of it.

In this chapter we shall concentrate on the three most frequently used types of functions, which are polynomials, trigonometric functions and the exponential function.

5.1 *Polynomials

Perhaps the most common type of function for use in almost every kind of application is polynomials, the main reason being that they can be simply evaluated. Polynomials can be formulated in different ways, and we begin by discussing the standard form.

5.1.1 Standard Polynomials

A polynomial of degree n has the general form

$$P(x) = a_n x^n + a_{n-1} x^{n-1} + \cdots + a_1 x + a_0, \tag{5.1}$$

where n is a non-negative integer and $a_n \neq 0$. Polynomials are easy to evaluate for any given number x, and this is one of the reasons for their popularity. In Chap. 18 we shall discuss some basic principles for computers. Here we just note that all computing must in the end be broken down to basic algebraic operations like addition,

B. Gustafsson, *Fundamentals of Scientific Computing*,
Texts in Computational Science and Engineering 8,
DOI 10.1007/978-3-642-19495-5_5, © Springer-Verlag Berlin Heidelberg 2011

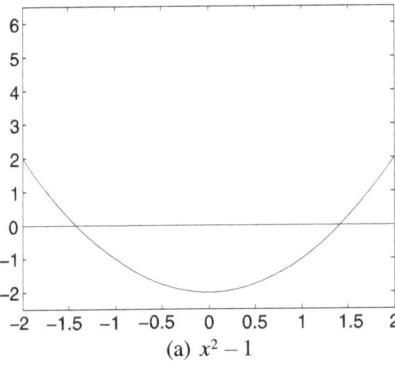

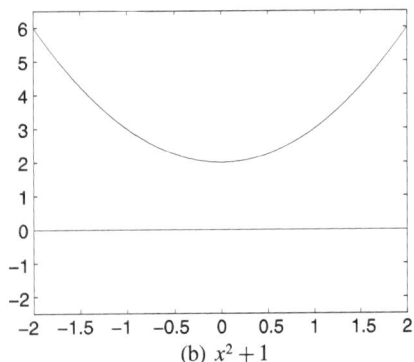

(a) $x^2 - 1$ (b) $x^2 + 1$

Fig. 5.1 Two polynomials

multiplication and division. The straightforward way of evaluating a polynomial at a certain point x is to compute the powers, multiply and add. For example,

$$a_3 x^3 + a_2 x^2 + a_1 x + a_0 = a_3 \cdot x \cdot x \cdot x + a_2 \cdot x \cdot x + a_1 \cdot x + a_0,$$

which shows that 6 multiplications and 3 additions are required. However, there is a better way of evaluation. We have

$$a_3 x^3 + a_2 x^2 + a_1 x + a_0 = \left((a_3 \cdot x + a_2) \cdot x + a_1 \right) \cdot x + a_0.$$

This formula is called *Horner's scheme*, and it requires only 3 multiplications and 3 additions. It is easy to see that a general polynomial of degree n requires n multiplications and n additions for evaluation at one point.

The *roots of a polynomial* $P(x)$, i.e., those points $x = x_j$ where $P(x) = 0$, play an important role in many respects. Consider the example $x^2 - 2 = 0$, which has the two solutions $x = \sqrt{2} = 1.414$ and $x = -\sqrt{2} = -1.414$. By looking at the graph of $P(x)$, we can see that there are no other roots. The curve cuts through the x-axis at two places and no more, as shown in Fig. 5.1(a).

Next we consider the equation $x^2 + 2 = 0$. The graph in Fig. 5.1(b) shows why there are no real solutions. The curve never cuts through the x-axis. However, by introducing complex numbers (see Sect. 3.2), there are roots of the polynomial also in this case. We write the equation as $z^2 + 2 = 0$, where $z = x + iy$ is now a complex number. We can solve the equation by expanding the square:

$$x^2 + 2ixy + i^2 y^2 + 2 = x^2 + 2ixy - y^2 + 2 = 0.$$

The real part and the imaginary part of the left hand side must be zero, which gives the system

$$x^2 - y^2 + 2 = 0,$$
$$2xy = 0.$$

From the second equation we see that either x or y must be zero, and we choose $x = 0$. Then it follows from the first equation that $y^2 = 2$, i.e., $y = \sqrt{2}$ or $y = -\sqrt{2}$.

The case $y = 0$ doesn't work, since it implies that $x^2 = -2$, which is impossible for a real number x. So, we have found the two solutions $z = i\sqrt{2}$ and $z = -i\sqrt{2}$.

The roots of a quadratic polynomial $z^2 + az + b = 0$ can be found from the general formula also in the complex case:

$$z_{1,2} = -\frac{a}{2} \pm \sqrt{\frac{a^2}{4} - b}. \tag{5.2}$$

We have discussed how to treat the square-root of a negative real number, but it remains to define what the square root of a general complex number is. We shall come back to this in Sect. 5.3.

It can be shown that a polynomial of degree n always has precisely n roots. However, it must be made clear how the roots should be counted. The standard formula for the solutions of the equation

$$x^2 - 2x + 1 = 0$$

gives

$$x_1 = 1 + \sqrt{1 - 1} = 1,$$
$$x_2 = 1 - \sqrt{1 - 1} = 1,$$

i.e., the polynomial has a *double root* $x = 1$. The polynomial $(x - 1)^n$ has the root $x = 1$ of multiplicity n. Therefore, the counting of the roots takes the multiplicity into account.

In general the roots are complex, even if the coefficients of the polynomial are real. The graph of the polynomial can be used only for determining the number of real roots (which may be useful enough). However, for real coefficients, it is easy to show that the complex roots come in pairs. If z is a complex root, then $\bar{z}$ is a root as well.

If the roots x_j are known, the polynomial can be written in the form

$$P(x) = c\big((x - x_n)(x - x_{n-1}) \cdots (x - x_1)\big),$$

where $c = a_n$ is a nonzero constant. If only one root $x = x_1$ is known, this form can be used to make the computation of the other roots easier. By dividing by the factor $(x - x_1)$ the degree of the polynomial is reduced one step. For example, by inspection we see that

$$P(x) = x^3 - 6x^2 + 11x - 6$$

has the root $x_1 = 1$, i.e.,

$$P(x) = (x^2 - 5x + 6)(x - 1).$$

(In general, the reduced degree polynomial is obtained by formal symbolic division.) Accordingly, the other two roots are obtained from the standard formula

$$x = \frac{5}{2} \pm \sqrt{\frac{25}{4} - 6},$$

i.e., $x_2 = 3$, $x_3 = 2$.

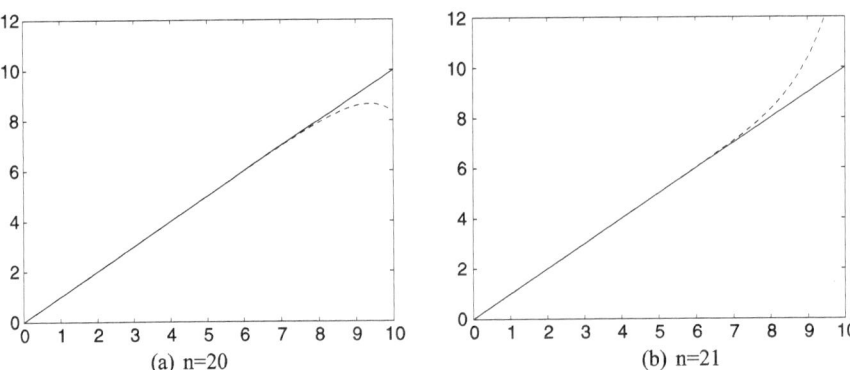

Fig. 5.2 Interpolation by a polynomial of degree n

Perhaps the most common use of polynomials is interpolation. Given $n+1$ points x_j with corresponding function values f_j, there is a unique polynomial $P(x)$, i.e., a unique set of coefficients $\{a_k\}_{k=0}^{n}$, such that $P(x_j) = f_j$ at all points x_j. The most common type of interpolation is linear interpolation, where the function values between two points x_0 and x_1 are approximated by a straight line. When increasing the number of given points, the degree of the polynomial goes up, and it seems like we would get a better approximation. This is true up to a certain point but, for sufficiently large degree n, the result may be an interpolating polynomial that has little to do with the true function $f(x)$. The reason is that the computation of the coefficients is very sensitive to perturbations.

For illustration, we consider the linear function $f(x) = x$ for $0 \leq x \leq 10$. An interpolation polynomial of degree n requires $n + 1$ interpolation points, and we choose the equidistant points $x_j = 10j/n$, $j = 0, 1, \ldots, n$. The coefficients a_k are determined by requiring that the polynomial is exact at the interpolation points x_j, which gives the system

$$a_0 + x_0 a_1 + x_0^2 a_2 + \cdots + x_0^n a_n = x_0,$$
$$a_0 + x_1 a_1 + x_1^2 a_2 + \cdots + x_1^n a_n = x_1,$$
$$\vdots$$
$$a_0 + x_n a_1 + x_n^2 a_2 + \cdots + x_n^n a_n = x_n.$$

The solution to this one is obvious. When a polynomial is interpolated by another one of the same or higher degree, we should get the given polynomial back, since the interpolation problem has a unique solution. In our case this means that the solution is zero coefficients a_k for all k except $k = 1$, where $a_1 = 1$. However, when solving the system by a MATLAB program, and plotting the result, we are in for a surprise when increasing n. Figure 5.2 shows the result for $n = 20$ and $n = 21$. The polynomial is completely wrong in the right part of the interval in both cases.

Let us now explain what the reason for the trouble is. We begin by introducing the concept of a basis for functions in analogy with vector basis discussed in Sect. 3.1.

For polynomials it is natural to define the powers x^j as basis functions. However, before doing so, we introduce a scalar product and a norm also for functions. For vectors we used sums of the products of the vector elements. When dealing with functions it is natural to go from sums to integrals, and we define the scalar product of two functions $f(x)$ and $g(x)$ defined in the interval $[0, 1]$ by

$$(f, g) = \int_0^1 f(x)g(x)\,dx.$$

Corresponding to the length of a vector we define the norm of a function as

$$\|f\| = \sqrt{(f, f)}.$$

It is convenient to normalize the basis functions such that the norm is one. We have

$$\int_0^1 (x^j)^2\,dx = \int_0^1 x^{2j}\,dx = \frac{1}{2j+1},$$

and we define the basis functions as

$$\phi_j(x) = \sqrt{2j+1}\,x^j, \quad j = 0, 1, \ldots, n.$$

Any polynomial of degree n can now be written as

$$P_n(x) = \sum_{j=0}^n b_j \phi_j(x), \tag{5.3}$$

where the new coefficients are $b_j = a_j/\sqrt{2j+1}$.

In Sect. 3.1 we demonstrated that the scalar product is a good measure of the degree of linear independence between two vectors. As is often the case in mathematics, we can apply the same machinery to a completely different area by making the proper definitions. In this case we go from vectors to polynomials. A scalar product value close to one for a pair of basis polynomials indicates almost linear dependence leading to a poor choice of basis. We pick two adjacent basis functions, which gives

$$(\phi_{j-1}, \phi_j) = \int_0^1 \sqrt{2j-1}\,x^{j-1}\sqrt{2j+1}\,x^j\,dx$$

$$= \frac{\sqrt{2j-1}\sqrt{2j+1}}{2j} = \sqrt{1 - 1/(4j^2)}.$$

Apparently, this number approaches one when j increases, and we can see that the two functions are becoming nearly linearly dependent already for quite moderate degree j. This is the reason for the ill conditioning of the interpolation problem. As a consequence, the system of equations for determining b_j (or a_j above) from the interpolation conditions becomes *ill conditioned*, which gives the computer trouble when solving it.

The lesson to be learned from this example is that, when working with polynomials in standard form, we should keep the degree low. However, if we really need higher degree, there is a way of getting around the difficulty demonstrated above. That is to use orthogonal polynomials, which will be introduced in Sect. 5.1.3.

Polynomials are well defined also in several dimensions. In 2D a second degree (quadratic) polynomial has the general form

$$a_{20}x^2 + a_{11}xy + a_{02}y^2 + a_{10}x + a_{01}y + a_{00}, \tag{5.4}$$

and a cubic polynomial has the form

$$a_{30}x^3 + a_{21}x^2y + a_{12}xy^2 + a_{03}y^3 + a_{20}x^2 + a_{11}xy + a_{02}y^2 + a_{10}x + a_{01}y + a_{00}. \tag{5.5}$$

The general form is

$$
\begin{aligned}
a_{pq}x^p y^q &+ a_{p,q-1}x^p y^{q-1} + \cdots + a_{p0}x^p \\
&+ a_{p-1,q}x^{p-1}y^q + a_{p-1,q-1}x^{p-1}y^{q-1} + \cdots + a_{p-1,0}x^{p-1} \\
&\vdots \\
&+ a_{0q}y^q + a_{0,q-1}y^{q-1} + \cdots + a_{00},
\end{aligned}
$$

and the generalization to higher dimensions should be clear from this. The coefficients are to be determined from given data, for example the function values at given interpolation points. The more coefficients there are, the more flexibility there is to adapt the polynomial to the data. The number of coefficients, i.e., the number of terms, is called the number of *degrees of freedom*.

There are also polynomials that are bilinear, biquadratic etc. A bilinear polynomial $P_1(x, y)$ is a first degree polynomial in each of the variables when the other variable is held constant, and similarly for the other types:

Bilinear: $P_1(x, y) = a_{11}xy + a_{10}x + a_{01}y + a_{00},$

Biquadratic: $P_2(x, y) = a_{22}x^2y^2 + a_{21}x^2y + a_{12}xy^2 + a_{20}x^2 + a_{11}xy + a_{02}y^2$
$$+ a_{10}x + a_{01}y + a_{00}.$$

These types of polynomials are frequently used in the piecewise polynomial version to be discussed in the next section, and in the finite element method in Chap. 11.

Exercise 5.1 The equation

$$x^3 - x^2 + x - 1 = 0$$

has the root $x_1 = 1$.

(a) Reduce the degree by taking out the factor $(x - 1)$ and compute the other two roots.
(b) Verify that there are two complex conjugate roots.
(c) Prove the general theorem: If the polynomial has real coefficients and a complex root z_0, then $\bar{z}_0$ is also a root.

Exercise 5.2 Write down the most general form of a cubic and a bicubic polynomial in two variables x and y. Which one has the highest number of degrees of freedom?

Exercise 5.3 A trilinear function of three variables x, y, z is linear in each variable when the other two are held constant. Write down the most general trilinear polynomial. What is the number of degrees of freedom?

Exercise 5.4 Write a program for evaluation of a polynomial by Horner's scheme, and verify that the same results are obtained by adding the terms according to the standard formulation.

5.1.2 Piecewise Polynomials

The difficulty encountered with the standard high degree polynomials and their almost linearly dependence has to do with their global character. The only way to increase the number of coefficients is to raise the degree, and it changes the function everywhere. One way to overcome the resulting ill-conditioning is to localize the polynomials and make them *piecewise polynomials*. Linear interpolation is the most common application of such functions. With given function values u_j at the points x_j, we get the interpolation function $v(x)$ as straight lines between each pair of neighboring points. The function $u(x)$ is approximated by a *piecewise linear* function.

The use of piecewise polynomials got an enormous boost with the introduction of finite element methods, see Chap. 11. Approximations of solutions to differential equations based on such functions are not only accurate, but they can also be computed in a very effective way.

Piecewise polynomials $v(x)$ can be written as a combination of basis functions $\phi_j(x)$

$$v(x) = \sum_{j=1}^{N} c_j \phi_j(x), \tag{5.6}$$

where $\phi_j(x)$ are now different from the standard polynomial case. But how should they be chosen? As an example we choose the interval $[0 \ 1]$, and allow for a nonuniform distribution of the grid points x_j (also called nodes) such that $x_{j+1} - x_j = \Delta x_j, \ j = 0, 1, \ldots, N - 1$ and $x_0 = 0, \ x_N = 1$. For convenience we assume that the function we want to interpolate vanishes at $x = 0$. With each node x_j we associate a function $\phi_j(x)$ that is defined by

$$\phi_j(x) = \begin{cases} (x - x_{j-1})/\Delta x_{j-1}, & x_{j-1} \le x < x_j, \\ (x_{j+1} - x)/\Delta x_j, & x_j \le x < x_{j+1}, \\ 0 & \text{otherwise.} \end{cases}$$

(Here we have introduced an extra node $x_{N+1} = 1 + \Delta x_{N-1}$.) These N basis functions are often called the *roof functions* or *hat functions*, and they are shown in Fig. 5.3. At each node x_j there is only one nonzero basis function $\phi_j(x)$. Therefore, the coefficients in the expansion (5.6) are easily obtained as $c_j = u_j$.

Fig. 5.3 Piecewise linear
basis functions

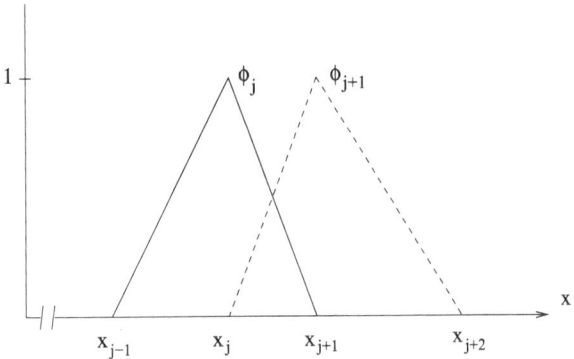

The basis functions and the form (5.6) may seem like an unnecessarily complicated way of describing the very simple procedure of drawing a straight line between given points (x_j, u_j). In each subinterval $[x_j\ x_{j+1}]$ the function value is simply obtained as

$$v(x) = u_j + \frac{x - x_j}{x_{j+1} - x_j}(u_{j+1} - u_j). \tag{5.7}$$

However, the introduction of the form (5.6) is for theoretical purposes. It allows us to generalize the analysis to higher degree polynomials as well as other types of interpolating functions.

Next we turn to piecewise quadratic polynomials, i.e., to functions that are second degree polynomials in each interval $I_j = [x_{j-1}, x_j]$, and continuous across nodes. Three values are required to determine the parabola in each subinterval, and we choose the midpoint as an extra node $x_{j-1/2}$. Since there is now an extra parameter, we need two basis functions $\phi_{j-1/2}$ and ϕ_j in each subinterval. The requirements are specified as follows:

$$\phi_j(x) = \begin{cases} 1 & \text{at } x = x_j \\ 0 & \text{at all other nodes} \\ \text{quadratic polynomial} & \text{in every } I_j \end{cases},$$

$$\phi_{j-1/2}(x) = \begin{cases} 1 & \text{at } x = x_{j-1/2} \\ 0 & \text{at all other nodes} \\ \text{quadratic polynomial} & \text{in every } I_j. \end{cases}$$

These functions are shown in Fig. 5.4.

There is a difference between the original nodes x_j and the extra nodes $x_{j-1/2}$. Any function

$$v = \sum_j (c_{j-1/2}\phi_{j-1/2} + c_j\phi_j)$$

has continuous derivatives of any order across $x = x_{j-1/2}$, but in general the derivative is discontinuous across $x = x_j$. Note that the basis function $\phi_{j-1/2}$ in the figure

Fig. 5.4 Piecewise quadratic basis functions

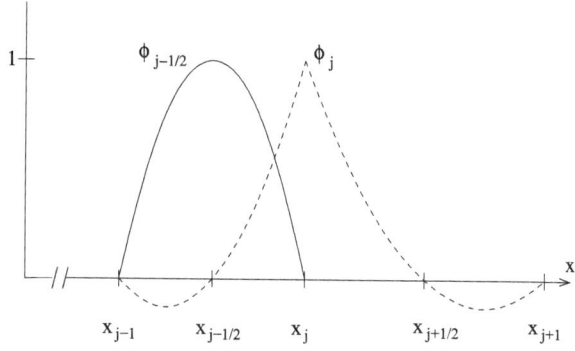

cannot be substituted by ϕ_j shifted a half-step leftwards. That would introduce a discontinuity in the derivative at $x = x_{j-1/2}$, and the function is no longer a quadratic polynomial in the whole interval I_{j-1} as required.

As an example, let us try interpolation for the function $e^{-50(x-0.5)^2}$ on the interval $0 \le x \le 1$ with n uniformly distributed given points. We know already that standard high order polynomials don't work well. Figure 5.5 shows the result for $n = 8$ and $n = 15$. Obviously, the result is bad, already for $n = 8$. With more points and higher degree polynomials, things become even worse in the sense that the maximal deviation increases.

When going to piecewise polynomials, the situation improves dramatically. Figure 5.6 shows the result for piecewise linear and piecewise quadratic functions and $n = 15$. The result is quite good considering the low degree polynomials.

When going to piecewise quadratic polynomials, the result can hardly be distinguished from the true solution, as shown in the right figure. Note that there are only a total of 15 points (x_j, u_j) given here as well. This means that the basis functions as shown in Fig. 5.4 are made broader such that the extra node with half-index is located at points x_j with integer index. The break points of the polynomials are located at $x_2, x_4, \ldots, x_{12}$, and there are only 7 different quadratic polynomials in total, each one being nonzero over two subintervals.

So far, the interpolation has been done to match the given function values u_j, which guarantees continuity. But the derivatives are usually discontinuous at the nodes. If the derivatives u'_j are available as given data as well, these can also be used when defining the interpolation polynomials. One of the most well known types is the *Hermite polynomials*. They have third degree in each interval with the four coefficients determined by u_{j-1} and u'_{j-1} at the left end point and u_j and u'_j at the right end point. As a consequence, not only is the interpolating function continuous everywhere, but also its derivatives. The effect of this is that the resulting function is smoother everywhere.

In the previous section we discussed polynomials in several dimensions and, as shown there, the generalization from 1D is quite straightforward. For piecewise polynomials an additional problem arises: how should we generalize the subinter-

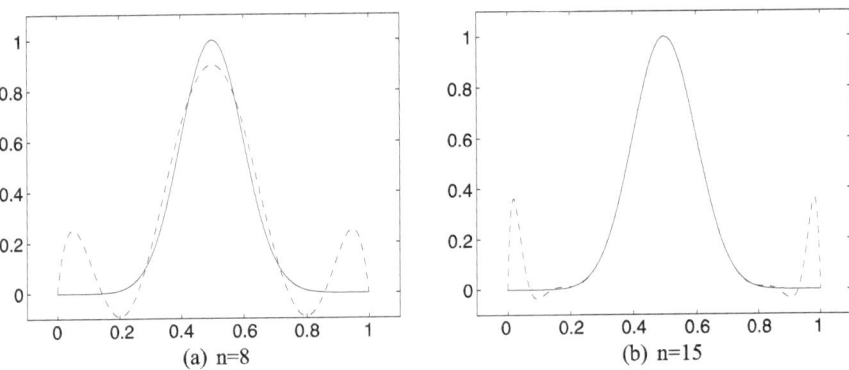

Fig. 5.5 Interpolation by standard polynomials (– –), true solution (—)

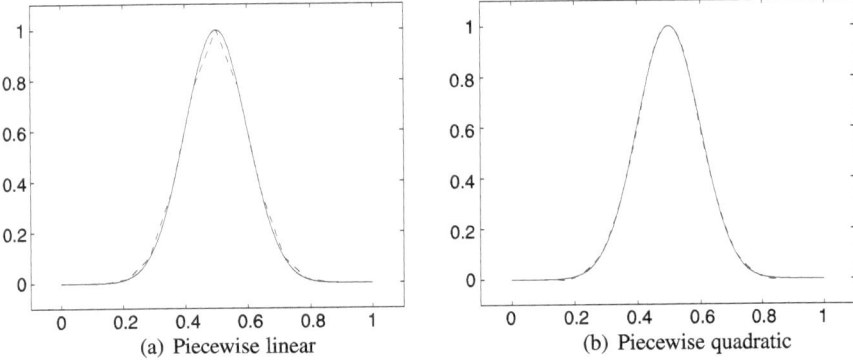

Fig. 5.6 Interpolation by piecewise polynomials, $n = 15$ (– –), true solution (—)

vals $[x_j, x_{j+1}]$? At a first glance, it seems that rectangles is the natural choice in 2D, with nodes at the corners. However, a first degree polynomial

$$a_{10}x + a_{01}y + a_{00}$$

has only 3 coefficients, i.e., we need 3 points to determine the interpolating polynomial. Consequently, there is an extra node in the rectangle that is not needed. This difficulty is avoided if we choose triangles instead with the three nodes at the corners. Geometrically we can think of the polynomial as a plane in 3D which is always uniquely determined by three points. Or think of table with three legs. It always stays steady, even on an uneven floor, while a four leg table does not.

But we shouldn't discount the rectangles as subdomains. If we add a fourth term to the linear polynomial we get a bilinear polynomial

$$a_{11}xy + a_{10}x + a_{01}y + a_{00},$$

which has 4 coefficients, and it can be used very naturally with rectangles. Figure 5.7 shows the two cases.

Fig. 5.7 Linear and bilinear functions

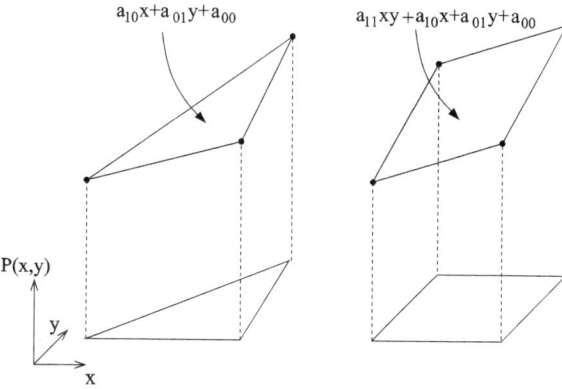

$a_{10}x + a_{01}y + a_{00}$ $a_{11}xy + a_{10}x + a_{01}y + a_{00}$

$P(x,y)$

Fig. 5.8 Nodes for quadratic and cubic polynomials

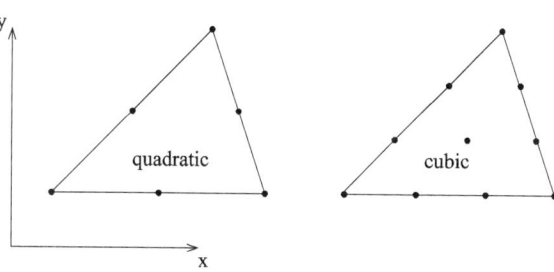

quadratic   cubic

Fig. 5.9 Two triangles with different polynomials

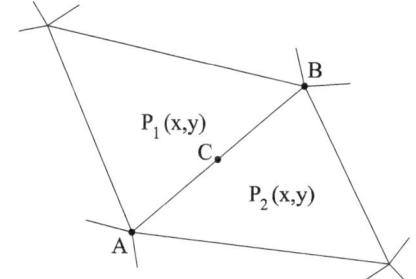

$P_1(x,y)$

$P_2(x,y)$

When going to second degree polynomials (5.4), triangles are again the natural choice. There are now 6 coefficients to be determined and so, if a node is added in the middle of each edge, we have the right number. For cubic polynomials (5.5), there are 10 coefficients. By adding two nodes at each edge, we get 9 nodes altogether, and we need one more. This one is chosen as the center of gravity of the triangle, see Fig. 5.8.

In 1D the interpolating polynomials are continuous across the nodes, since we are requiring that the polynomials on each side equal the given value. In 2D the continuity doesn't follow immediately. Two polynomials have the same value at a common node, but do they agree along the whole edge? Indeed they do.

Consider the two triangles in Fig. 5.9 with polynomials $P_1(x, y)$ and $P_2(x, y)$ respectively.

Fig. 5.10 Nodes for Hermite
polynomials

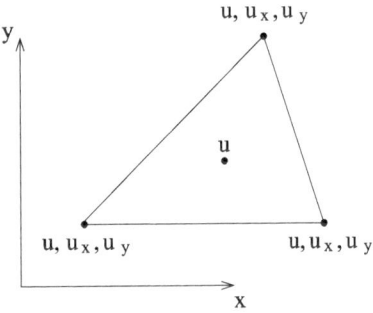

We assume first that the polynomials are linear:

$$P_1(x, y) = a_1 x + a_2 y + a_3,$$
$$P_2(x, y) = b_1 x + b_2 y + b_3.$$

The common edge AB is described by a linear function $y = \alpha x + \beta$, and therefore we can write the two polynomials at the edge as

$$P_1(x, y) = a_1 x + a_2(\alpha x + \beta) + a_3 = c_1 x + c_2,$$
$$c_1 = a_1 + \alpha a_2, \ c_2 = \beta a_2 + a_3,$$
$$P_2(x, y) = b_1 x + b_2(\alpha x + \beta) + b_3 = d_1 x + d_2,$$
$$d_1 = b_1 + \alpha b_2, \ d_2 = \beta b_2 + b_3.$$

Since they are both linear functions of the variable x, they are uniquely determined by their common values at the end points A and B. Hence, they are identical on the whole edge AB.

Turning to quadratic polynomials, we use the same argument. The linear form of the common edge is substituted into the two second degree polynomials resulting in two quadratic polynomials in the single variable x. They are uniquely determined by the three common nodes A, B, C, and we can draw the same conclusion as for the linear case: they are identical along the whole edge AB. The same property can be shown for cubic polynomials with 4 nodes along each triangle edge.

The cubic Hermite polynomials introduced above in 1D can be generalized to several dimensions as well. In 2D, there are 10 coefficients to be determined. By prescribing also the two partial derivatives at the corners, we have 9 values altogether, and we need one more. This one is chosen as the function value at the center of gravity, see Fig. 5.10. (The derivatives $\partial u / \partial x$ and $\partial u / \partial y$ are denoted by u_x and u_y.)

In 3D there are even more coefficients for each type of polynomial. Here we just note that a linear function $ax + by + cz + d$ has four coefficients, and a tetrahedron with the nodes in the corners is the right type of subdomain.

Exercise 5.5 Derive the analytical form of the piecewise quadratic basis functions in Fig. 5.4.

Exercise 5.6 Let $u_{1/2}, u_1, u_{3/2}, \ldots, u_N, u_{N+1/2}$ be $2N + 1$ function values given at the points $x_{1/2}, x_1, x_{3/2}, \ldots, x_N, x_{N+1/2}$. What are the coefficients c_j and $c_{j+1/2}$ in the interpolating function

$$v(x) = \sum_{j=1}^{N} c_j \phi_j(x) + \sum_{j=0}^{N} c_{j+1/2} \phi_{j+1/2}(x)?$$

Exercise 5.7 Write a program for piecewise linear interpolation by using (5.7) and for piecewise quadratic interpolation by using the coefficients derived in Exercise 5.6. Compare the results for the same number of given data points.

Exercise 5.8 Assume that two cubic polynomials are identical at four points on a common triangle edge. Prove that they are identical along the whole side.

5.1.3 Orthogonal Polynomials

We have seen in the previous section that higher degree polynomials become more and more linearly dependent in the sense that the scalar product becomes closer to one (after normalization such that $\|P_j\| = 1$).

Let us now see how we can construct an orthogonal polynomial basis. Consider the linear polynomial $x + 1$ on the interval $[0, 1]$. With the basis functions 1 and x, we get the scalar product

$$(1, x) = \int_0^1 1 \cdot x \, dx = \frac{1}{2}, \tag{5.8}$$

showing that they are not orthogonal. Let us now write the polynomial in the different form

$$\frac{3}{2} + \frac{1}{2}(2x - 1) \equiv \frac{3}{2} P_0(x) + \frac{1}{2} P_1(x),$$

where

$$P_0(x) = 1, \qquad P_1(x) = 2x - 1.$$

We can see this as a switch from the basis functions 1 and x to the new ones 1 and $2x - 1$. The orthogonality condition is now fulfilled:

$$(P_0, P_1) = \int_0^1 (2x - 1) \, dx = 0.$$

We have *orthogonalized* the polynomial basis. This procedure can be continued to any degree of the polynomials. The general class of orthogonal polynomials is obtained by requiring that each polynomial is orthogonal to all others except itself:

$$\int_0^1 P_m(x) P_n(x) \, dx = 0 \quad \text{if } m \neq n.$$

By using the Kronecker δ-function (see Sect. 3.1), the orthogonality condition can be written in the form

$$\int_0^1 P_m(x)P_n(x)\,dx = c_n\delta_{mn},$$

where c_n is a nonzero parameter that depends on n. This leads to a variant of the Legendre polynomials that will be treated below.

By introducing a weight function in the integrals, and also changing the x-interval, the orthogonality property can be defined in a more general way:

$$(f,g) \equiv \int_a^b f(x)g(x)w(x)\,dx = 0.$$

Here $w(x)$ is a positive function that is chosen to put more weight in certain parts of the integration interval. The left end point a can be $-\infty$ as well, and b can be ∞.

There are two types of orthogonal polynomials that are of particular interest when solving partial differential equations. These are the Chebyshev and the Legendre polynomials, and we shall describe these in Chap. 7.

5.2 *Trigonometric Functions

In most areas of science, engineering and daily life, one is using degrees for measuring angles, with 360 degrees corresponding to a full circle. However, in mathematics it is convenient to use another unit for measuring angles, and we shall begin this section by discussing that. Figure 5.11 shows a circle with radius one. Corresponding to the angle θ degrees, there is a piece x of the circle, and the length of x is easy to compute. We know that the circle has a full circumference 2π, and obviously we have $x = 2\pi\theta/360$. This unit of measuring angles is called *radian*, and for the full circle we have $0 \le x < 2\pi$. Actually, we can extend the definition such that x can take any value. The second rotation around the circle corresponds to $2\pi \le x \le 4\pi$, and so on. Negative values of x correspond to rotation in the other direction, i.e., clockwise. Consider now the triangle shown in the right part of Fig. 5.11 that corresponds to the angle x in the circle. The longest side in this triangle has length one, and the lower right corner has the angle 90 degrees, i.e., $\pi/2$ radians. If x is the angle of the lower left corner, the length of the vertical and horizontal sides are denoted by $\sin x$ and $\cos x$ respectively. These are the basic trigonometric functions, and they are *periodic*. This means that if we add any multiple $2n\pi$ to a certain angle x, where n is an integer, we get the same value back such that $\sin(x + 2n\pi) = \sin x$, and similarly for cos. Furthermore, the trigonometric functions are well defined for negative angles. For example, the radial line corresponding to the angle x in the circle can as well be defined by the angle $-(2\pi - x) = x - 2\pi$. This agrees with the periodicity property, which tells us that $\sin(x - 2\pi) = \sin x$.

There are many formulas that are convenient to use when dealing with trigonometric functions. Here we just note that by the famous Pythagoras' Theorem, we

Fig. 5.11 *sin* and *cos* defined by triangles, *x* measured in radians

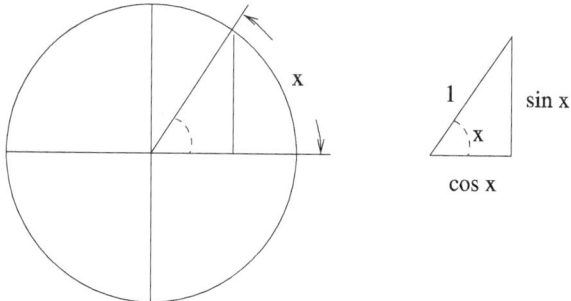

have $(\sin x)^2 + (\cos x)^2 = 1$ for all angles x, showing that $|\sin x| \leq 1$ and $|\cos x| \leq 1$. Furthermore, we have the remarkably simple differentiation rules

$$\frac{d}{dx}\sin(ax) = a\cos(ax),$$

$$\frac{d}{dx}\cos(ax) = -a\sin(ax).$$

It follows from this that the functions $\sin(ax)$ and $\cos(ax)$ both satisfy the differential equation

$$\frac{d^2u}{dx^2} + a^2u = 0.$$

One can define the trigonometric functions in other ways, one of them being by infinite series

$$\sin x = x - \frac{x^3}{3!} + \frac{x^5}{5!} - \cdots,$$

$$\cos x = 1 - \frac{x^2}{2!} + \frac{x^4}{4!} - \cdots$$

(5.9)

that follows by Taylor series expansion. We saw in Sect. 4.1.2 that one has to be careful when dealing with infinite sums, but here one can show that they both converge for all values of x. At a first glance it may be surprising that convergence doesn't require a bound on $|x|$, since the powers x^j are enormous for large $|x|$ and j. However, the factorials $j!$ in the denominators grow even faster for increasing j, and convergence is a fact.

Trigonometric functions show up in many different situations that are not connected to geometric problems and triangles. If we think of x as time instead of an angle and change notation $x \rightarrow t$, the functions $\sin(at)$ and $\cos(at)$ are well defined for all times t. They are *oscillating functions*, i.e., they oscillate between 1 and -1 for increasing time. If the constant a is large, the oscillations are fast, and we have a high frequency function. In acoustics, the functions describe small pressure variations, and large values of a correspond to high tones. Figure 5.12 shows the function $\sin t + 0.2\sin(20t)$, i.e., there is a high frequency tone on top of a base tone, where the high frequency component has a small amplitude.

Fig. 5.12 The function $\sin t + 0.2 \sin(20t)$

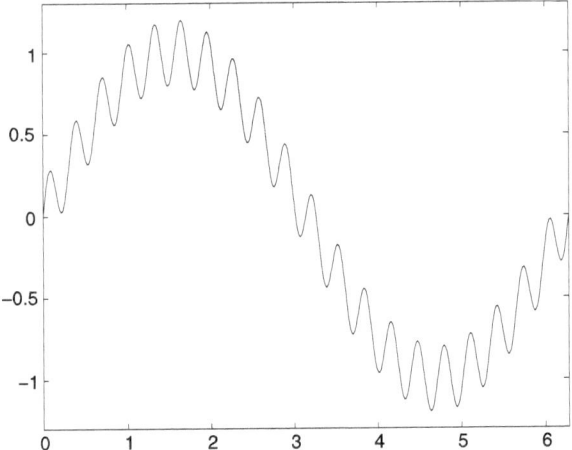

The third basic trigonometric function is

$$\tan x = \frac{\sin x}{\cos x}, \quad x \neq \pm \left(\frac{\pi}{2} + 2\pi n \right), \ n = 0, 1, \ldots.$$

The geometric interpretation is the quotient between the vertical and horizontal triangle sides in Fig. 5.11.

The trigonometric functions have their corresponding inverse functions. For the sine function one may ask the question: For a given value x between -1 and 1, what is the angle θ in radians such that $\sin \theta = x$? There is not a unique answer to this question, since we can add any multiple of 2π to θ and get the same value x back. Furthermore, $\sin \theta$ takes on each value twice during a full 2π rotation. To obtain uniqueness, we make the choice $-\pi/2 \leq \theta \leq \pi/2$, and call the function $\arcsin x$. It is natural to call it the inverse of the sine function, and therefore one sometimes uses the notation $\sin^{-1} x$. In words, the definition can be expressed as: "$\arcsin x$ is the angle θ in radians between $-\pi/2$ and $\pi/2$ such that $\sin \theta = x$".

The other trigonometric functions have their inverses as well, and we have

$$\theta = \arcsin x = \sin^{-1} x, \quad -1 \leq x \leq 1, \quad -\frac{\pi}{2} \leq \theta \leq \frac{\pi}{2},$$

$$\theta = \arccos x = \cos^{-1} x, \quad -1 \leq x \leq 1, \ 0 \leq \theta \leq \pi,$$

$$\theta = \arctan x = \tan^{-1} x, \quad -\infty < x < \infty, \quad -\frac{\pi}{2} < \theta < \frac{\pi}{2}.$$

The first two are shown in Fig. 5.13.

There are many trigonometric formulas that are handy to use in calculus. A few of them are listed in Appendix A.3.

Exercise 5.9 Use the Pythagoras' Theorem to prove $\sin^2 x + \cos^2 x = 1$.

Exercise 5.10 Use the geometric definition in Fig. 5.11 of the trigonometric functions to prove

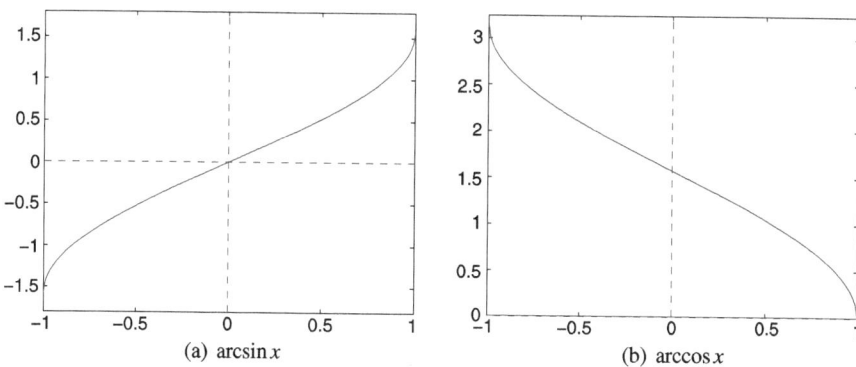

Fig. 5.13 The arcsin- and arccos-functions

(a) $\sin(-x) = -\sin x$
(b) $\cos(-x) = \cos x$
(c) $\sin(\pi - x) = \sin x$
(d) $\cos(\pi - x) = -\cos x$

Exercise 5.11 The series expansions of $\sin x$ and $\cos x$ can be used to derive the identity $\sin 2x = 2 \sin x \cos x$. Check that the two expansions agree up to order x^3.

5.3 **The Exponential Function

A very special function is obtained if we require that the rate of change is proportional to the function itself. If the proportionality constant is one, this leads to the differential equation

$$\frac{du}{dx} = u.$$

One way of finding the solution is to use the Taylor expansion discussed in Sect. 4.2. We expand around the point $x = 0$, and by using the differential equation we get

$$\frac{du}{dx} = u, \qquad \frac{d^2u}{dx^2} = \frac{du}{dx} = u, \qquad \frac{du^3}{dx^3} = \frac{d^2u}{dx^2} = u, \qquad \ldots,$$

i.e., all derivatives equal the function itself. With $u(0) = 1$, the Taylor series becomes

$$u(x) = 1 + x + \frac{x^2}{2} + \frac{x^3}{2 \cdot 3} + \cdots,$$

One can show that the series converges for all x, and it is an entire function. We call it the *exponential function* defined by

$$e^x = \sum_{n=0}^{\infty} \frac{x^n}{n!}. \tag{5.10}$$

(By definition we have $x^0 = 1$ and $0! = 1$.) The infinite series converges rapidly in the sense that, for reasonable values of x, a few terms give an almost correct result. This may seem even more surprising than in the case of trigonometric functions, since all terms have the same sign for $x > 0$. The explanation is again the enormous growth of the denominators $n!$ for increasing n.

The constant e is a number with remarkable properties. It is a so called irrational number $e = 2.71828\ldots$ with infinitely many decimals, and it emerges just about everywhere in mathematics. By using the chain rule for differentiation (see Appendix A.1), we get

$$\frac{d}{dx}e^{cx} = ce^{cx}$$

for any constant c.

The exponential function grows very quickly. For example, $e^5 \approx 148$, and $e^{10} \approx 22026$. The concept *exponentially growing* is often used in many different connections, even outside science. It is understood as a function b^x for any positive number $b > 1$. Since we can always find a number c such that $e^c = b$, we get back to the exponential function by $b^x = (e^c)^x = e^{cx}$, where c is the growth coefficient. If c is positive, the exponential function grows for increasing x; if c is negative, it decreases.

As another example of the rapid exponential growth, we take the population growth in the World. In rough numbers, the World population has grown from 6.7 bill. in 2008 to 6.8 bill. in 2009, which is a 1.5% increase. This yearly growth may not look very strong. However, let us make the somewhat unlikely assumption that the population p will grow at the same rate during the coming century. In 2010 there will be $6.8 \cdot 1.015$ bill. people, in 2011 $6.8 \cdot 1.015 \cdot 1.015$ bill. etc. The general evolution is described by the formula

$$p(t) = 6.8 \cdot 1.015^t,$$

where t is the elapsed time after 2009. At the end of this century there will be $p(90) = 26.0$ bill. people! In many countries there are politicians who argue that we need a stronger population growth in order to take care of the increasing number of elderly people. If we assume that this idea will take hold in the whole world such that the yearly growth is doubled to 3%, then the population at the end of the century will be $p(90) = 6.8 \cdot 1.03^{90} = 97.2$ bill.

If $x = e^a$, the exponent a is called the *natural logarithm* of x, and is denoted by $\ln x$ for any positive number x. (Negative numbers e^a cannot be achieved for any real number a.) It is the number that should be put in the exponent in order to recover the original number x. Obviously it satisfies

$$x = e^{\ln x}.$$

The natural algorithm is a special case of the general logarithm which is defined for any positive *base b*, and we write it as $\log_b x$. It satisfies the relation

$$x = b^{\log_b x}.$$

The choice $b = 10$ is the base used in daily life giving rise to the decimal system, and we have for example $\log_{10} 1000000 = 6$. When it comes to digital computers, the base is chosen as $b = 2$, and we have for example $\log_2 64 = 6$.

Sometimes one talks about *logarithmic growth*, which means a function $f(x) = c \log_b x$, where b is a constant. In contrast to exponential growth, it grows much more slowly than x. With the base $b = 10$ and $c = 1$, a million-fold increase in x from 10 to 10 million corresponds to a 6-fold increase on the logarithmic scale. An example is the Richter scale used for classifying earthquakes. It is the 10-logarithm of the local magnitude of the displacement at the center, which means that the seemingly modest increase from 7 to 8 on the Richter scale corresponds to a ten-fold increase in magnitude. Actually, the increase is even worse. The more relevant measure of an earthquake is the released energy, and it is proportional to the power 1.5 of the magnitude. Consequently, we should compare $(10^7)^{1.5}$ to $(10^8)^{1.5}$, which is a factor 32.

The exponential function can be generalized to complex arguments x. This sounds like an unnecessary generalization if we have numerical solution of differential equations with real solutions in mind. But we shall see that on the contrary, it becomes a central part of Fourier analysis, which is a powerful tool for analysis of numerical methods.

We consider first the function $e^{i\theta}$, where θ is a real number. By formally substituting $i\theta$ for θ in the Taylor expansion, we get

$$e^{i\theta} = 1 - \frac{\theta^2}{2!} + \frac{\theta^4}{4!} - \cdots + i\left(\theta - \frac{\theta^3}{3!} + \frac{\theta^5}{5!} + \cdots\right).$$

What a surprise! By comparing to (5.9) with θ substituted by x, the two series are identified as the trigonometric functions $\cos\theta$ and $\sin\theta$, and we know that they exist for all θ. The relation is

$$e^{i\theta} = \cos\theta + i\sin\theta$$

with the angle θ measured in radians. By using basic trigonometric formulas, we also have

$$e^{-i\theta} = \cos(-\theta) + i\sin(-\theta) = \cos\theta - i\sin\theta.$$

By combining the two equations, we get explicit formulas for the trigonometric functions in terms of the exponential function:

$$\cos\theta = \frac{e^{i\theta} + e^{-i\theta}}{2}, \qquad \sin\theta = \frac{e^{i\theta} - e^{-i\theta}}{2i}.$$

The absolute value is by definition

$$|e^{i\theta}| = \sqrt{(\cos\theta)^2 + (\sin\theta)^2} = 1,$$

i.e., when θ goes from 0 to 2π, the unit circle is described.

If $z = x + i\theta$ is a general complex number, e^z is well defined through its Taylor expansion. A convenient representation is obtained by

$$e^z = e^{x+i\theta} = e^x e^{i\theta} = r(\cos\theta + i\sin\theta),$$

where $r = e^x$. When θ goes from 0 to 2π, a circle with radius r and centered at the origin is described.

A particularly beautiful example of mathematics is obtained for the special case $\theta = \pi$, for which

$$e^{\pi i} = -1.$$

It seems like magic that the irrational numbers e and π and the imaginary number i can be combined in such a simple formula. But once we have accepted the Taylor series expansion and the identity $e^{i\theta} = \cos\theta + i\sin\theta$ which follows from it, the relation is immediately obtained.

We shall see later that numerical methods often use trigonometric functions for approximation of certain functions. Also here, it seems that the introduction of complex numbers is a complication, when actually dealing with real functions. However, it turns out that it often leads to a simplification.

In Sect. 3.2 we introduced complex numbers, and here we have now also introduced a complex *function* of a complex variable. This is by no means a simple matter, and leads to the theory of *analytic functions*. This is not the place to go deeper into this theory, but we shall still use the exponential function e^z, where z is a complex variable. The derivative of this function exists to any order, and just as for real variables we have

$$\frac{d^p(e^z)}{dz^p} = e^z, \quad p = 0, 1, 2, \ldots.$$

The differentiation rules for all elementary functions apply also for complex arguments. When it comes to differential equations, the introduction of the exponential function and the very simple rule for differentiation

$$\frac{d}{dz}e^{iz} = ie^{iz} \tag{5.11}$$

simplifies the analysis. Even if the solution of the differential equation is actually real, the analysis is done in complex space, and the imaginary part is simply disregarded in the end.

The exponential function with complex arguments is indeed very useful when representing complex numbers in general, and elementary functions of them. Let us first discuss the square-root of a complex number z. A quadratic equation

$$z^2 + 2az + b = 0$$

with complex coefficients a and b has the solutions

$$z_{1,2} = -a \pm \sqrt{a^2 - b},$$

which is formally identical to the real case. But what do we mean by the square-root? It is convenient to write a complex number as

$$z = re^{i\theta}$$

as demonstrated above. The angle θ is not uniquely defined. Clearly there is no need to use more than one period of length 2π, since it covers the whole 360 degree

Fig. 5.14 Multiplication and
division of complex numbers

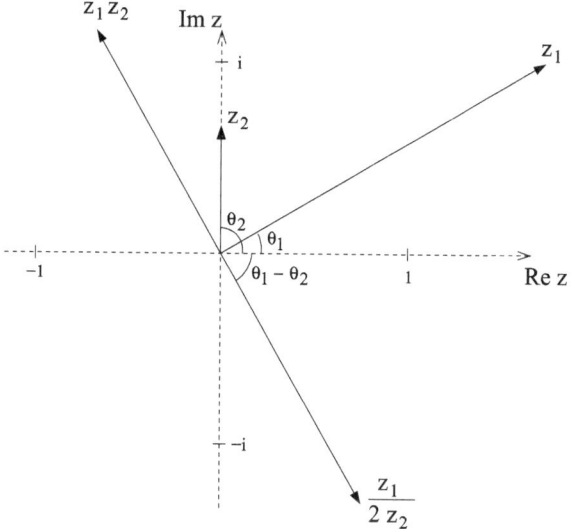

circle. But still there is a choice. We could for example use $0 \leq \theta < 2\pi$ as well as $-\pi < \theta \leq \pi$. We choose the latter one and define

$$\sqrt{z} = \sqrt{r}(e^{i\theta})^{1/2} = \sqrt{r}e^{i\theta/2},$$

where we know that $-\pi/2 < \theta/2 \leq \pi/2$. This defines the *principle branch* of the square-root, which is consistent with $\sqrt{1} = 1$ (not with $\sqrt{1} = -1$). We say that there is a cut along the negative real axis.

Multiplication and division can also be conveniently handled by the exponential representation. With

$$z_1 = r_1 e^{i\theta_1}, \quad -\pi < \theta_1 \leq \pi,$$
$$z_2 = r_2 e^{i\theta_2}, \quad -\pi < \theta_2 \leq \pi,$$

we have

$$z_1 z_2 = r_1 r_2 e^{i(\theta_1 + \theta_2)},$$
$$\frac{z_1}{z_2} = \frac{r_1}{r_2} e^{i(\theta_1 - \theta_2)}.$$

Figure 5.14 illustrates these rules in the complex plane, where the numbers are represented by vectors in the complex plane.

Figure 5.15 shows the square and the square-root of two numbers z_1 and z_2, where $z_2 = \overline{z_1}$.

Finally we solve the equation

$$z^N = 1,$$

where N is a positive integer. According to the general rule, there are N roots corresponding to N different representations in the form $1^{1/N}$. Here we are not looking

Fig. 5.15 The square and
square-root of complex
numbers

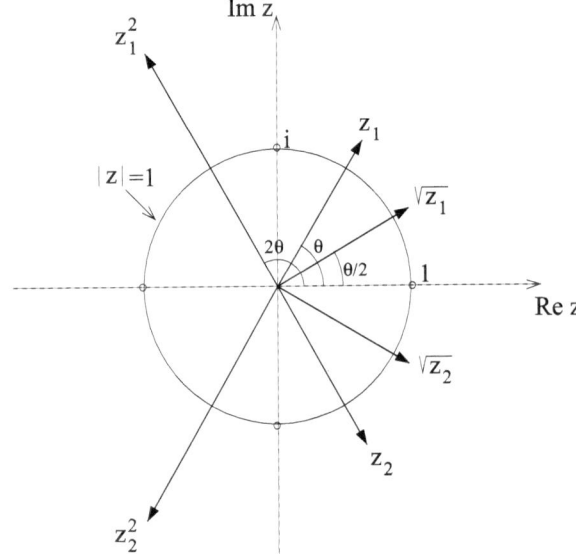

for a unique solution, but rather for all possible different solutions. By using the N
equivalent exponential representations

$$1 = e^{i2\pi n}, \quad n = 0, 1, \ldots, N - 1,$$

we obtain the solutions

$$z_{n+1} = e^{i2\pi n/N}, \quad n = 0, 1, \ldots, N - 1. \tag{5.12}$$

These are called the *roots of unity*, and are evenly spread around the unit circle as
shown in Fig. 5.16. Note that any attempt to introduce another root by letting $n = N$
will fail (as it should do), since $z_{N+1} = e^{i2\pi N/N} = 1$ coincides with z_1.

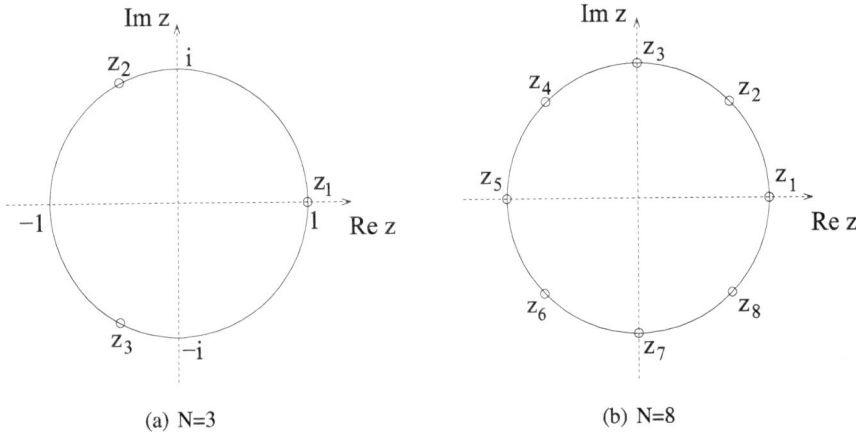

Fig. 5.16 The unit roots

Exercise 5.12 Use a computer program to demonstrate that for any fixed integer p and constant a one can always find a number x_0 such that $e^x > ax^p$ for $x > x_0$.

Exercise 5.13 Use a computer program to demonstrate that for any fixed negative integer p one can always find a number x_0 such that $e^{-x} < x^p$ for $x > x_0$.

Exercise 5.14 Prove that $\sqrt{2}\,\mathrm{Re}(e^{i\pi/4}) = 1$ by using the geometric representation of complex numbers.

Exercise 5.15 Find all roots of the equation $z^N - 1 = 0$ for $N = 2, 4, 8$ by using the geometric representation.

Part II
Fundamentals in Numerical Analysis

Chapter 6
The Fourier Transform

One of the main tools in mathematical and numerical analysis is the use of various types of *transforms*. A function $f(x)$ is normally represented through its value at any given point x. But there are other ways of representing it. It is often possible to express the function in terms of certain basis functions $\phi_k(x)$, such that

$$f(x) = \sum_{k=0}^{\infty} c_k \phi_k(x). \tag{6.1}$$

With all basis functions $\phi_k(x)$ known, the coefficients c_k characterize the function completely. Sometimes it is more natural to represent the function as an integral

$$f(x) = \int_a^b c(\xi)\phi(\xi, x)\, d\xi, \tag{6.2}$$

where a may be $-\infty$ and b may be ∞. Here the function $c(\xi)$ is the new representation, and it is a generalization of the sequence of numbers $\{c_k\}$ above. The operation that changes the representation from the original one to the set of coefficients c_k or the function $c(\xi)$ is called a *transform* of the function. For computation and approximation of functions, infinite series of the type (6.1) are made finite. It turns out that such finite series play a fundamental role in numerical analysis, and in particular for the solution of partial differential equations.

The most well known transform is the *Fourier transform* invented by the French mathematician Joseph Fourier (1768–1830). It exists in both versions above with either a series or an integral representation. Here we shall concentrate on the series version, since it is the most natural basis for numerical approximation. The basis functions are e^{ikx}, which are closely connected to the trigonometric functions $\cos kx$ and $\sin kx$ (see Sect. 5.3). The integer k is called the *wave number*. If x stands for time, k is called the *frequency*, and a high frequency means that the function is oscillating rapidly as a function of x.

The Fourier transform has an amazing potential when it comes to numerical solution methods. There are two fundamental but different applications for it. The first one is as a tool for the *analysis* of numerical methods. As we shall see, it makes it possible to find out how a certain numerical method behaves when it comes to

B. Gustafsson, *Fundamentals of Scientific Computing*,
Texts in Computational Science and Engineering 8,
DOI 10.1007/978-3-642-19495-5_6, © Springer-Verlag Berlin Heidelberg 2011

such fundamental properties as stability, accuracy and convergence. The transform
works almost like magic in the sense that complicated relations between coupled
state variables are transferred to much simpler relations in the transformed space,
where they can be easily analyzed before putting the algorithm on the computer.

The second use of the Fourier transform is as a direct tool in the actual *computation* of the approximate solution. It acts as the central part of the solution algorithm
for pseudospectral methods, as we shall see in Chap. 12. The success here is partly
due to the invention of the fast Fourier transform that was made in the 1960s.

We begin by discussing the classic Fourier series in the first section, and then the
discrete Fourier transform in the next section.

6.1 Fourier Series

Here we shall discuss the case with periodic functions $f(x)$, i.e., functions that
satisfy $f(x+2\pi) = f(x)$ for all x. Under certain conditions on $f(x)$ one can prove
that it can be represented as an infinite series

$$f(x) = \sum_{k=-\infty}^{\infty} \hat{f}(k)e^{ikx}. \tag{6.3}$$

The coefficients $\hat{f}(k)$ represent the amplitude for the wave number k, and they are
often called the *spectrum* of $f(x)$. For certain purposes, the spectrum tells a lot more
than a graph of the function itself.

Next we define the scalar product of two functions $f(x)$ and $g(x)$ as

$$(f, g) = \frac{1}{2\pi} \int_0^{2\pi} \overline{f}(x)g(x) \, dx. \tag{6.4}$$

Sometimes we use the notation $(f(\cdot), g(\cdot))$ to indicate that the functions f and g
have an argument. (The scalar product is a number, not a function, and therefore the
variable x is substituted by a dot.) Since

$$\frac{1}{2\pi} \int_0^{2\pi} e^{-ikx} e^{i\nu x} \, dx = \frac{1}{2\pi} \int_0^{2\pi} e^{i(\nu-k)x} \, dx = \begin{cases} 1 & \text{for } k = \nu, \\ 0 & \text{for } k \neq \nu, \end{cases}$$

the basis functions e^{ikx} are orthonormal. This is the best of all situations both for
vectors and for functions, see Chaps. 3 and 5. Not only is the sensitivity of the
coefficients $\hat{f}(k)$ to perturbations in data minimal, the coefficients are also easily
obtained for a given function $f(x)$. After multiplying by $e^{i\nu x}$ and integrating the
Fourier series above, all terms except one vanish, and we get

$$\hat{f}(k) = (e^{ik\cdot}, f(\cdot)) = \frac{1}{2\pi} \int_0^{2\pi} e^{-ikx} f(x) \, dx. \tag{6.5}$$

It may look like a complication to introduce complex basis functions even in the
case where $f(x)$ is real, but the computations are actually simpler in this way. To see
what happens when $f(x)$ is real, we choose the example $f(x) = \sin x$. By using well

known trigonometric formulas for the cos- and sin-functions (see Appendix A.3), we get

$$\hat{f}(k) = \frac{1}{2\pi} \int_0^{2\pi} e^{-ikx} \sin x \, dx = \frac{1}{2\pi} \int_0^{2\pi} (\cos kx - i \sin kx) \sin x \, dx$$

$$= \frac{1}{2\pi} \int_0^{2\pi} \frac{1}{2}\left(\sin((1+k)x) + \sin((1-k)x)\right) dx$$

$$+ \frac{i}{2\pi} \int_0^{2\pi} \frac{1}{2}\left(\cos((1+k)x) - \cos((1-k)x)\right) dx.$$

If $k \neq -1$ and $k \neq 1$, then the integrals are zero. For the exceptional values we have

$$\hat{f}(-1) = \frac{i}{2}, \qquad \hat{f}(1) = -\frac{i}{2},$$

which gives the series

$$f(x) = \hat{f}(-1)e^{-ix} + \hat{f}(1)e^{ix} = \frac{i}{2}(\cos x - i \sin x) - \frac{i}{2}(\cos x + i \sin x) = \sin x.$$

The coefficients are complex but such that the imaginary parts of the complex functions cancel each other in the sum.

The general rule for real functions $f(x)$ is obtained by observing that

$$\hat{f}(-k) = \frac{1}{2\pi} \int_0^{2\pi} e^{ikx} f(x) \, dx = \overline{\hat{f}(k)}.$$

Therefore, if we match each wave number with its negative counterpart, we get with $\hat{f}(k) = a + bi$

$$\hat{f}(k)e^{ikx} + \hat{f}(-k)e^{-ikx}$$

$$= (a + bi)(\cos kx + i \sin kx) + (a - bi)\left(\cos(-kx) + i \sin(-kx)\right)$$

$$= 2a \cos kx - 2b \sin kx.$$

Therefore we can write a real function in the form

$$f(x) = \sum_{k=0}^{\infty} c_k \cos kx + \sum_{k=1}^{\infty} s_k \sin kx,$$

where

$$c_k = 2 \operatorname{Re} \hat{f}(k), \quad k = 0, 1, \ldots,$$

$$s_k = -2 \operatorname{Im} \hat{f}(k), \quad k = 0, 1, \ldots.$$

Here $\hat{f}(k)$ are the complex Fourier transform coefficients.

The conclusion is that by going to complex arithmetic even for real functions, the algebra becomes simpler due to the nice properties of the exponential function, and in the end all imaginary parts cancel.

Several variants of Fourier series can be used for some special cases where the function is not periodic. The functions $\sin kx$ are 2π-periodic, but they can also be used as basis functions on the interval $[0, \pi]$ for approximation of functions that are zero at the boundaries. Such expansions are used for a few examples in Chap. 12, where the solution of differential equations by spectral methods is discussed.

For practical computations there can be only a finite number of terms in the series. For convenience we assume that N is an even number, and define the approximation

$$f(x) \approx f_N(x) = \sum_{k=-N/2}^{N/2} \hat{f}(k)e^{ikx}.$$

Higher wave numbers k correspond to faster oscillations, which allows for better representation of functions with variation on a small scale. Therefore we expect smaller errors with more terms in the approximating series, and the convergence properties as $N \to \infty$ become interesting. We would like to have a situation where the coefficients $\hat{f}(k)$ decrease rapidly for increasing $|N|$, such that $f_N(x)$ represents $f(x)$ very well for small N. The convergence theory for Fourier series is well developed, and as a general rule the series converges faster for smoother functions. This is hardly a surprise. A smooth function $f(x)$ has a small high frequency part, and the error will be small if the higher terms are left out.

In applications one is often dealing with discontinuous functions, and they can certainly not be classified as smooth. As an example we take the function

$$f(x) = \begin{cases} 1 & \text{for } 0 < x \leq \pi, \\ 0 & \text{for } \pi < x \leq 2\pi. \end{cases}$$

The integrals defining the Fourier coefficients are easy to compute, and we have

$$\hat{f}(k) = \begin{cases} \frac{1}{2} & \text{for } k = 0, \\ \frac{1}{|k|\pi} \sin(|k|x) & \text{for } k \text{ odd}, \\ 0 & \text{for } k \text{ even}. \end{cases}$$

Figure 6.1 shows $f_N(x)$ for $N = 40$ and $N = 160$.

More terms give a better approximation, but there are still evident oscillations around the discontinuities. (Note that the periodicity assumption causes a discontinuity at $x = 0$ and $x = 2\pi$ as well.) It can be proven that no matter how many terms are included in the Fourier series, there will always be an "overshoot" and an "undershoot" very close to the discontinuity. On the other hand, the approximation catches the location of the discontinuity very well, and for some analytical purposes this might be sufficient.

The oscillating behavior shown here is known as the *Gibbs phenomenon* after the American physicist Josiah Willard Gibbs (1839–1903). Actually, others discovered and analyzed this phenomenon at greater depth, but it was still associated with Gibbs' work. Much work has been devoted lately to numerical techniques for eliminating this unwanted effect. It has been fairly successful for one-dimensional problems, but in several dimensions the Gibbs phenomenon still causes severe difficulties.

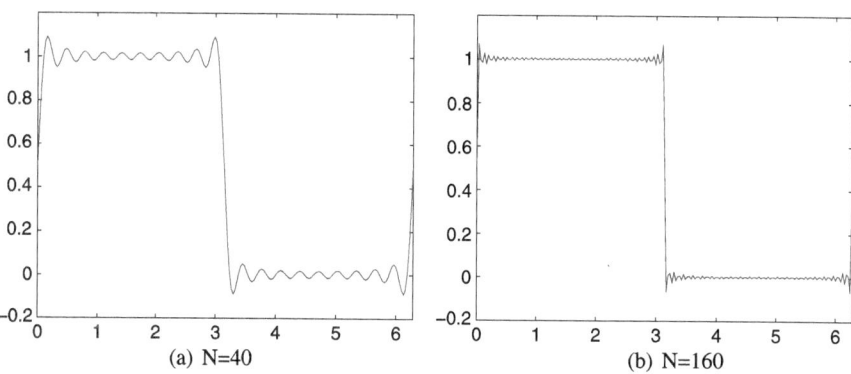

(a) N=40   (b) N=160

Fig. 6.1 Fourier series for a discontinuous function

In this presentation we have assumed that the functions are periodic. If instead we are dealing with functions that are defined on the whole real axis $-\infty < x < \infty$, the Fourier representation goes over from a series to an integral

$$f(x) = \int_{-\infty}^{\infty} \hat{f}(\xi) e^{i\xi x}\, d\xi,$$

where it is assumed that $f(x)$ is such that the integral exists. The function $\hat{f}(\xi)$ is obtained as a generalized integral as well:

$$\hat{f}(\xi) = \frac{1}{2\pi} \int_{-\infty}^{\infty} e^{-i\xi x} f(x)\, dx.$$

In computational mathematics it is not very natural to deal with infinite intervals, and therefore we shall mainly use the Fourier series representation in this book.

Fourier analysis is a powerful tool when it comes to differential equations, in particular for partial differential equations. The main reason for this is the remarkable property of differential operators acting on the exponential function. With a function $f(x)$ given as a Fourier series (6.3), we have

$$\frac{df}{dx} = \frac{d}{dx} \sum_{k=-\infty}^{\infty} \hat{f}(k) e^{ikx} = \sum_{k=-\infty}^{\infty} \hat{g}(k) e^{ikx} = g(x), \qquad \hat{g}(k) = ik\hat{f}(k).$$

(Here it is assumed that the differentiated sum converges, which is not always true.) The coefficients in Fourier space for g are obtained by a simple multiplication by ik of the original coefficients. This property becomes essential when analyzing PDE. For the equation

$$\frac{\partial u}{\partial t} = \lambda \frac{\partial u}{\partial x} \tag{6.6}$$

we write the solution as a Fourier series in space

$$u(x,t) = \sum_{k=-\infty}^{\infty} \hat{u}(k,t) e^{ikx},$$

and plug it into the differential equation

$$\sum_{k=-\infty}^{\infty} \frac{\partial \hat{u}}{\partial t}(k,t)e^{ikx} = \lambda \sum_{k=-\infty}^{\infty} ik\hat{u}(k,t)e^{ikx}.$$

Since the functions e^{ikx} are linearly independent, the coefficients in the two sums must be the same:

$$\frac{\partial \hat{u}}{\partial t}(k,t) = \lambda ik\hat{u}(k,t), \quad k = 0, \pm 1, \pm 2, \dots.$$

The PDE has been reduced to a series of scalar ODE, and this new form is called the *Fourier transform* of the partial differential equation (6.6). The ODE is trivially solved, and we have

$$\hat{u}(k,t) = e^{\lambda ikt}\hat{u}(k,0), \quad k = 0, \pm 1, \pm 2, \dots.$$

The scalar λ determines the growth of the Fourier coefficients with time. We have

$$|\hat{u}(k,t)| = |e^{\lambda ikt}||\hat{u}(k,0)| = e^{-\operatorname{Im}(\lambda)kt}|\hat{u}(k,0)|, \quad k = 0, \pm 1, \pm 2, \dots.$$

If $\operatorname{Im}(\lambda) \neq 0$, there is a growth for those coefficients $\hat{u}(k,t)$ corresponding to the wave numbers k with sign opposite to the one for $\operatorname{Im}(\lambda)$. Furthermore, since there is no bound on the wave numbers, the exponential growth is unbounded as well. Even if we are considering a fixed point in time, the relative growth $|\hat{u}(k,t)|/|\hat{u}(k,0)|$ becomes more and more severe for increasing $|k|$. This is typical for ill posed problems. For our simple equation we obviously need real numbers λ.

A more precise statement about the growth in physical space is obtained through *Parseval's relation* which connects the magnitude of the solutions in physical space and Fourier space. A convenient measure of the solution is the norm based on the scalar product defined in (6.4):

$$\|u(\cdot,t)\|^2 = \frac{1}{2\pi}\int_0^{2\pi} |u(x,t)|^2\,dx.$$

By using the orthogonality of the basis functions, it easy to show that

$$\|u(\cdot,t)\|^2 = \sum_{k=-\infty}^{\infty} |\hat{u}(k,t)|^2.$$

If the Fourier coefficients do not grow, then the norm of the solution doesn't grow:

$$\|u(\cdot,t)\|^2 = \sum_{k=-\infty}^{\infty} |\hat{u}(k,t)|^2 \leq \sum_{k=-\infty}^{\infty} |\hat{u}(k,0)|^2 = \|u(\cdot,0)\|^2.$$

The PDE treated here is very simple, but the Fourier technique is applicable to much more general equations. Since it transforms complicated differential operators to multiplication by scalars, it is a very powerful tool for analysis. We shall come back to this type of analysis in later chapters when discussing various applications.

Exercise 6.1 A series $\sum_{k=0}^{\infty} 1/k^p$ converges if and only if $p > 1$. Give a sufficient condition on the Fourier coefficients $\hat{f}(k)$ of a periodic function $f(x)$ such that $d^2 f/dx^2$ exists for all x.

6.2 The Discrete Fourier Transform

The truncation of the Fourier series to a finite number of terms could be a basis for numerical approximation. But the computation of the coefficients requires some sort of numerical method for computing the integrals. This means that only a finite number of function values $f(x_j)$ are used, and we can think of the truncated Fourier transform as an operation connecting a finite set of point values $f(x_j)$ to a finite set of Fourier coefficients. By choosing different integration formulas, we arrive at different discrete transforms.

When dealing with discrete functions, the scalar product defined by an integral should be substituted by another scalar product defined by a sum. It is essential to retain the orthogonality of the basis functions also with the new scalar product.

We shall derive the discrete version of the Fourier transform by using the interpolation property as the fundamental concept.

We are assuming that $f(x)$ is a 2π-periodic function and that its values $f_j = f(x_j)$ are known at the grid points $x_j = j\Delta x$, $j = 0, 1, \ldots, N$ where N is an even number and $(N+1)\Delta x = 2\pi$. (The assumption of even N is only for convenience, see Exercise 6.5.) From the known values we want to construct an interpolation function

$$g(x) = \sum_{k=-N/2}^{N/2} c_k e^{ikx}, \tag{6.7}$$

which agrees with $f(x)$ at the grid points. The interpolation conditions are

$$f_j = \sum_{k=-N/2}^{N/2} c_k e^{ikx_j}, \quad j = 0, 1, \ldots, N. \tag{6.8}$$

For the discrete representation we are dealing with sums instead of integrals. However, the beautiful generality of mathematics allows for the use of the same machinery as for the continuous Fourier transform. The discrete function values f_j can be considered as the components in a vector $\mathbf{f}$. We can then define a scalar product and a norm for vectors $\mathbf{f}$ and $\mathbf{g}$ by

$$(\mathbf{f}, \mathbf{g}) = \frac{1}{2\pi} \sum_{j=0}^{N} \overline{f}_j g_j \Delta x,$$

$$\|\mathbf{f}\|^2 = (\mathbf{f}, \mathbf{f}).$$

The $N+1$ basis vectors are

$$\mathbf{e}_k = \begin{bmatrix} e^{ikx_0} \\ e^{ikx_1} \\ \vdots \\ e^{ikx_N} \end{bmatrix}, \quad k = -N/2, -N/2+1, \ldots, N/2,$$

and we shall show that they are orthonormal. If $m \neq k$, we have by using the summation formula for geometric series (see Sect. 4.1.2)

$$
(\mathbf{e}_m, \mathbf{e}_k) = \frac{1}{2\pi} \sum_{j=0}^{N} e^{-imx_j} e^{ikx_j} \Delta x = \frac{\Delta x}{2\pi} \sum_{j=0}^{N} e^{i(k-m)j\Delta x}
$$

$$
= \frac{\Delta x}{2\pi} \frac{1 - e^{i(k-m)(N+1)\Delta x}}{1 - e^{i(k-m)\Delta x}} = \frac{\Delta x}{2\pi} \frac{1 - e^{i(k-m)2\pi}}{1 - e^{i(k-m)\Delta x}} = 0.
$$

If $m = k$, we get

$$
(\mathbf{e}_m, \mathbf{e}_k) = \frac{1}{2\pi}(N+1)\Delta x = 1,
$$

and the proof of orthonormality is complete.

The solution of the interpolation problem now follows easily. We simply multiply (6.8) by e^{-imx_j} and sum it over j, i.e., we take the scalar product of (6.8) with $\mathbf{e}_m$. The result is

$$
c_k = (\mathbf{e}_k, \mathbf{f}),
$$

or equivalently

$$
c_k = \frac{1}{2\pi} \sum_{j=0}^{N} f_j e^{-ikx_j} \Delta x, \quad k = -N/2, -N/2+1, \ldots, N/2. \tag{6.9}
$$

This is called the *Discrete Fourier Transform (DFT)* and (6.8) is its inverse. Because of the close relation to the trigonometric functions, the procedure (6.7), (6.9) is called *trigonometric interpolation*.

Obviously, the DFT is closely related to the Fourier transform from the previous section. In fact, the sums occurring in (6.9) are special approximations of the integrals in (6.5). We have simply picked the left function value in each interval $[x_j, x_{j+1}]$ and multiplied it by Δx to get the area of the corresponding integral, just as for the Riemann sum illustrated in Fig. 2.11.

The discrete version of the Fourier transform is used as a tool for analyzing difference methods for time dependent problems. Just as for the continuous transform for differential equations we need a connection between the norm of the grid function and the norm of its Fourier coefficients. It is given by the *discrete Parseval's relation*:

The coefficients c_k defined in (6.9) of the Fourier series (6.8) satisfy

$$
\sum_{k=-N/2}^{N/2} |c_k|^2 = \frac{1}{2\pi} \sum_{j=0}^{N} |f_j|^2 \Delta x. \tag{6.10}
$$

This relation follows easily by the orthogonality of the basis grid functions, and we shall demonstrate in Sect. 10.3 how to use it when estimating the solution of the difference scheme.

The condition that N should be an even number was introduced only for the purpose of getting a nice symmetric formula for the Fourier series, and it can be removed. In fact the Fourier series can be defined by any sum of the form

$$g(x) = \sum_{k=M}^{M+N} c_k e^{ikx},$$

if $N+1$ interpolation points are given, but the common choice is $M = 0$. However, we cannot use more than $N+1$ basis functions, since there are at most $N+1$ linearly independent vectors $\mathbf{e}_k$ containing the $N+1$ elements e^{ikx_j}.

There is another way of considering the basis functions e^{ikx_j} for the DFT. Recall the unit roots z_j defined in (5.12). After a renumbering we have

$$e^{ikx_j} = e^{ikj\Delta x} = e^{ik2\pi j/(N+1)} = z_k^j, \quad k = -N/2, -N/2+1, \ldots, N/2,$$

which shows that the elements of the basis vectors are the powers of the unit roots:

$$\begin{bmatrix} 1 \\ z_{-N/2} \\ z_{-N/2}^2 \\ \vdots \\ z_{-N/2}^N \end{bmatrix}, \quad \begin{bmatrix} 1 \\ z_{-N/2+1} \\ z_{-N/2+1}^2 \\ \vdots \\ z_{-N/2+1}^N \end{bmatrix}, \quad \ldots, \quad \begin{bmatrix} 1 \\ z_{N/2} \\ z_{N/2}^2 \\ \vdots \\ z_{N/2}^N \end{bmatrix}.$$

Here we have assumed that $N+1$ is odd, which means that

$$z_{N/2} = e^{i\frac{N}{2}\Delta x} = e^{i(\pi - \frac{\Delta x}{2})},$$
$$z_{-N/2} = e^{-i\frac{N}{2}\Delta x} = e^{i(\pi + \frac{\Delta x}{2})},$$

i.e., these two elements are located symmetrically on the unit circle on each side of the point $z = -1$. If we try to add an extra wave number $k = N/2 + 1$, we get the new element

$$z_{N/2+1} = e^{i(\frac{N}{2}+1)\Delta x} = e^{i(\pi + \frac{\Delta x}{2})} = z_{-N/2},$$

i.e., we hit an already existing unit root. This shows what we have already concluded: The maximal number of basis vectors is $N+1$.

We can interpret the DFT and its inverse as operations by matrices on vectors. With the vectors

$$\mathbf{f} = \begin{bmatrix} f_0 \\ f_1 \\ \vdots \\ f_N \end{bmatrix}, \quad \mathbf{c} = \begin{bmatrix} c_{-N/2} \\ c_{-N/2+1} \\ \vdots \\ c_{N/2} \end{bmatrix}$$

and the matrix

$$F = \frac{1}{2\pi} \begin{bmatrix} \mathbf{e}_{-N/2}^* \\ \mathbf{e}_{-N/2+1}^* \\ \vdots \\ \mathbf{e}_{N/2}^* \end{bmatrix}, \tag{6.11}$$

(with * denoting the transposed complex conjugate form) it is easily checked that
the DFT can be expressed as

$$\mathbf{c} = F\mathbf{f},$$
$$\mathbf{f} = 2\pi F^*\mathbf{c}.$$

Apparently the inverse transform is $F^{-1} = 2\pi F^*$, i.e., when disregarding the normalization constant 2π, F is a unitary matrix. The inverse DFT is simply an expansion of the vector $\mathbf{f}$ in terms of the basis vectors $\{\mathbf{e}_k\}$.

   The algorithm for computation of the DFT has a special history. Each coefficient
c_k requires approximately $2N$ arithmetic operations (on complex numbers) according to the formula (6.9) and, since there are $N + 1$ coefficients, the total operation
count is approximately $2N^2$ (N-independent terms are not counted). In applications
N is often quite large and, if the transform is to be carried out again and again in
some central part of an algorithm, the computation may quickly get out of hand.
Therefore, when the *Fast Fourier Transform* (*FFT*) was invented by James Coo-
ley and John Tukey in 1965, it was an enormous step forward. The FFT produces
the same numbers as the DFT, but it does so by using a special fast algorithm.
Some people think that it is the most significant achievement ever for computational
mathematics. The FFT is a clever reorganization of the arithmetic, such that the total
number of operations goes down to something that is proportional to $N \log N$. This
is true also for the inverse DFT (6.8). The original algorithm was developed for the
case that $N = 2^p$, where p is an integer. But this restriction was later removed, and
present day FFT's allow for any number N.

   The FFT provides a spectral analysis of the function $f(x)$ as does the Fourier
transform. Each coefficient c_k represents the amplitude corresponding to the wave
number k. Consider for example the function in Fig. 5.12 which consists of two
sine-waves. The result of the FFT is zero values for all coefficients except four,
which are

$$c_1 = -0.5i,$$
$$c_{-1} = 0.5i,$$
$$c_{20} = -0.1i,$$
$$c_{-20} = 0.1i.$$

Just as for the (continuous) Fourier transform in the previous section, we compute
the coefficients in the trigonometric polynomial and obtain $\sin x + 0.2 \sin 20x$.

   The problem treated here is an example of a spectral analysis of a given function,
and the result is often presented by a graph that shows the amplitude for the different
wave numbers. Figure 6.2 shows the graph for our case. It is a very simple one, but
we present it because such figures are often shown in just about every application
area dealing with spectral analysis.

   A more realistic situation is when the signal is polluted by noise. The left part
of Fig. 6.3 shows such a function, and it is not that easy to distinguish the typical
wave numbers just by looking at it. However, the FFT provides the sine-coefficients
shown in the right part of the same figure, and it clearly shows the peaks at $k = 40$
and $k = 100$.

Fig. 6.2 The discrete Fourier
transform for
$\sin x + 0.2 \sin 20x$

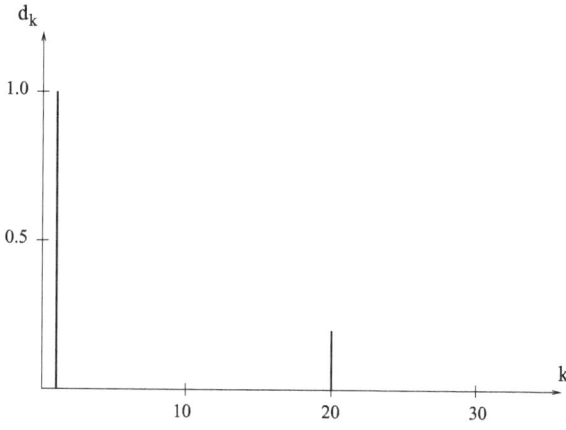

In fact, the function is

$$f(x) = \sin 40x + 0.2 \sin 100x + 0.5r(x)$$

sampled at 501 points, where $r(x)$ is a random function uniformly distributed in the
interval $[0, 1]$. The lower wave number $k = 40$ could have been identified from the
graph of $f(x)$, but hardly the higher wave number $k = 100$.

The discrete Fourier transform is well defined for any number sequence $\{f_j\}$, but
the interpolation property holds of course only if f_j represents the function value at
the point $x_j = j\Delta x$. The other assumption we have been using is periodicity. But
how does the transform know that the function is periodic when we don't provide the
value f_{N+1} at the right end point $x_{N+1} = 2\pi$? Well, it does not. A natural condition
is that f_N is close to f_0, since the assumption is that $f_{N+1} = f_0$. If this condition
doesn't hold, we pay for it by a very poor representation of $f(x)$ for x-values in
between the interpolation points. Figure 6.4 shows the interpolated function $g(x)$
obtained with the FFT acting on $f(x) = x$ with $N = 21$.

It interpolates at the grid points, but is a very bad approximation in between.
In particular we note that it is aiming for what it thinks is a perfect interpolation
$g(2\pi) = f(0) = 0$ at $x = 2\pi$.

The FFT is one of the most frequently used tools in just about every application area. The idea with a transform in general is to represent a certain function
with something other than its values at grid points. The identification of dominating frequencies k reveals more than by simply looking at the plot of the function
$f(x)$. Figure 6.3 could be an application in acoustics, where the two frequencies are
identified by the discrete Fourier transform.

In the previous section we discussed how Fourier analysis can be applied to
partial differential equations in order to gain insight into the behavior of the solutions. For approximations based on discretization, it is quite natural to perform
the analysis by using the discrete Fourier transform. Indeed it is a very powerful
and common analysis tool also in that case. In Sect. 10.3 we shall come back to this
issue.

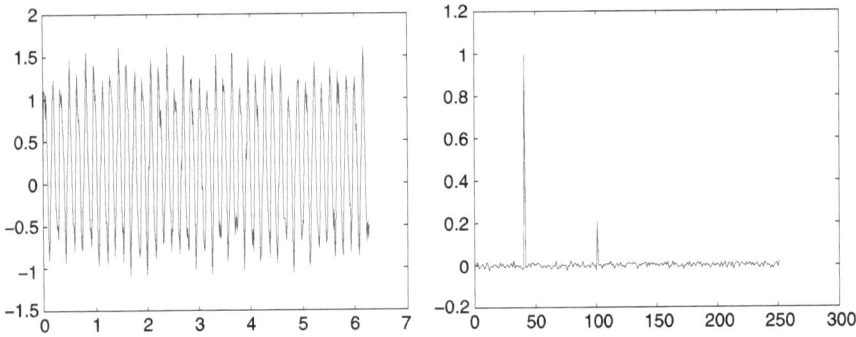

Fig. 6.3   The function $\sin 40x + \sin 100x + r(x)$ and its discrete Fourier transform

Fig. 6.4 Interpolating
function $g(x)$ for $f(x) = x$

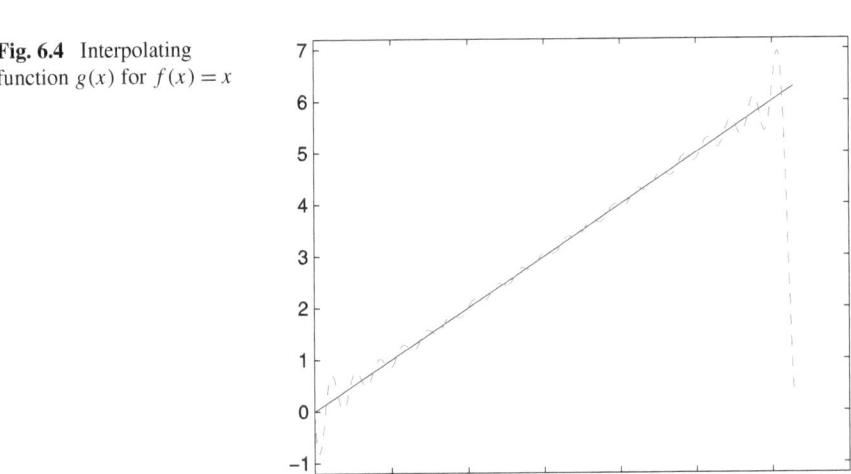

Exercise 6.2 Find the coefficients in the expansion $\cos 2x + \sin 2x = \sum_k c_k e^{ikx}$.

Exercise 6.3 Prove Parseval's relation (6.10) by using the orthonormality of the basis grid functions.

Exercise 6.4 Prove that the function $\tilde{g}(x) = \sum_{k=-N/2+n}^{N/2+n} c_k e^{ikx}$ is identical to $g(x)$ in (6.7) for any integer n.

Exercise 6.5 Prove that there is a unique Fourier interpolating function $g(x) = \sum_{k=1}^{M} c_k e^{ikx}$ also when there is an odd number of given points. Use the same steps in the proof as above when arriving at (6.9).

Exercise 6.6 Write a program that computes the interpolating function (6.7) by using the formula (6.9). Use it for a real set of numbers $\{f_j\}_0^N$, and check that $\mathrm{Re}(g(x_j)) = f_j$. Is $\mathrm{Im}(g(x))$ zero for all x?

Exercise 6.7 Use the program in Exercise 6.6 for experimenting with various sets of data. For example, find the monthly temperature data over a number of years for a city of your choice, and use the FFT (6.9) for identifying the one year periodicity. As an alternative, use the MATLAB function *fft(x)* for the discrete Fourier transform of a vector x (see Sect. 18.2).

Chapter 7
Polynomial Expansions

Fourier series are not well suited to represent nonperiodic functions. Other basis functions are needed, and polynomials are a possible choice since they are easy to handle computationally. Resolution of functions with fast variation require polynomials of high degree. However, it was demonstrated in Sect. 5.1 that standard polynomials with x^n as basis functions are no good for higher degrees, and that orthogonal polynomials are a better alternative. Various types of such polynomials were developed a long time ago, and they played a role in the development of classic applied mathematics and approximation theory. Later they became more obscure as a result of the introduction of piecewise polynomials that are more convenient for finite element methods, as we shall see. However, during the last decades there has been a remarkable renascence for orthogonal polynomials when it comes to numerical solution of PDE by spectral methods, as we shall see in Chap. 12. This is why we introduce them here.

7.1 Orthogonal Polynomial Expansions

In analogy with Fourier series, we can under certain conditions represent a given function as an infinite orthogonal polynomial expansion

$$f(x) = \sum_{n=0}^{\infty} a_n P_n(x).$$

Here it is assumed that $\{P_n(x)\}$ is a sequence of orthogonal polynomials with the scalar product

$$(f, g) = \int_a^b f(x)g(x)w(x)\,dx. \tag{7.1}$$

B. Gustafsson, *Fundamentals of Scientific Computing*,
Texts in Computational Science and Engineering 8,
DOI 10.1007/978-3-642-19495-5_7, © Springer-Verlag Berlin Heidelberg 2011

It is also assumed that the sum converges for all x with $a \leq x \leq b$. After multiplication of $f(x)$ by each $P_n(x)$ and integrating, the coefficients a_n are obtained from the explicit formula

$$a_n = \frac{(P_n, f)}{\| P_n \|^2}, \quad n = 0, 1, \ldots.$$

Only for very simple model cases can the integrals be evaluated by analytical means. In general, we must use some sort of numerical integration. This means that a finite number of discrete points x_j are involved, and we have actually a transformation between a finite number sequence $\{f_j = f(x_j)\}$ and a finite number of coefficients $\{a_j\}$. This is a discrete transform, just as we saw for the Fourier case above. The question is how the discrete points should be chosen for different kinds of orthogonal polynomials. Again we shall see that we can use classical results associated with well known mathematicians who were active long ago.

There are two types of orthogonal polynomials that are of particular interest when solving partial differential equations. These are the Chebyshev and Legendre polynomials, and we shall describe these in the next two sections.

7.2 Chebyshev Polynomials

The end points of the interval and the weight function determine the scalar product (7.1) and the type of orthogonal polynomial. With $a = -1$, $b = 1$ and $w(x) = 1/\sqrt{1 - x^2}$, we have the scalar product

$$(f, g) = \int_{-1}^{1} \frac{f(x)g(x)}{\sqrt{1 - x^2}} \, dx,$$

which leads to the classic *Chebyshev polynomials* $T_n(x)$ named after the Russian mathematician Pafnuty Chebyshev (1821–1894). Obviously the weight function puts more weight on both ends of the interval. At a first glance it looks like we are in trouble here, since the weight function is not defined at the end points $x = -1$ and $x = 1$. Do we have to work solely with functions that tend to zero as x approaches the end points? Indeed not. Here we have a case where the integral exists for any bounded function $f(x)g(x)$ despite the singularity. For the simple case $f(x)g(x) = 1$ we can actually compute the integral analytically. In Appendix A.1 we find the differentiation rule

$$\frac{d}{dx} \arcsin x = \frac{1}{\sqrt{1 - x^2}},$$

where $\arcsin x$ is the inverse of the sine function, see Sect. 5.2. We get

$$\int_{-1}^{1} \frac{1}{\sqrt{1 - x^2}} \, dx = \arcsin 1 - \arcsin(-1) = \frac{\pi}{2} + \frac{\pi}{2} = \pi.$$

The orthogonality relations for the Chebyshev polynomials are

$$\int_{-1}^{1} T_m(x) T_n(x) \frac{1}{\sqrt{1 - x^2}} \, dx = \begin{cases} 0 & \text{if } m \neq n, \\ \pi & \text{if } m = n = 0, \\ \pi/2 & \text{if } m = n \neq 0. \end{cases}$$

There is no need to find the explicit form of each basis function for evaluation of the Chebyshev polynomials. Just like Horner's scheme for standard polynomials, there is a simple recursive formula well suited for the computer:

$$\begin{aligned} T_0(x) &= 1, \\ T_1(x) &= x, \\ T_{n+1}(x) &= 2x\,T_n(x) - T_{n-1}(x), \quad n = 1, 2, \dots. \end{aligned} \tag{7.2}$$

There is more nice mathematics that applies to the Chebyshev polynomials. It can be shown that they satisfy the differential equation

$$\frac{d}{dx}\left(\sqrt{1-x^2}\frac{dT_n(x)}{dx}\right) + \frac{n^2}{\sqrt{1-x^2}}T_n(x) = 0, \quad n = 0, 1, 2, \dots \tag{7.3}$$

for $-1 \le x \le 1$, and the boundary conditions are

$$T_n(-1) = (-1)^n, \qquad T_n(1) = 1.$$

Since the function $\cos\xi$ has its range between -1 and 1, the variable transformation $x = \cos\xi$ is natural. The new function

$$\tilde{T}_n(\xi) = T_n(\cos\xi)$$

is then defined in the interval $0 \le \xi \le \pi$, and one can show that it satisfies the differential equation

$$\frac{d^2\tilde{T}_n(\xi)}{d\xi^2} + n^2\tilde{T}_n(\xi) = 0, \quad n = 0, 1, 2, \dots. \tag{7.4}$$

But this equation is easy to solve. With the boundary conditions

$$\tilde{T}_n(0) = 1, \qquad \tilde{T}_n(\pi) = (-1)^n,$$

the solution is

$$\tilde{T}_n(\xi) = \cos(n\xi), \qquad n = 0, 1, \dots.$$

This means that we have an explicit form of the original polynomial:

$$T_n(x) = \cos(n \arccos x), \quad n = 0, 1, \dots, \ -1 \le x \le 1.$$

This formula is not normally used for direct computation, but it has great significance for algorithms involving Chebyshev polynomials, as we shall see. Another consequence of this identity is that the inequality $|T_n(x)| \le 1$ for all n is directly verified.

As for the Fourier case, the formulas for interpolation are closely connected to the choice of numerical integration also for the Chebyshev polynomials. We choose the interpolation points as

$$x_n = \cos\left(\frac{n\pi}{N}\right), \quad n = 0, 1, \dots, N.$$

These points are called the *Gauss–Lobatto* points, and they were originally derived for obtaining accurate numerical integration formulas. The interpolation problem is

$$g(x) = \sum_{n=0}^{N} a_n T_n(x),$$

$$g(x_n) = f(x_n), \quad n = 0, 1, \ldots, N.$$

(Note that the x_n-points are numbered backwards such that $x_{n+1} < x_n$.) One can prove that the coefficients are given by

$$a_n = \frac{1}{c_n} \sum_{j=0}^{N} \frac{1}{c_j} f(x_j) T_n(x_j) \Delta x, \quad n = 0, 1, \ldots, N,$$

$$c_n = \begin{cases} 1, & n = 1, 2, \ldots, N-1 \\ 2, & n = 0, N, \end{cases} \tag{7.5}$$

where $\Delta x = 2/N$. Looking back at the cosine expression for T_n, we note that

$$T_n(x_j) = \cos(n \arccos x_j) = \cos\left(n \arccos\left(\cos\frac{j\pi}{N}\right)\right) = \cos\frac{n j \pi}{N}.$$

This gives the final transform

$$a_n = \frac{1}{c_n} \sum_{j=0}^{N} \frac{1}{c_j} f(x_j) \cos\frac{j n \pi}{N} \Delta x, \quad n = 0, 1, \ldots, N. \tag{7.6}$$

(Recall that the x_j-points are numbered backwards!) This is a remarkable formula. Starting out with a somewhat strange weight function and orthogonal polynomials, we ended up with a cosine transform! This one can be computed by using the fast Fourier transform.

We summarize the whole procedure.

1. Choose the Gauss–Lobatto points $x_j = \cos(j\pi/N)$, $j = 0, 1, \ldots, N$, for a discrete representation of a given function $f(x)$.
2. Compute the coefficients a_n defined in (7.6) by using a fast discrete cosine transform.
3. The function

$$g(x) = \sum_{n=0}^{N} a_n T_n(x),$$

interpolates $f(x)$ at the points x_j, $j = 0, 1, \ldots, N$, and is a good approximation of $f(x)$ for all x with $-1 \le x \le 1$.

The interpolation points have a denser distribution close to the boundaries, which corresponds to the structure of the polynomials. Figure 7.1 shows the Chebyshev polynomial of degree 40. It is oscillatory with a faster variation near the boundaries, which means that it is able to catch features in a function on a smaller scale there.

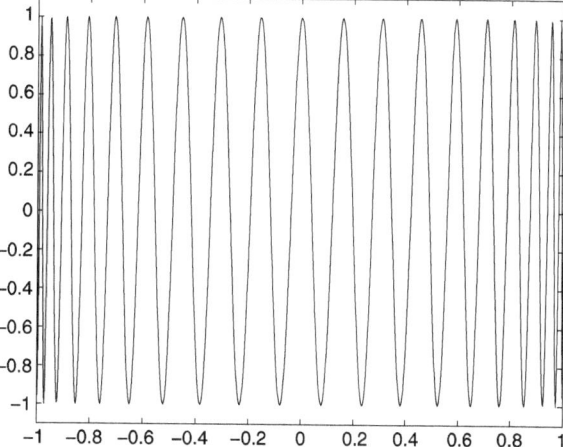

Fig. 7.1 Chebyshev polynomial $T_{40}(x)$

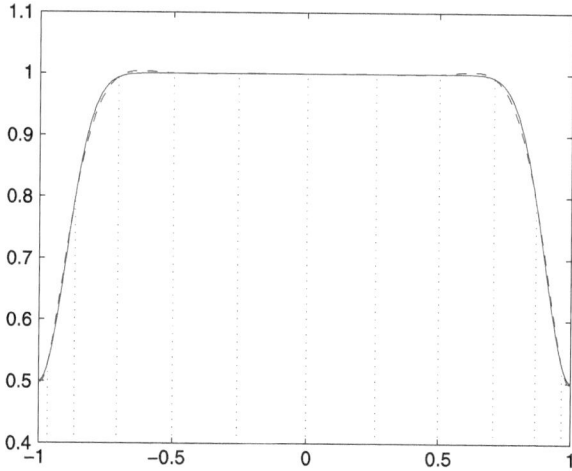

Fig. 7.2 Chebyshev interpolation, $N = 12$ (−−). Original function (—)

In general the clustering of the points is a disadvantage, since it does not take into account the location of the small scale variations in the function that is to be interpolated. However, for certain applications it may be an advantage. For example, in fluid dynamics, there are often boundary layers, i.e., a strong variation of the state variables near solid walls. In such cases, the Chebyshev distribution may give an almost optimal representation. Figure 7.2 shows a constructed example with boundary layers at both boundaries. The function is

$$f(x) = 0.5\left(1 - e^{-50(x+1)^2} + 1 - e^{-50(x-1)^2}\right),$$

and the figure shows the Chebyshev interpolation polynomial of degree $N = 12$. The dotted vertical lines are located at the Gauss–Lobatto points. The approximating function is remarkably good with only 13 interpolation points.

Exercise 7.1 Write a program for evaluation of $T_n(x)$ according to (7.2). Define a sequence $\{f(x_j)\}$, and compute the coefficients a_n first by using (7.5), and then by (7.6). Check that the results are identical.

Exercise 7.2 Prove that $\tilde{T}_n(\xi)$ satisfies (7.4) if $T_n(x)$ satisfies (7.3).

7.3 Legendre Polynomials

With the same x-interval as for the Chebyshev polynomials but weight function $w(x) = 1$, the scalar product is

$$(f, g) = \int_{-1}^{1} f(x) g(x) \, dx.$$

The corresponding orthogonal polynomials satisfy

$$\int_{-1}^{1} P_m(x) P_n(x) \, dx = \frac{2}{2n+1} \delta_{mn},$$

and they are called the *Legendre polynomials* after the French mathematician Adrien Marie Legendre (1752–1833). There is also a recursive formula:

$$P_0(x) = 1,$$
$$P_1(x) = x,$$
$$P_{n+1}(x) = \frac{2n+1}{n+1} x P_n(x) - \frac{n}{n+1} P_{n-1}(x), \quad n = 1, 2, \ldots.$$

The Gauss–Lobatto points for the Legendre polynomials are the $N+1$ roots of the polynomial

$$(1 - x^2) \frac{d P_N(x)}{dx} = 0.$$

The interpolation polynomial

$$g(x) = \sum_{n=0}^{N} a_n P_n(x)$$

which equals $f(x)$ at the Gauss–Lobatto points, is obtained with the coefficients

$$a_n = \frac{1}{c_n} \sum_{n=0}^{N} f(x_j) P_n(x_j) \Delta x, \quad n = 0, 1, \ldots, N,$$

$$c_n = \begin{cases} 2/(2n+1), & n = 0, 1, \ldots, N-1, \\ 2/N, & n = N, \end{cases}$$

where $\Delta x = 2/N$. This is the *discrete Legendre transform*.

Any interval $[a, b]$ on the x-axis can of course be transformed to the interval $[-1, 1]$ that is used for the two classes of polynomials described here. We can also go the other way around from $[-1, 1]$ to $[a, b]$. As an example we take the Legendre polynomials and transform them to the interval $[0, 1]$. This is done by the change of variable $x = 2\xi - 1$. The new polynomial is denoted by $\tilde{P}_n(\xi) = P_n(2\xi - 1)$, and the following relations hold:

$$\tilde{P}_0(\xi) = 1,$$
$$\tilde{P}_1(\xi) = 2\xi - 1,$$
$$\tilde{P}_{n+1}(\xi) = \frac{2n+1}{n+1}(2\xi - 1)\tilde{P}_n(\xi) - \frac{n}{n+1}\tilde{P}_{n-1}(\xi), \quad n = 1, 2, \ldots,$$
$$\int_0^1 \tilde{P}_m(\xi)\tilde{P}_n(\xi)\,d\xi = \frac{1}{2n+1}\delta_{mn}.$$

These polynomials are called the *shifted Legendre polynomials*.

Chapter 8
Least Square Problems

Almost all kinds of approximation are based on the principle that a complicated function is represented by a simpler function that is easy to evaluate. This simpler function has a certain number of parameters that are determined such that the approximation error is small in some sense. The finite element method for differential equations is based on this idea. In order to understand the principles, we discuss in this chapter the simpler problem of approximating a given function without any differential equations involved.

One question is how to measure the error. A very natural measure is the maximal deviation from the true function, but it turns out that the theory becomes simpler if the square of the error is integrated over the interval of interest. This leads to the *least square method*.

8.1 Vector Approximations

We begin by considering a simple example. Assume that we want to approximate the vector $\mathbf{u} = [x, y, z]$ in the 3-dimensional space by another vector $\mathbf{v} = [a, b, 0]$ in 2-dimensional space. It seems reasonable to choose $a = x$ and $b = y$, but we shall derive the solution in a strict way, which can be generalized to more complicated problems. We repeat first the basic concepts that are required. The scalar product between two vectors $\mathbf{u}_1$ and $\mathbf{u}_2$ is

$$(\mathbf{u}_1, \mathbf{u}_2) = x_1 x_2 + y_1 y_2 + z_1 z_2,$$

and the length of a vector $\mathbf{u}$ is

$$\|\mathbf{u}\| = \sqrt{(\mathbf{u}, \mathbf{u})} = \sqrt{x^2 + y^2 + z^2}.$$

Two vectors $\mathbf{u}_1$ and $\mathbf{u}_2$ are orthogonal if $(\mathbf{u}_1, \mathbf{u}_2) = 0$.

B. Gustafsson, *Fundamentals of Scientific Computing*,
Texts in Computational Science and Engineering 8,
DOI 10.1007/978-3-642-19495-5_8, © Springer-Verlag Berlin Heidelberg 2011

The approximation criterion is to minimize the length of the error vector. The basis vectors are chosen as the unit vectors

$$\mathbf{w}_1 = \begin{bmatrix} 1 \\ 0 \\ 0 \end{bmatrix}, \qquad \mathbf{w}_2 = \begin{bmatrix} 0 \\ 1 \\ 0 \end{bmatrix},$$

such that $\mathbf{v}$ can be written in the form

$$\mathbf{v} = a\mathbf{w}_1 + b\mathbf{w}_2. \tag{8.1}$$

We try to find the parameters a and b such that

$$\|\mathbf{u} - \mathbf{v}\|^2 = (x - a)^2 + (y - b)^2 + z^2$$

is as small as possible (minimizing the square of the length gives the same result as minimizing the length). Since all three terms are positive, the minimum is trivially obtained as

$$x - a = 0,$$
$$y - b = 0,$$

which can also be written in the more complicated form

$$\big(\mathbf{u} - (a\mathbf{w}_1 + b\mathbf{w}_2), \mathbf{w}_1\big) = 0,$$
$$\big(\mathbf{u} - (a\mathbf{w}_1 + b\mathbf{w}_2), \mathbf{w}_2\big) = 0.$$

This form shows that the error vector is orthogonal to the two basis vectors in the approximating subspace. Since every vector $\mathbf{v}$ in the two-dimensional (x, y)-space can be written as $\mathbf{v} = a_1\mathbf{w}_1 + a_2\mathbf{w}_2$, it follows that the error vector is orthogonal to all vectors in the 2D x/y-space, i.e., it is orthogonal to the whole 2D space. The solution $a = x$, $b = y$ is called the *projection* of $\mathbf{u}$ on the 2D-space, and it is illustrated in Fig. 8.1.

Let us next take a look at the general case. We want to approximate the N-dimensional vector $\mathbf{u}$ by

$$\mathbf{v} = \sum_{j=1}^{M} a_j\mathbf{w}_j, \tag{8.2}$$

where $\mathbf{w}_j$ are basis vectors, and $M < N$. How should the coefficients a_j be chosen such that $\|\mathbf{u} - \mathbf{v}\|$ is minimal? To find out, we assume that $\mathbf{v}$ is the optimal solution, and perturb it by a vector $\varepsilon\mathbf{d}$, where $|\varepsilon|$ is a small number. This means that

$$\|\mathbf{u} - \mathbf{v}\|^2 \le \|\mathbf{u} - \mathbf{v} - \varepsilon\mathbf{d}\|^2 = \|\mathbf{u} - \mathbf{v}\|^2 - 2\varepsilon(\mathbf{u} - \mathbf{v}, \mathbf{d}) + \varepsilon^2\|\mathbf{d}\|^2.$$

The parameter ε is arbitrary and, by choosing it sufficiently small in magnitude, the second term will dominate. Furthermore, by choosing the right sign of ε, this term will be negative. This leads to the conclusion that the condition for a minimum is

$$(\mathbf{u} - \mathbf{v}, \mathbf{d}) = 0 \tag{8.3}$$

for all vectors $\mathbf{d}$ in the admissible vector space spanned by the basis vectors $\mathbf{w}_j$. Therefore we can write the minimization condition as

$$(\mathbf{v}, \mathbf{w}_k) = (\mathbf{u}, \mathbf{w}_k), \quad k = 1, 2, \ldots, M,$$

Fig. 8.1 Projection of a 3D-vector **u** on **v** in 2D-space

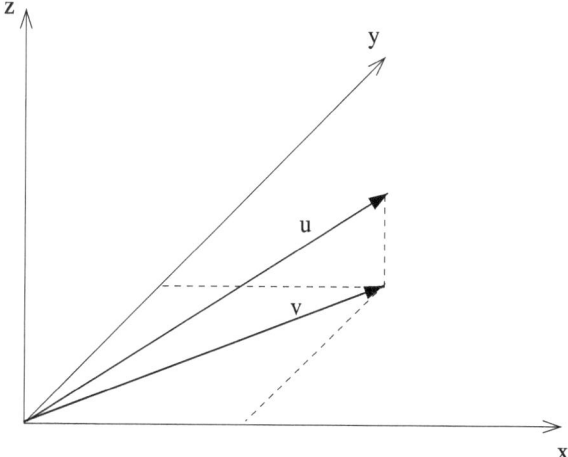

or equivalently

$$\sum_{j=1}^{M}(\mathbf{w}_j, \mathbf{w}_k)a_j = (\mathbf{u}, \mathbf{w}_k), \quad k = 1, 2, \ldots, M. \tag{8.4}$$

This condition is fundamental and, as we shall see, it shows up again when dealing with functions instead of vectors, and also when solving differential equations.

8.2 The Continuous Case

In this section we shall consider approximations of functions $f(x)$ defined everywhere in the interval $[0, 1]$. If $f(x)$ is a complicated function to evaluate, it is an advantage to approximate it with another simpler function $g(x)$. We introduced orthogonal polynomials $P_j(x)$ in Sect. 5.1.3 satisfying the condition

$$(P_j, P_k) = \int_0^1 P_j(x)P_k(x)\,dx = 0, \quad j \neq k.$$

These are the shifted Legendre polynomials $P_j(x)$ introduced in Sect. 7.3, and we use them as basis functions for the approximation

$$g(x) = \sum_{j=0}^{n} a_j P_j(x).$$

The least square problem is formulated just as above by the requirement that the norm of the error

$$\|f - g\|^2 = \int_0^1 \left(f(x) - \sum_{j=0}^{n} a_j P_j(x)\right)^2 dx$$

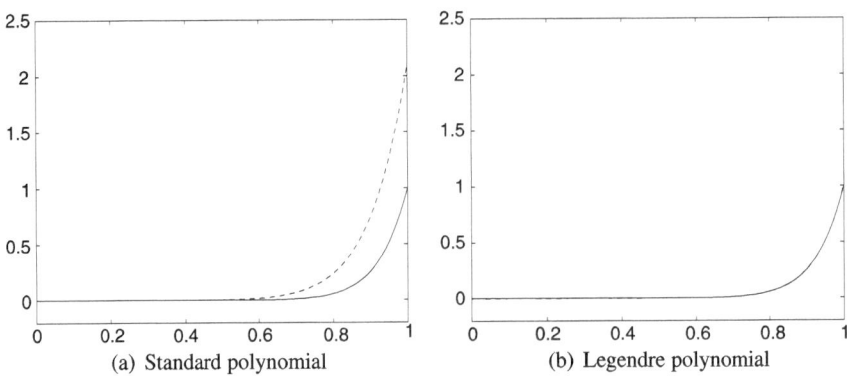

Fig. 8.2 Approximation $(--)$ of x^{13} $(—)$ with polynomials of degree 13

is as small as possible. This problem was treated in the previous section for vectors, and the same arguments can be applied here as well. The scalar product and norm can be handled in precisely the same way with the definition for either vectors or continuous functions. This leads to the condition of orthogonality between the error and the subspace, i.e.,

$$(f - g, P_k) = 0, \quad k = 0, 1, \ldots, n. \tag{8.5}$$

The linear system for determining the coefficients a_j is

$$\sum_{j=0}^{n} (P_j, P_k) a_j = (f, P_k), \quad k = 0, 1, \ldots, n,$$

which corresponds precisely to (8.4) for the vector case. The scalar products on the left hand side are known integrals, and for $j \neq k$ they are all zero. From the formulas in Sect. 7.3 it follows that

$$a_j = (2j + 1)(f, P_j), \quad j = 0, 1, \ldots, n.$$

Here, the integrals $(f, P_j) = \int_0^1 f(x) P_j(x) \, dx$ may have to be computed numerically.

With standard polynomials, the system for the coefficients a_j would be dense and, furthermore, ill conditioned. As a test we choose $f(x) = x^n$, and go all the way to degree n with the approximating polynomials. That means that we should get exactly x^n back as an approximation. But with standard polynomials, we get into trouble already for quite low degree. Figure 8.2(a) shows the result for $n = 13$, and it is way off the target. Actually, the MATLAB system gives a warning "Matrix is close to singular", and the computer cannot do any better. On the other hand, the Legendre approximation shown in Fig. 8.2(b) is exact. Indeed, it will stay exact for any reasonable degree $n \geq 13$.

The machinery for the least square method is well defined by simply specifying the basis functions and the scalar product. With orthogonal polynomials the x-interval and the weight function determine everything. For example,

$$(f, g) = \int_{-1}^{1} f(x)g(x)\frac{1}{\sqrt{1 - x^2}}\,dx$$

gives the Chebyshev polynomials discussed in Sect. 7.2.

8.3   The Discrete Case

The least square method is probably best known for the discrete case, i.e., the function values $f(x)$ are known only at a number of discrete points $x = x_j$. These values are usually measured, and measuring errors are present. Even if the underlying basis functions are known from some theoretical considerations, the measuring errors are reason enough to find a new and smoother function that approximates the registered values. Interpolation is not a good alternative, since it forces the constructed function to pass exactly through the measured points.

We assume that there are $N + 1$ uniformly distributed points

$$x_j = j\Delta x, \quad j = 0, 1, \ldots, N,$$

with $(N + 1)\Delta x = 1$. In this case the scalar product between two sets of function values $\{f_j\}$ and $\{g_j\}$ is defined by the sum

$$(f, g) = \sum_{j=0}^{N} f_j g_j \Delta x,$$

and the *norm* by

$$\|f\| = \sqrt{(f, f)}.$$

With our choice of Δx we get the proper normalization such that the norm of the discretized function $f(x) = 1$ is $\|f\| = 1$. This means that x_N is not the end point $x = 1$ but rather $x = 1 - \Delta x$.

With basis functions $P_j(x)$, the least square approximation

$$g(x) = \sum_{j=0}^{n} a_j P_j(x)$$

is obtained exactly as in the continuous case just by substituting the integrals by sums. (Actually we are back to the vector case treated in Sect. 8.1.) The orthogonality condition

$$(f - g, P_k) = 0, \quad k = 0, 1, \ldots, n,$$

for the error gives the system

$$\sum_{j=0}^{n} (P_j, P_k)a_j = (f, P_k), \quad k = 0, 1, \ldots, n. \tag{8.6}$$

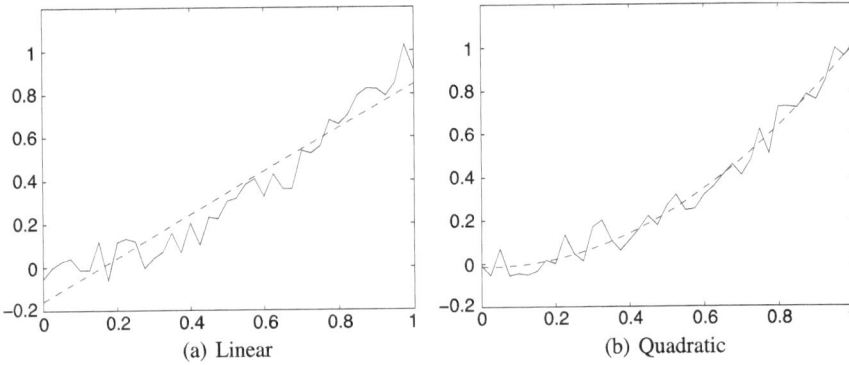

Fig. 8.3 Least square approximation $(--)$, true discrete function $(—)$

for the coefficients a_j. Standard polynomials can be used only for very low degrees. On the other hand, it is a very common choice for many applications. In fact, considered as a statistical technique, the basis functions $\{1,\ x\}$ lead to so-called linear regression analysis of a statistical sample. Figure 8.3(a) shows the straight line which is obtained for the discrete function

$$f_j = x_j^2 + 0.2\,\text{rand}(x_j),$$

where $\text{rand}(x)$ is a uniformly distributed random function in the interval $-0.5 \le x \le 0.5$. Figure 8.3(b) shows the least square second degree polynomial approximation for the same type of function. (The random numbers have the same statistical distribution, but are individually different.)

Let us now take a look at a more realistic example from real life. The discussion concerning the climate change is intense, the main question being if there really is a correlation between the greenhouse gases in the atmosphere and the global temperature. The underlying mathematical model is very complicated, and this is not the place to provide any new proofs in either direction. However, we shall perform an experiment that shows that one has to be very careful with any firm conclusions.

Since 1979, The National Space Science & Technology Center publishes monthly measurements of the global average temperature at the Earth surface over the years. Figure 8.4 shows these measurements including 2009, i.e., 372 readings $T(j)$ in all.

There are quite wild oscillations, and it is not easy to correlate these to any particular activities either from the Sun or from human and other activities on the Earth. When talking about clear trends, we want to see smoother curves with the irregular perturbations taken out. Both scientists and laymen have a preference for straight lines or slowly varying functions like low degree polynomials, where conclusions are easy to make. It has of course not been verified at all that there is any underlying polynomial form of the temperature as a function of time but for lack of any better knowledge of the type of functions, we can use them anyway. But the human eye is not sufficient to uniquely define the approximation even if the basis functions are given. Let us see what kind of approximation the least square method produces for our example.

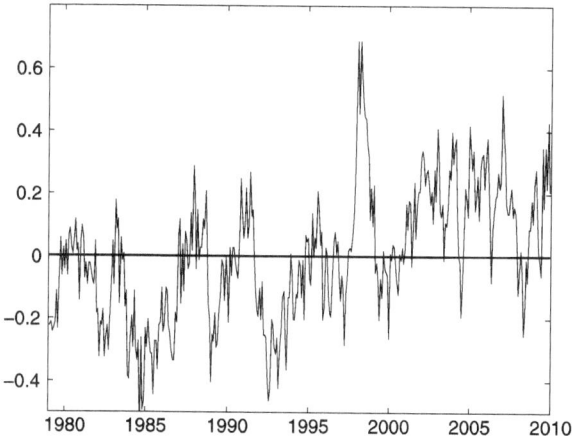

Fig. 8.4 Global temperature at the Earth surface 1979–2009

The most basic question is whether or not there is any increase in the global temperature at all. The common type of analysis is based on linear functions, and the first subfigure of Fig. 8.5 shows a clear rise of the temperature as a function of time. If we next go to second degree polynomials, we get the second figure, which shows an even more pronounced positive trend during the last years (positive in a mathematical sense, negative in view of the effects on human society). In principle, higher degree polynomials should be able to pick up the functional behavior of the data more accurately. The next two subfigures show the result with 3rd and 4th degree polynomials. They both show decreasing temperature at the end of the measured period. Somebody who is arguing that we should worry about temperature increase can use degree 1 or 2 of the polynomials, while the opponent saying that there is nothing to worry about can use degree 3 or 4. Clearly these least square approximations provide no solid base for either one to draw any certain conclusions. Something to think about when taking part in the current climate change debate!

For higher degree polynomials, the least square procedure will fail to give any reasonable results also for the discrete case. However, just as for the continuous case, polynomials can be constructed that are orthogonal with the discrete scalar product, resulting in a well conditioned computational problem. We discussed the basic principles for orthogonal polynomials in Sect. 5.1.3, and in Chap. 7.

We have discussed polynomial approximation here. Of course any other functions can be used for least square approximations, as long as we have a good set of basis functions $\phi_j(x)$. The coefficients a_j in the approximation $g(x) = \sum a_j \phi_j(x)$ are obtained by solving (8.6) with P_j, P_k substituted by ϕ_j, ϕ_k. For periodic functions we have the Fourier expansions with trigonometric basis functions. Since these functions are orthogonal in both the continuous and discrete cases with a uniform point distribution, it is an ideal choice. Trigonometric functions as well as Chebyshev and Legendre polynomials for the solution of differential equations are further discussed in Chap. 12.

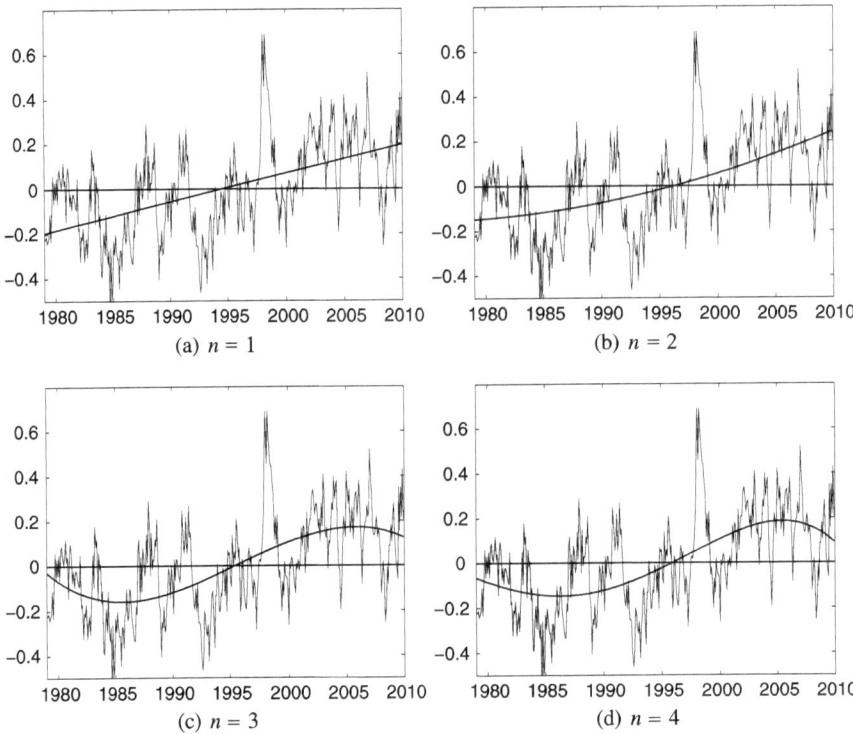

Fig. 8.5 Least square approximation with n degree polynomials $(--)$, true discrete function $(—)$

We end this chapter by mentioning the closely connected problem of finding a solution of an overdetermined linear system of equations

$$A\mathbf{x} = \mathbf{b},$$

where A is a rectangular $m \times n$ matrix with $m > n$. In general there is of course no solution to this problem, but it makes sense to define a least square solution $\mathbf{x}^*$ defined by

$$\|A\mathbf{x}^* - \mathbf{b}\|^2 = \min_{\mathbf{x}} \|A\mathbf{x} - \mathbf{b}\|^2.$$

Assuming that $\mathbf{x}^*$ is the minimizing vector, we perturb it by a vector $\varepsilon\mathbf{x}$, where $|\varepsilon|$ is a small scalar. We have

$$\|A\mathbf{x}^* - \mathbf{b}\|^2 \leq \|A(\mathbf{x}^* + \varepsilon\mathbf{x}) - \mathbf{b}\|^2 = \|A\mathbf{x}^* - \mathbf{b}\|^2 + 2\varepsilon(A\mathbf{x}^* - \mathbf{b}, A\mathbf{x}) + \mathcal{O}(\varepsilon^2)$$

for all vectors $\mathbf{x}$. Since $|\varepsilon|$ is arbitrary small, the condition for a minimum is

$$(A\mathbf{x}^* - \mathbf{b}, A\mathbf{x}) = 0,$$

which is a slightly different form compared to the condition (8.3) for the pure approximation problem of a given vector. It can be modified further to the condition

$$\left(A^T(A\mathbf{x}^* - \mathbf{b}), \mathbf{x}\right) = 0.$$

We have here an N-dimensional vector $A^T(A\mathbf{x}^* - \mathbf{b})$ that is orthogonal to all vectors in an N-dimensional space, and this is impossible only for a nonzero vector. Hence, the final condition for a minimum is

$$A^T A\mathbf{x}^* = A^T \mathbf{b}.$$

This system is called the *normal equations*.

It can be shown that, if the column vectors of A are linearly independent, then the matrix $A^T A$ is nonsingular, and the solution vector $\mathbf{x}^*$ is uniquely determined. We write it as

$$\mathbf{x}^* = A^+\mathbf{b},$$

where the matrix A^+ is called the *pseudo-inverse* of A. If A is a square matrix, the pseudo-inverse is identical to the true inverse.

It can be shown that the normal equations are ill conditioned already for moderate sized n, just as the systems arising for approximation with standard polynomials discussed above. Accordingly, the numerical solution methods must be designed by using some sort of orthogonalization procedure.

Exercise 8.1 MATLAB has a function *polyfit*(x, f, n) that returns the coefficients in the least square polynomial of degree n for the point set (x_j, f_j). Use this function to plot the least square approximation for various n and data sets of your own choice (for example the stock market index the last month).

Part III
Numerical Methods for Differential Equations

Chapter 9
Numerical Methods for Differential Equations

In this chapter we shall discuss some basic considerations when solving differential equations numerically.

9.1 Basic Principles

In Chap. 2 we demonstrated different types of differential equations. It is convenient to distinguish between the independent variables $\mathbf{x} = [x_1, x_2, \ldots, x_d]$ for space variables and t for time. Here d can be any number, but often with $1 \leq d \leq 3$ as in physical space. We introduce the abstract notation $\mathscr{P}(\partial\mathbf{x}, \partial t)$ for a differential operator. By this we mean that $\mathscr{P}\mathbf{u}$ is an expression that contains the partial derivatives of $\mathbf{u}$ with respect to the space and time coordinates. There are essentially three types of problems for the unknown (vector) variable $\mathbf{u}$ depending on where and how the known data are given:

- Initial value problems (IVP)
- Boundary value problems (BVP)
- Initial-boundary value problems (IBVP)

In abstract form they are characterized as follows.

- **Initial value problem:** $-\infty < x_j < \infty$, $t \geq 0$

$$\mathscr{P}(\partial\mathbf{x}, \partial t)\mathbf{u} = \mathbf{F}(\mathbf{x}, t),$$
$$\mathbf{u}(\mathbf{x}, 0) = \mathbf{f}(\mathbf{x}).$$

- **Boundary value problem:** Domain Ω, boundary $\partial\Omega$

$$\mathscr{P}(\partial\mathbf{x})\mathbf{u} = \mathbf{F}(\mathbf{x}), \quad \mathbf{x} \in \Omega,$$
$$\mathbf{u}(\mathbf{x}) = \mathbf{g}(\mathbf{x}), \quad \mathbf{x} \in \partial\Omega.$$

B. Gustafsson, *Fundamentals of Scientific Computing*,
Texts in Computational Science and Engineering 8,
DOI 10.1007/978-3-642-19495-5_9, © Springer-Verlag Berlin Heidelberg 2011

- **Initial-boundary value problem:** $\mathbf{x} \in \Omega$, $t \geq 0$

$$\mathscr{P}(\partial\mathbf{x}, \partial t)\mathbf{u} = \mathbf{F}(\mathbf{x}, t), \quad \mathbf{x} \in \Omega,$$
$$\mathbf{u}(\mathbf{x}, t) = \mathbf{g}(\mathbf{x}, t), \quad \mathbf{x} \in \partial\Omega,$$
$$\mathbf{u}(\mathbf{x}, 0) = \mathbf{f}(\mathbf{x}), \quad \mathbf{x} \in \Omega.$$

Here $\mathbf{F}$ represents a given (vector) function which is independent of $\mathbf{u}$, and it is often called a *forcing function*. We have introduced the compact notation $\partial\mathbf{x}$ when several space variables x_j are involved. For example, for the simplest form of the heat equation in two space dimensions, the differential operator $\mathscr{P}$ is

$$\mathscr{P}(\partial\mathbf{x}, \partial t) \equiv \frac{\partial}{\partial t} - \frac{\partial^2}{\partial x_1^2} - \frac{\partial^2}{\partial x_2^2}.$$

The independent variables $\mathbf{x}$ and t can of course stand for any type of variable other than space and time. Furthermore, the initial condition for $t = 0$ may contain time derivatives as well if the differential equation contains higher order time derivatives. Likewise, the boundary conditions may contain space derivatives as well.

Figure 9.1 shows the three types of problems.

There may be derivatives of different order in the differential equation, and those of highest order are the most significant ones when it comes to the fundamental properties. We consider a second order scalar PDE in two space dimensions:

$$a\frac{\partial^2 u}{\partial x^2} + 2b\frac{\partial^2 u}{\partial x \partial y} + c\frac{\partial^2 u}{\partial y^2} = 0,$$

where a, b and c are given constants. It is classified as one of three different types:

- Hyperbolic
- Parabolic
- Elliptic

They are fundamentally different, and it is important to classify a certain PDE correctly before trying to solve it. Hyperbolic and parabolic PDE can be solved as IVP or IBVP, while elliptic PDE are solved as BVP.

The classification is based on the Fourier transform. We demonstrated in Sect. 6.1 how it is applied to a PDE. Here we use the notation ξ and η for the wave numbers, i.e., the Fourier modes have the form $\hat{u}e^{i(\xi x + \eta y)}$. The Fourier transform of the differential equation is

$$-\left(a\xi^2 + 2b\xi\eta + c\eta^2\right)\hat{u} = 0.$$

The properties of the differential equation are closely connected to the properties of the polynomial which is multiplying the Fourier coefficient, and we are back to classification of polynomials, which is classic mathematics associated with the geometrical intersection between planes and cones. A nontrivial solution $\hat{u} \neq 0$ requires

$$a\xi^2 + 2b\xi\eta + c\eta^2 = 0. \tag{9.1}$$

This equation describes an ellipse, a parabola or a hyperbola in the ξ, η plane depending on the sign of the expression $ac - b^2$. In accordance with this, the PDE is called

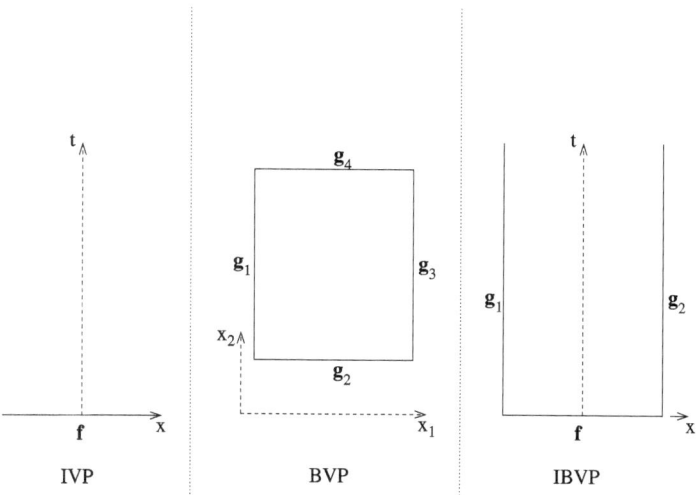

Fig. 9.1 The three types of problems for differential equations

- elliptic if $ac - b^2 > 0$,
- parabolic if $ac - b^2 = 0$,
- hyperbolic if $ac - b^2 < 0$.

An elliptic equation must be solved as a boundary value problem, while parabolic and hyperbolic equations are solved as initial or initial-boundary value problems. Trying the elliptic Laplace equation $\partial^2 u / \partial x^2 + \partial^2 u / \partial y^2 = 0$ as an initial value problem with the "time-like" variable y, (9.1) implies $\eta = \pm i\xi$. Accordingly, the solution has the form

$$u(x, y) = \alpha e^{-\xi y + i\xi x} + \beta e^{\xi y + i\xi x} \tag{9.2}$$

for some constants α and β. Even if the initial condition is such that the solution is nongrowing in the beginning, the inherent "explosive" exponential growth for one of the terms will be activated by the slightest perturbation. We have here another example of an ill posed problem.

For the hyperbolic wave equation $\partial^2 u / \partial x^2 - \partial^2 u / \partial y^2 = 0$, we get in the same way the solution

$$u(x, y) = \alpha e^{i\xi(y+x)} + \beta e^{i\xi(-y+x)}. \tag{9.3}$$

Here we have two components that are both bounded for any time, showing that it can be solved as an initial value problem. Hyperbolic PDE are also well defined with only first order derivatives involved. In Chap. 15, we shall consider such systems further.

In the parabolic case, the lower order derivatives play a more fundamental role. The parabolic heat equation $\partial^2 u / \partial x^2 - \partial u / \partial y = 0$ has already been treated with y as a time-like variable, and we shall come back to it again in Chap. 16.

In order to simplify the presentation, we shall most of the time discuss problems in one space dimension (or no space variable at all for ODE initial value problems), and scalar differential equations, i.e., the vector **u** has only one element u. Another simplification is the case where $\mathscr{P}(\partial\mathbf{x}, \partial t)$ is linear, i.e., it satisfies

$$\mathscr{P}(au + v) = a\mathscr{P}u + \mathscr{P}v$$

for any constant a. Unfortunately real life is often nonlinear, and we have to deal with nonlinear differential equations as well. The generalization and implementation of numerical methods may not be much more complicated for such problems, but a strict analysis of the properties certainly is.

A pure initial value problem has no boundaries at all, but a computation in practice must of course be limited to a finite domain. It is a special difficulty to handle the artificial boundaries that are introduced in this way, and we shall touch upon this issue in Chap. 15.

A special type of initial value problems is the case with periodic solutions in space such that $u(x, t) = u(x + p, t)$ for all x and t and some constant p. The computation is then limited to one period and, for convenient use of trigonometric functions in the analysis, it is natural to transform the period length to 2π. The domain in space can be chosen arbitrarily, but $0 \le x \le 2\pi$ is a convenient choice. There are no boundary conditions at the two boundaries $x = 0$ and $x = 2\pi$. For the computation we just use the fact that the function and all its derivatives are identical at the two boundaries.

The principle of discretization was introduced in Chap. 2. For difference methods it means that the unknown state variables are represented by their values at the grid points $\{x_j\}$. The approximation of the derivatives is obtained by going back to the definition of derivatives, but stopping short of carrying out the limit process to the end. In this way we get relations between function values that are located closely together. For initial value problems, where a certain state is known at a certain time t_0, this disretization procedure gives rise to a step-by-step algorithm which makes it possible to obtain the solution at any later time $t > t_0$. There is no condition at the end of the interval that needs to be matched by the solution. For a boundary value problem, where the state variables are known at the whole boundary of the domain, there is a coupling between the variables across the whole domain. Consequently, a large system of equations for the unknowns at the grid points must be solved.

The second principle for approximation is the representation of the solution as a finite sum

$$v(x) = \sum_{j=0}^{N} a_j \phi_j(x),$$

where the functions $\phi_j(x)$ are known basis functions that are chosen in such a way that the coefficients a_j can be computed rapidly. This means that the solution is well defined everywhere, but there is still a discretization procedure involved. The finite sum cannot satisfy the differential equation everywhere and, consequently, some relaxation has to be introduced.

One principle is the *Galerkin method*, named after the Russian mathematician Boris Grigoryevich Galerkin (1871–1945). The differential equation $\mathscr{P}u = F$ is transferred to a weaker form by multiplying with the basis functions and integrating. For a boundary value problem we get

$$\int_{\Omega} \big(\mathscr{P}(\partial x) v(x) - F(x) \big) \phi_j(x)\, dx = 0, \quad j = 0, 1, \ldots, N.$$

The discretization is here realized by choosing N as a finite number. This method is very similar to the condition (8.5) for the least square approximation problem in the sense that the error in the differential equation is orthogonal to the subspace which is spanned by the basis functions ϕ_j. For the finite element method, the approximating functions are piecewise polynomials. These are defined by their values at certain grid points or nodes, so also here the discretization introduces a computational grid.

Another principle is obtained by relaxing the conditions on the approximate solution v, so that it satisfies the differential equation only at certain grid-points x_j. This type of discretization is called *collocation*. The basis functions in the sum can then be any functions, but of course they must be chosen such that computations become simple and fast. Also here we have a computational grid consisting of the *collocation points* x_j.

For a time dependent problem in one space dimension, the solution has the form

$$u(x, t) = \sum_{j=0}^{N} a_j(t) \phi_j(x).$$

The Galerkin method can still be applied, but now we are left with a system of time dependent differential equations for the coefficients $a_j(t)$. This is what is called the method of lines, and it requires a numerical method for solution in time, usually chosen as a finite difference method. The finite element discretization in space gives its name to the whole method.

The collocation method can be applied also here. For a given set of collocation points x_j, the approximation $v(x, t)$ is differentiated with respect to x, and then represented at the grid points. Again we get a system of ordinary differential equations for $v(x_j, t)$, that is usually solved by finite difference approximation. This is called a *pseudo-spectral method*. The basis functions are usually taken as trigonometric functions or orthogonal polynomials.

All three methods are associated with computational grids which represent the discretization. We distinguish between the two main types: *structured grids* and *unstructured grids*. There are precise definitions of these concepts, but here we simply think of structured grids that can be mapped to rectangular grids in 2D or grids consisting of parallelepipeds in 3D. Different grid points can be identified by a pair or triple of indices in 2D and 3D respectively:

$$u_{jk}, \quad j = 0, 1, \ldots, N_x, \ k = 0, 1, \ldots, N_y,$$
$$u_{jkl}, \quad j = 0, 1, \ldots, N_x, \ k = 0, 1, \ldots, N_y, \ l = 0, 1, \ldots, N_z.$$

(The numbering may of course be chosen differently.) In 1D, a grid is always structured in this sense, but we distinguish between *uniform* and *nonuniform* grids. In the first case the step size is constant, such that $x_j = j \Delta x$. In several dimensions, the step size is constant in each coordinate direction. No such restriction applies for nonuniform grids.

Finite difference and spectral methods are both associated with structured grids, and we shall discuss these further in Chaps. 10 and 12. They can be constructed also for unstructured grids, but then it is much more complicated to keep up the accuracy. Finite element methods on the other hand adapt to unstructured grids much more easily. That is part of the reason for their success in applications. But one should keep in mind that algorithms based on methods on structured grids are much faster, the reason being that the complicated data structures for keeping track of the grid points are not necessary.

In the following chapters we shall give a description of the three types of numerical methods for differential equations.

Whatever method we are using, there is the question of how fine the computational grid should be. We talk about this in general terms as *resolution*. In the next section we shall discuss this concept in the most fundamental sense, i.e., the necessary number of points for representing a certain function.

9.2 Resolution

Solution procedures based on discretization lead to computational problems with increasing size for decreasing step size. Obviously there is a lower limit on the step size even by the largest available computers. On the other hand, the solution may contain small-scale features that are significant for the understanding of a certain physical process.

We consider a simple model example. The function

$$u(x) = e^{-800(x-0.2)^2}$$

represents a pulse with its peak at $x = 0.2$. Let us now perturb this function with an oscillating sin-function

$$u(x) = e^{-800(x-0.2)^2} + 0.05 \sin 2\pi k x, \tag{9.4}$$

where k is the wave number. If k is large, the function oscillates quickly as a function of x, and a good representation of this function requires a small step size Δx. If the computer is given $u(x)$ only at the grid points, there is no way that it can know about the oscillations between the grid points if the grid is too coarse. Figure 9.2 shows how the graphic software treats the function for $k = 50$ and four different step sizes $\Delta x = 1/N$.

The first curve is the correct one. The remaining three are the result of a resolution with approximately 100 points, which is not good enough. By changing the step size with just 1%, the results differ completely. For $N = 100$, the grid points fall exactly on the zeros of the oscillating function, and we obtain erroneously the exact original unperturbed function.

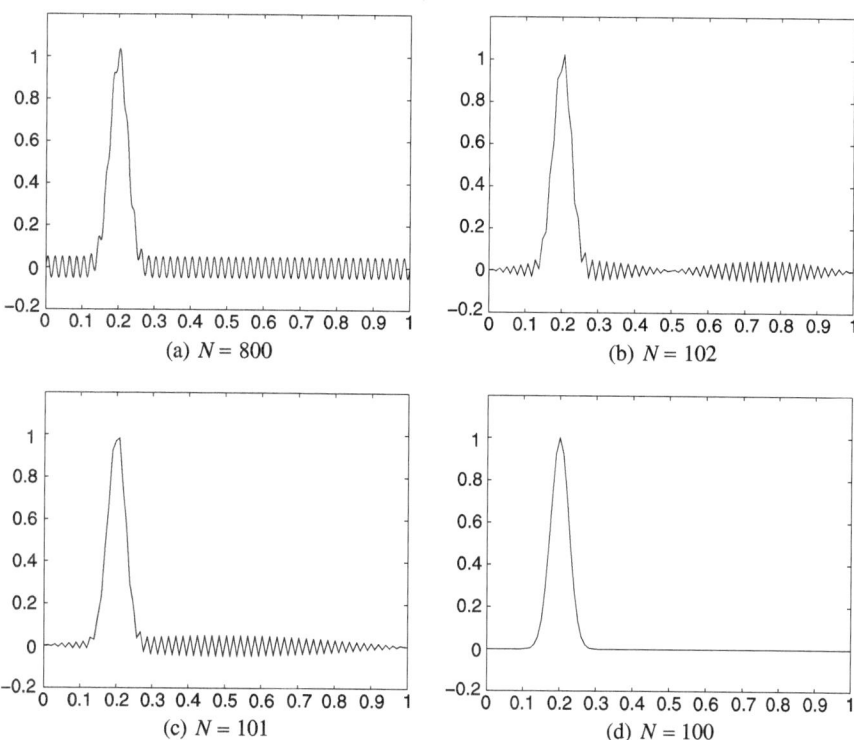

Fig. 9.2 Representation of the function (9.4) for different step sizes

This is an illustration of the obvious fact that representation of the finest structure of a certain state variable requires sufficiently many grid points. On the other hand, in many applications it is good enough to know the main features of the solution. For instance, in our example with the transport equation (2.18), where the initial function simply travels with the speed c, the position of the main pulse at different points in time may be the main interest. A good solver of the initial value problem achieves this even if the grid is not fine enough to resolve the oscillating part, but rather use the clean profile in Fig. 9.2(d).

There are other cases where a computation with poor resolution may still be useful. For instance, in our example the integral of the solution is not affected by the perturbation:

$$\int_0^1 u(x)\,dx = \int_0^1 (e^{-800(x-0.2)^2} + 0.05\sin 2\pi k x)\,dx$$

$$= \int_0^1 e^{-800(x-0.2)^2}\,dx = \sqrt{\pi/800}.$$

For the discrete case the integral is replaced by a finite sum, and one can show that also this one is essentially unaffected by the sine function. Furthermore, the integral doesn't change when the pulse is moving across the domain as long as it is not near the boundaries:

$$\int_0^1 u(x)\,dx = \int_0^1 u(x-ct)\,dx = \sqrt{\pi/800}.$$

Hence, even if we start with the poorly resolved function, the approximated integral of the solution will be kept almost constant. In many applications we have a similar situation. In such a case the problem is solved without the small scale features resolved, but a certain quantity of interest is still obtained with sufficient accuracy. We shall discuss this further in Sect. 17.4.

However, a word of warning is appropriate here. In real life problems, it is very difficult to analyze the effects of small-scale variations on the results. It is certainly much safer to make sure that the grid is fine enough for the smallest features of the solution, even if it requires large computational resources.

Chapter 10
Finite Difference Methods

Finite difference methods were used centuries ago, long before computers were available. As we have seen in Chap. 2, these methods arise quite naturally by going back to the definition of derivatives, and just stopping short of taking the limit as the step size tends to zero. We shall first discuss ordinary differential equations and then partial differential equations.

10.1 Ordinary Differential Equations

Consider an initial value problems of the type

$$\frac{du}{dt} = g(u),$$
$$u(0) = f,$$

(10.1)

where $g(u)$ is a general function of the solution u itself, and f is a known value. The simple method (2.9) discussed in Sect. 2.2.2 generalizes to

$$u_{n+1} = u_n + g(u_n)\Delta t, \quad n = 0, 1, \ldots,$$
$$u_0 = f$$

for our equation. The method is called the *Euler method* after Leonhard Euler, and it provides a very simple formula for implementation on a computer. When u_n is known, the value $g(u_n)$ and the right hand side can be computed. However, depending on the properties of $g(u)$, it may be necessary to use other forms of discretizations. For example, a better centering of $g(u)$ in the interval $[t_n, t_{n+1}]$ would improve the result. By taking the average of the end points we get the so called *trapezoidal rule*

$$u_{n+1} = u_n + \frac{g(u_n) + g(u_{n+1})}{2}\Delta t, \quad n = 0, 1, \ldots,$$
$$u_0 = f.$$

B. Gustafsson, *Fundamentals of Scientific Computing*,
Texts in Computational Science and Engineering 8,
DOI 10.1007/978-3-642-19495-5_10, © Springer-Verlag Berlin Heidelberg 2011

However, there is one complication here. In order to compute u_{n+1} at each step, we must solve a nonlinear equation

$$u_{n+1} - \frac{g(u_{n+1})}{2} \Delta t = u_n + \frac{g(u_n)}{2} \Delta t.$$

When the new time level is involved in this way, we have an *implicit method* in contrast to *explicit methods*. Except for very simple functions $g(u)$, we must use numerical methods for solving the nonlinear equation at each step. This is a typical situation in computational mathematics. In order to obtain more accurate numerical solutions, we may have to design more complicated numerical methods. However, we must make sure that the increased manual effort in construction and programming, results in a faster solution procedure on the computer for obtaining a certain accuracy.

Let us now study the linear problem

$$\frac{du}{dt} = -u, \tag{10.2}$$
$$u(0) = 1.$$

It has the exponentially decreasing solution $u(t) = e^{-t}$, and there is of course no need to use a numerical method. But we do that anyway to illustrate some interesting phenomena. The Euler method is

$$u_{n+1} = u_n - u_n \Delta t, \quad n = 0, 1, \ldots,$$

and we try the different time steps $\Delta t = 2.1$ and $\Delta t = 1.9$. The result is shown in Fig. 10.1 together with the true solution. Clearly, the numerical solution is completely wrong. Furthermore, for the larger time step, the amplitude of the solution is growing, and it will never approach zero which the true solution will do for large t.

It is easy to see why things are going wrong. The scheme can be written as

$$u_{n+1} = (1 - \Delta t)u_n, \quad n = 0, 1, \ldots.$$

The number sequence $\{u_n\}$ is nonincreasing for increasing n if $|1 - \Delta t| \leq 1$, i.e., if $\Delta t \leq 2$. We call the method stable if this condition is satisfied. For $1 < \Delta t \leq 2$ the solution will be oscillating, but at least it will not take off without bounds.

Let us next modify the scheme such that the right hand side $-u$ of the differential equation is taken as $-u_{n+1}$ in the interval $[t_n, t_{n+1}]$. Then the method becomes

$$u_{n+1} = u_n - u_{n+1} \Delta t, \quad n = 0, 1, \ldots,$$

or equivalently

$$u_{n+1} = \frac{1}{1 + \Delta t} u_n, \quad n = 0, 1, \ldots.$$

This is called the *Euler backward* method. We run the same cases as above, and the result is shown Fig. 10.2. The oscillatory behavior is gone and, despite the very large time steps, the solution looks reasonable for all t.

The behavior is again easy to explain. The number multiplying u_n satisfies the stability condition $|1/(1 + \Delta t)| \leq 1$ for all Δt, and we call the method *unconditionally stable*.

Fig. 10.1 Solution of (10.2) by the Euler method, $\Delta t = 2.1\ (--)\ \Delta t = 1.9\ (-\cdot)$, true solution (—)

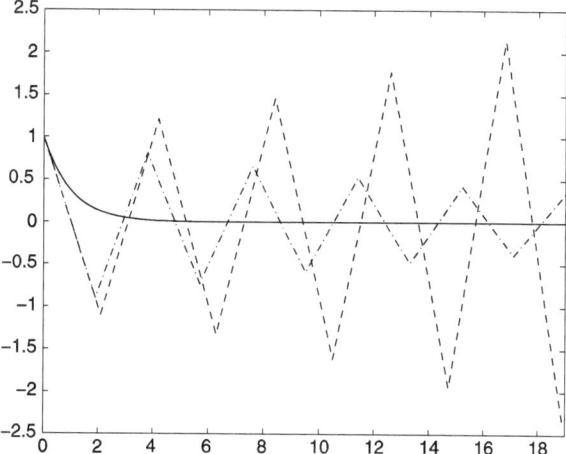

Fig. 10.2 Solution of (10.2) by the Euler backward method, $\Delta t = 2.1\ (--)\ \Delta t = 1.9\ (-\cdot)$, true solution (—)

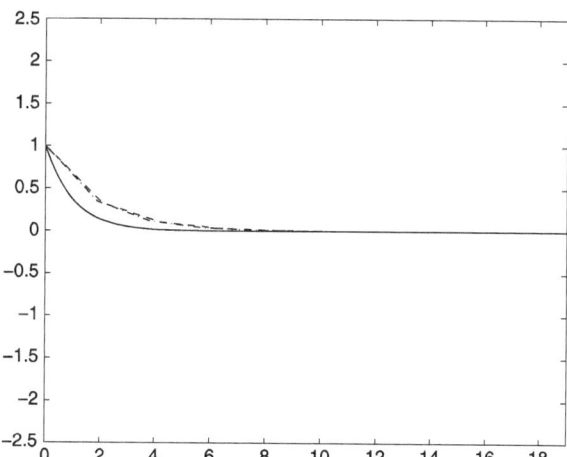

It seems that it should not make much difference if we choose to approximate the right hand side $-u$ of (10.2) by the value $-u_n$ at one end of the interval or by $-u_{n+1}$ at the other end. But obviously it does.

For systems of ODE, the methods for scalar equations can be generalized by simply switching to vector notation. For example, the Euler backward method for the differential equation

$$\frac{d\mathbf{u}}{dt} = g(t, \mathbf{u})$$

is

$$\mathbf{u}_{n+1} = \mathbf{u}_n + \mathbf{g}(t_{n+1}, \mathbf{u}_{n+1})\Delta t.$$

This is simple enough to write down, but what does it take to solve it? The unknown vector is $\mathbf{u}_{n+1}$, and it is determined by the vector equation

$$\mathbf{u}_{n+1} - \mathbf{g}(t_{n+1}, \mathbf{u}_{n+1})\Delta t = \mathbf{u}_n. \tag{10.3}$$

This is a nonlinear system of N equations for the N unknown elements in $\mathbf{u}_{n+1}$. It seems like a hopeless task to solve such a system for each time step, but it is not. We shall discuss iterative solution methods for it in Chap. 13.

The analysis of a system of ODE is much harder compared to a scalar ODE, but there are effective tools to simplify the analysis. We take a linear system

$$\frac{d\mathbf{u}}{dt} = A\mathbf{u},$$

where A is an $N \times N$ matrix. Assuming that A has N linearly independent eigenvectors, we let T be the matrix that takes A to diagonal form (see Sect. 3.4):

$$T^{-1}AT = \Lambda = \text{diag}(\lambda_1, \lambda_2, \ldots, \lambda_N).$$

We now multiply the differential equation from the left by T^{-1}. Since T does not depend on t, and $T^{-1}T = I$, the differential equation can be written as

$$\frac{d(T^{-1}\mathbf{u})}{dt} = T^{-1}ATT^{-1}\mathbf{u},$$

or with $\mathbf{v} = T^{-1}\mathbf{u}$.

$$\frac{d\mathbf{v}}{dt} = \Lambda\mathbf{v}.$$

But this is a set of scalar ODE that are independent of each other, and the analysis has become considerably simpler.

As an example we take a case where all the eigenvalues λ_j are real and negative. This means that

$$|v_j(t)| = |e^{\lambda_j t}v_j(0)| \leq |v_j(0)|, \quad j = 1, 2, \ldots, N,$$

and obviously we have

$$\|\mathbf{v}(t)\| \leq \|\mathbf{v}(0)\|$$

for the vector norm. For the original ODE we get

$$\|\mathbf{u}(t)\| = \|T\mathbf{v}(t)\| \leq \|T\|\|\mathbf{v}(t)\| \leq \|T\|\|\mathbf{v}(0)\| \leq \|T\|\|T^{-1}\|\|\mathbf{u}(0)\|.$$

For the original system there may be an increase of the norm, but an a priori bound is known, and it is independent of t. The bound

$$\text{cond}(T) = \|T\|\|T^{-1}\|$$

is called the *condition number* of the matrix T, and it is going to show up again when discussing linear systems of algebraic equations in Chap. 14. Here we conclude that a system of ODE becomes sensitive to perturbations, and therefore harder to solve, when the eigenvectors of the coefficient matrix A are almost linearly dependent.

Let us now analyze the Euler backward method for the same system

$$\mathbf{u}_{n+1} = \mathbf{u}_n + A\mathbf{u}_{n+1}\Delta t.$$

After the same type of transformation as we used for the differential equation, we get

$$\mathbf{v}_{n+1} = \mathbf{v}_n + \Lambda \mathbf{v}_{n+1}\Delta t.$$

From the scalar analysis above we know that each component of $\mathbf{v}$ is a nonincreasing sequence for increasing n. Obviously this leads to the inequality

$$\|\mathbf{v}_n\| \le \|\mathbf{v}_0\|$$

for the vector norm. For the original scheme we get in the same way as for the ODE system

$$\|\mathbf{u}_n\| = \|T\mathbf{v}_n\| \le \|T\|\|\mathbf{v}_n\| \le \|T\|\|\mathbf{v}_0\| \le \|T\|\|T^{-1}\|\|\mathbf{u}_0\| = \mathrm{cond}(T)\|\mathbf{u}_0\|.$$

The bound is identical to the one for the ODE system.

The conclusion from this exercise is that the eigenvalue analysis is very powerful. It shows that when analyzing a certain difference method for a system of ODE, we gain much knowledge by analyzing how it works for a scalar equation

$$\frac{du}{dt} = \lambda u,$$

which goes under the name the *test equation*. In Sect. 6.1 this equation was discussed briefly as a result of a Fourier transformed PDE. The number λ is there the Fourier transform of a differential operator in space, and its location in the complex plane is essential for the properties of the original PDE. The solution is

$$u(t) = e^{\lambda t}u(0),$$

and we note that it is nonincreasing with time if and only if $\mathrm{Re}\,\lambda \le 0$.

But λ may as well be the discrete Fourier transform of a difference operator in space, and in that case the solution of the test equation tells something about the semidiscrete approximation.

As another example of discretization in time, we apply the trapezoidal rule

$$u_{n+1} = u_n + \frac{\lambda\Delta t}{2}(u_n + u_{n+1}),$$

or equivalently

$$u_{n+1} = \frac{1 + \lambda\Delta t/2}{1 - \lambda\Delta t/2}u_n.$$

For negative λ, the sequence $\{u_n\}$ is nonincreasing, and we have an unconditionally stable scheme.

The parameters λ and Δt will always occur as $\mu = \lambda\Delta t$ in the right hand side for any consistent one-step scheme for the test equation, and the general form is

$$u_{n+1} = z(\mu)u_n, \tag{10.4}$$

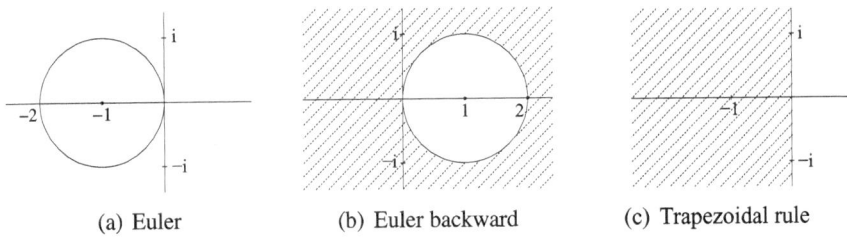

(a) Euler   (b) Euler backward   (c) Trapezoidal rule

Fig. 10.3 Stability domains

where the *amplification factor* $z(\mu)$ is a scalar function of μ. From a stability point of view the interesting question is for what values of μ do we have $|z(\mu)| \leq 1$. We recall that the eigenvalues of a matrix may be complex even if the matrix is real, and it is therefore necessary to consider complex μ. We make the formal definition:

The *stability domain* for a difference method (10.4) is the set $S(\mu)$ in the complex plane which satisfies $|z(\mu)| \leq 1$.

The (shaded) stability domains for the Euler, Euler backward and trapezoidal method are shown in Fig. 10.3.

Since $\mathrm{Re}(\mu) = \mathrm{Re}(\lambda \Delta t) \leq 0$ if and only if $\mathrm{Re}\,\lambda \leq 0$, we note that the trapezoidal rule is the only method that is stable for exactly those values of λ where the true solution is nonincreasing. The Euler method has further stability restrictions, while the Euler backward method is "overstable", i.e., it is stable also for certain λ where the true solution grows.

A few warnings concerning the test equation are appropriate. The assumption of a full set of eigenvectors is not always fulfilled, and then a scalar ODE doesn't tell it all. Secondly, the matrix A may depend on t, and the diagonalization does not go through that easily.

Even worse is of course a nonlinear ODE. In that case one can linearize the equation, which we shall sketch for the ODE

$$\frac{du}{dt} = g(u), \tag{10.5}$$

where $g(u)$ is a nonlinear function of u. We make a small perturbation $u \rightarrow u + v$, where $|v|$ is small, and plug it into the differential equation. A Taylor expansion gives

$$\frac{du}{dt} + \frac{dv}{dt} = g(u + v) = g(u) + \frac{dg}{du}(u)v + \mathcal{O}(|v|^2).$$

When using the differential equation (10.5) and neglecting the square terms, we get the equation

$$\frac{dv}{dt} = \frac{dg}{du}(u)v.$$

If we now assume that the function $u = u(t)$ is known, we have a linear differential equation for $v(t)$. In a real application we do not of course know u, since that is the solution we want to compute. But we may know for example that the derivative

dg/du is negative, which brings us back to the example above. By understanding how the linear problem behaves, we know how a small perturbation of the solution to the original problem develops with time. In the same way we gain knowledge about the original difference scheme by studying the corresponding linear difference scheme.

The procedure described here is known as *linearization* of the equation, and is a very common analysis tool. If we know that there is a bound on small perturbations when time increases, the computation can be done with more confidence.

A fundamental question is of course how accurate the numerical solution is. As an example, we consider the Euler method. The first question is how well the difference scheme approximates the differential equation, and the answer is obtained by substituting the true solution $u(t)$ of the differential equation into the difference scheme. Since it cannot be expected to satisfy this scheme exactly, we have

$$u(t_{n+1}) = u(t_n) + \Delta t g\big(u(t_n)\big) + R,$$

and the question is how big is the remainder R? The Taylor expansion gives

$$u(t_{n+1}) = u(t_n) + \Delta t \frac{du}{dt}(t_n) + \frac{\Delta t^2}{2}\frac{d^2 u}{dt^2}(t_n) + \mathcal{O}(\Delta t^3),$$

and by using the differential equation $du/dt = g$ we get

$$u(t_{n+1}) = u(t_n) + \Delta t g(t_n) + \frac{\Delta t^2}{2}\frac{d^2 u}{dt^2}(t_n) + \mathcal{O}(\Delta t^3).$$

Since we are dealing with differential equations, it is natural to normalize the equation by dividing by Δt:

$$\frac{u(t_{n+1}) - u(t_n)}{\Delta t} = g(t_n) + \frac{\Delta t}{2}\frac{d^2 u}{dt^2}(t_n) + \mathcal{O}(\Delta t^2).$$

By letting Δt tend to zero, we recover the differential equation in the limit. The error for finite but small Δt

$$T(\Delta t) = \frac{\Delta t}{2}\frac{d^2 u}{dt^2}(t_n) + \mathcal{O}(\Delta t^2) = \mathcal{O}(\Delta t)$$

is called the *truncation error*. (The error $R = \mathcal{O}(\Delta t^2)$ defined above is called the *local truncation error*.)

It is important to distinguish between the truncation error on one hand, describing the error in the approximation of the *differential equation*, and the error $u_n - u(t_n)$ in the approximate *solution* on the other hand. It can be shown that they are of the same order under the important condition that the difference scheme is stable in a certain sense. We shall not go into those details here. For linear ODE the analysis is not very difficult. For the Euler and Euler backward schemes one can show that the error is of the order $\mathcal{O}(\Delta t)$, while it is $\mathcal{O}(\Delta t^2)$ for the trapezoidal rule.

In general, if $T(\Delta t) = \mathcal{O}(\Delta t^p)$ with $p > 0$, then the difference scheme is *consistent*, and we say that the difference scheme has *order of accuracy p*. If we also have $|u_n - u(t_n)| = \mathcal{O}(\Delta t^p)$ with $p > 0$, then the numerical solution *converges* to the true solution as $\Delta t \to 0$, i.e., for any fixed time $t = T$ we have

$$\lim_{\Delta t \to 0} |u_{T/\Delta t} - u(T)| = 0.$$

We say that the difference scheme is *convergent*.

In practical computations one can of course never reach the limit $\Delta t = 0$. However, the theoretical concept of convergence is still fundamental. If a certain computation gives a result that is not accurate enough, we would like to get a more accurate result if the computation is repeated with a smaller time step. This can be expected with a convergent difference scheme.

The examples we have discussed so far have order of accuracy one or two. The difference methods used in practice are often of higher order. There are essentially two ways of achieving this. One is to aim for *one-step methods* where only one time level t_n is used for computing u_{n+1}. This requires several stages in the computation, and we arrive at the large class of *Runge–Kutta methods*, named after the German mathematicians Carl Runge (1856–1927) and Martin Wilhelm Kutta (1867–1944). The most common method is the fourth order version

$$k_1 = g(u_n),$$

$$k_2 = g\left(u_n + \frac{\Delta t}{2}k_1\right),$$

$$k_3 = g\left(u_n + \frac{\Delta t}{2}k_2\right),$$

$$k_4 = g(u_n + \Delta t k_3),$$

$$u_{n+1} = u_n + \frac{\Delta t}{6}(k_1 + 2k_2 + 2k_3 + k_4).$$

It may seem like a strange formula, but the simple test equation $du/dt = \lambda u$ indicates how it is derived. For this equation we have

$$\frac{du}{dt} = \lambda u, \qquad \frac{d^2 u}{dt^2} = \lambda^2 u, \qquad \frac{d^3 u}{dt^3} = \lambda^3 u, \qquad \frac{d^4 u}{dt^4} = \lambda^4 u,$$

and by Taylor expansion

$$u(t + \Delta t) = u(t) + \Delta t \lambda u(t) + \frac{\Delta t^2}{2}\lambda^2 u(t)$$

$$+ \frac{\Delta t^3}{6}\lambda^3 u(t) + \frac{\Delta t^4}{24}\lambda^4 u(t) + \mathcal{O}(\Delta t^5).$$

The Runge–Kutta method for our equation is

$$k_1 = \lambda u_n,$$

$$k_2 = \left(\lambda + \frac{\Delta t}{2}\lambda^2\right)u_n,$$

$$k_3 = \left(\lambda + \frac{\Delta t}{2}\lambda^2 + \frac{\Delta t^2}{4}\lambda^3\right)u_n,$$

$$k_4 = \left(\lambda + \Delta t \lambda^2 + \frac{\Delta t^2}{2}\lambda^3 + \frac{\Delta t^3}{4}\lambda^4\right)u_n,$$

$$u_{n+1} = \left(1 + \Delta t \lambda + \frac{\Delta t^2}{2}\lambda^2 + \frac{\Delta t^3}{6}\lambda^3 + \frac{\Delta t^4}{24}\lambda^4\right)u_n,$$

Fig. 10.4 Stability domains
for Runge–Kutta methods

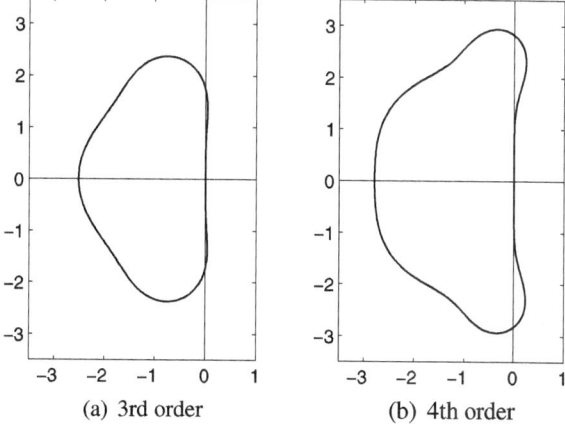

(a) 3rd order   (b) 4th order

i.e., exactly the Taylor expansion above. After dividing by Δt, we find that the truncation error is $\mathcal{O}(\Delta t^4)$ as it should be. After a little more work for the general nonlinear differential equation, the result is the same. The Runge–Kutta method presented here has fourth order accuracy.

Note that it is a one-step method in the sense that the solution u_n at only one time level is required in order to compute u_{n+1}. But there are several *stages* in the computational procedure.

The stability domain $S(\mu)$ is obtained by finding the values of μ for which

$$|z(\mu)| = \left| 1 + \mu + \frac{\mu^2}{2} + \frac{\mu^3}{6} + \frac{\mu^4}{24} \right| \le 1.$$

In Fig. 10.4 $S(\mu)$ is shown for both the third and fourth order Runge–Kutta methods. In the third order case, $z(\mu)$ has the same expansion as in the fourth order case, except for the last term, which is not present.

Runge–Kutta type methods of very high order have been derived over the years. They all have the same structure as above, but the number of stages grows with higher order of accuracy (more expressions k_j to be stored).

The method above is explicit, but there are also implicit Runge–Kutta methods. They have the same overall structure but, in the formula for k_j, the same quantity k_j occurs also in the right hand side in the argument v of $g(v)$. This requires the solution of a nonlinear equation at each stage for each step $u_n \rightarrow u_{n+1}$. The advantage is that the stability properties improve.

Another way of constructing high order methods is to involve more than two time levels when advancing the solution one step. The simplest such method is the second order accurate *leap-frog method*

$$u_{n+1} = u_{n-1} + 2\Delta t g(u_n).$$

It is a special case of a *linear multistep method*, and it requires two initial values to get started. If u_0 is a given initial value for the differential equation, we need also u_1. That value must be computed by a one-step method.

A general linear multistep method has the form

$$\alpha_m u_{n+m} + \alpha_{m-1} u_{n+m-1} + \cdots + \alpha_0 u_n$$
$$= \Delta t g(\beta_m u_{n+m} + \beta_{m-1} u_{n+m-1} + \cdots + \beta_0 u_0).$$

The word "linear" in the name for this class refers to the fact that $g(u)$ occurs in a linear way in the formula (not as $g^2(u)$ for example), and has nothing to do with the type of function $g(u)$, which may very well be nonlinear. One can also use the different form

$$\alpha_m u_{n+m} + \alpha_{m-1} u_{n+m-1} + \cdots + \alpha_0 u_n$$
$$= \Delta t \big(\beta_m g(u_{n+m}) + \beta_{m-1} g(u_{n+m-1}) + \cdots + \beta_0 g(u_0) \big),$$

which has the same order of accuracy. There is a significant flexibility in the choice of coefficients α_j and β_j. For a given order of accuracy, there are many ways of choosing the coefficients. If β_{n+m} is nonzero, the method is implicit but, if we want to keep the simpler explicit structure obtained with $\beta_{n+m} = 0$ while keeping the order of accuracy, we have to add more time levels at the other end.

The leap-frog method above has a symmetric and simple structure, and it is tempting to generalize it to higher order. By Taylor expansion it is easy to show that

$$\frac{du}{dt}(t) = \frac{1}{\Delta t} \left(-\frac{1}{12} u(t + 2\Delta t) + \frac{2}{3} u(t + \Delta t) - \frac{2}{3} u(t - \Delta t) + \frac{1}{12} u(t - 2\Delta t) \right)$$
$$+ \mathcal{O}(\Delta t^4), \tag{10.6}$$

which leads to the simple fourth order method

$$-\frac{1}{12} u_{n+4} + \frac{2}{3} u_{n+3} - \frac{2}{3} u_{n+1} + \frac{1}{12} u_n = \Delta t g(u_{n+2}). \tag{10.7}$$

In order to find out about the stability domain for the test equation, $g(u_{n+2})$ is replaced by λu_{n+2}. It is easy to determine when a given one-step method is stable as we saw above, but here we encounter a new difficulty. When more time levels are involved, how do we analyze stability? We write a general difference equation as

$$c_m u_{n+m} + c_{m-1} u_{n+m-1} + \cdots + c_0 u_n = 0.$$

The key to the analysis is the roots z_j of the *characteristic equation*

$$c_m z^m + c_{m-1} z^{m-1} + \cdots + c_0 = 0,$$

which is formally obtained by substituting $u_n = z^n$ and then dividing by z^n. If all the roots are distinct, then the general solution has the form

$$u_n = a_1 z_1^n + a_2 z_2^n + \cdots + a_m z_m^n,$$

where the constants a_j are determined by the m initial conditions. For stability we require that the solution has no growing component, and obviously the condition is $|z_j| \le 1$ for all j. If there is a double root z_1, the form of the solution is

$$u_n = (a_1 + a_2 n) z_1^n + a_3 z_3^n + \cdots + a_m z_m^n.$$

If $|z_1| = 1$, the solution will grow without bound when n increases. If on the other hand $|z_1| < 1$, then the component $a_2 n z_1^n$ will grow initially as n increases, but then it will decrease. This means that the solution stays bounded by a constant K which is independent of n. If there is a root with multiplicity higher than two, the polynomial multiplying it will be of higher degree, but the conclusion is again that the solution stays bounded independent of n if $|z_1| < 1$.

For the test equation, the roots will be functions of $\mu = \lambda \Delta t$. The definition of the stability domain for linear multistep methods is:

$$S = \{\mu : \text{all roots } z_j(\mu) \text{ satisfy } |z_j(\mu)| \le 1, \text{ multiple roots satisfy } |z_j(\mu)| < 1\}.$$

Let us now go back to the leap-frog method. The characteristic equation is

$$z^2 - 2\mu z - 1 = 0$$

with the roots

$$z_{1,2} = \mu \pm \sqrt{\mu^2 + 1}.$$

It is easily shown that the stability domain is just the line segment

$$\{\mu : \text{Re}\,\mu = 0, \ |\text{Im}\,\mu| < 1\}$$

on the imaginary axis. However, it is not as bad as it looks. Many problems are such that the coefficient matrix of the linearized system of ODE has purely imaginary eigenvalues. A simple example is

$$\frac{du}{dt} = v, \qquad \frac{dv}{dt} = -u,$$

which can be written as

$$\frac{d\mathbf{u}}{dt} = A\mathbf{u}, \qquad \mathbf{u} = \begin{bmatrix} u \\ v \end{bmatrix}, \qquad A = \begin{bmatrix} 0 & 1 \\ -1 & 0 \end{bmatrix} \tag{10.8}$$

The eigenvalues of A are given by

$$\lambda^2 + 1 = 0,$$

i.e., $\lambda_{1,2} = \pm i$. Accordingly, the leap-frog scheme

$$\mathbf{u}_{n+1} = \mathbf{u}_{n-1} + 2\Delta t A \mathbf{u}_n.$$

is stable for $\Delta t < 1$.

Next we take a look at the fourth order method (10.7) derived above. A minimal requirement is that it should be stable for $\mu = 0$ corresponding to a constant solution u independent of time. For the fourth order method above, it turns out that one of the roots is $z_1 = 7.873$ showing that the method is completely useless. The same conclusion follows for any symmetric method of the same type with accuracy $6, 8, \ldots$. Here we have a case where increasing the formal order of accuracy has a negative effect. The stability is a key concept that always must be kept in mind.

Fig. 10.5 Solution of (10.9)
by the Euler method,
$\Delta t = 0.001$ $(--)$,
$\Delta t = 0.01025$ $(—)$

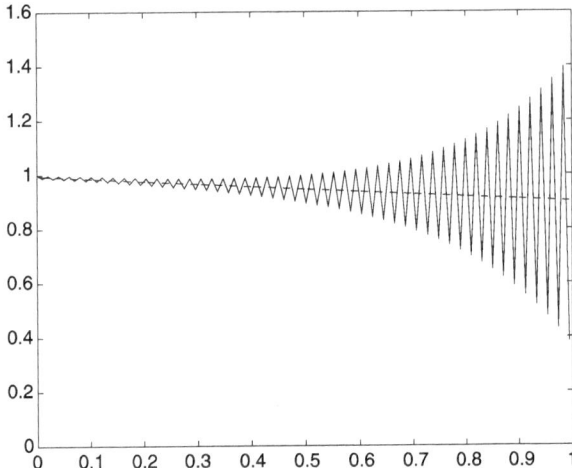

Finally we shall discuss an important class of ODE that is quite common in applications. Consider the initial value problem

$$\frac{d}{dt}\begin{bmatrix} u \\ v \end{bmatrix} = \begin{bmatrix} -100 & 99.9 \\ 99.9 & -100 \end{bmatrix}\begin{bmatrix} u \\ v \end{bmatrix},$$

$$u(0) = 1,$$

$$v(0) = 0.99.$$

(10.9)

We apply the Euler method with the two different time-steps $\Delta t = 0.001$ and $\Delta t = 0.01025$, and the result is shown in Fig. 10.5 for $u(t)$. The dashed curve with the shorter time step is smooth, and as we will see below, it is close to the true solution. The solution with the larger time step goes completely wrong. Obviously we have again a case with an instability, but there is actually a new observation to be made here.

We go back to the scalar problem (10.2) with the true solution shown in Fig. 10.2. In this case the stability limit is not a severe restriction, since the time steps have to be relatively small anyway in order to resolve the solution properly at the beginning of the time interval. Actually, a time step of the order $\Delta t = 0.1$ seems quite reasonable when looking at the graph, and this is far below the stability limit $\Delta t = 2$. The solution of (10.9) is very smooth, and a time step of the order $\Delta t = 0.1$ is certainly enough to get a good resolution. What is the reason for the severe time restriction?

The coefficient matrix has the two eigenvalues $\lambda_1 = -0.1$ and $\lambda_2 = -199.9$, i.e., they are far apart. Such a system of ODE is called *stiff*, and it is very common in many different types of applications. One of them occurs in chemistry when dealing with a number of components where the chemical reactions take place on different time scales. Another application is obtained when discretizing partial differential equations as we shall see later on.

With the new variables $\phi = (u + v)/2$ and $\psi = (u - v)/2$ in (10.9) we get

$$\frac{d\phi}{dt} = -0.1\phi, \qquad \frac{d\psi}{dt} = -199.9\psi,$$

$$\phi(0) = 0.995, \qquad \psi(0) = 0.005,$$

which is a direct way of diagonalizing the system. The function $\psi(t) = 0.005e^{-199.9t}$ is almost zero all the time, and the other function $\phi(t) = 0.995e^{-0.1t}$ is very smooth. By going back to the original variables $u = \phi + \psi$ and $v = \phi - \psi$, we see that they are very smooth as well. However, the crucial point is that the eigenvalue $\lambda_2 \approx -200$, entering in the form of e^{-200t}, is present all the time, even if it is annihilated for the true solution by the choice of initial values. The discrete solution introduces perturbations triggering the "parasitic" solution, which cannot be handled by the Euler scheme if the time steps are not extremely small.

In the example above, a good approximation would be to assume from the beginning that $u - v = 0$ all the time. For the general case, such a type of assumption leads to a new type of system. Let $\mathbf{u}(t)$ and $\mathbf{v}(t)$ be two vector functions with m and n components respectively. A general *differential-algebraic system* has the form

$$\frac{d\mathbf{u}}{dt} = \mathbf{f}(\mathbf{u}, \mathbf{v}),$$

$$\mathbf{g}(\mathbf{u}, \mathbf{v}) = \mathbf{0}.$$

Here the vector $\mathbf{f}$ has m components, and $\mathbf{g}$ has n components.

Differential-algebraic systems are limits of stiff systems and can be handled as such when solving them numerically, but in general special methods are used.

Stiff systems and differential-algebraic systems have been studied extensively, and very effective numerical methods are available today.

Modern variants of finite difference methods work with variable step size and even variable order of accuracy. The algorithm contains various types of sensors that are estimating the error. If a local fast change of the solution occurs, the solver works hard to reduce the time step in order to produce a well resolved solution.

There are many ODE-solvers on the market today. The MATLAB system has at least seven solvers, mainly divided into stiff and nonstiff classes. The *ode45-*function is the standard solver, using a Runge–Kutta method.

Exercise 10.1 When analyzing the linear ODE $du/dt = \lambda(t)u$, it is of interest to know the limit $\max_t |\lambda(t)|$. Derive the linearized form of the nonlinear ODE

$$\frac{du}{dt} = \frac{1}{u^2 + 1}, \tag{10.10}$$

and determine the limit of the coefficient that corresponds to $\lambda(t)$.

Exercise 10.2 Consider the initial value problem (10.2) with the solution $u(t) = e^{-t}$. Prove that the difference scheme

$$u_{n+1} = u_{n-1} - 2\Delta t u_n$$

is unstable for any Δt, while

$$u_{n+1} = u_{n-1} - \Delta t \, (u_{n+1} + u_{n-1})$$

is stable for all Δt.

Exercise 10.3 Use the MATLAB ODE-solvers *ode45* and *ode23* (see Sect. 18.2) for solving (10.10) with the initial value $u(0) = 0$. Compare the choice of time steps for the two methods.

10.2 Partial Differential Equations

Let us now turn to partial differential equations and the so-called transport equation (2.18) with $c = 1$. We introduce the two-dimensional grid $(x_j, t_n) = (j \Delta x, n \Delta t)$ and the grid function u_j^n as an approximation of $u(j \Delta x, n \Delta t)$, see Fig. 10.6.

Note that we have switched notation from the previous section by changing the subscript n indicating time level t_n, to a superscript. In this way it is easier to distinguished from the subscript j indicating the grid point in space. One has to be careful though, not to confuse the superscript n with the power notation.

At a certain point (x_j, t_n) we substitute

$$\frac{\partial u}{\partial x} \rightarrow \frac{u_{j+1}^n - u_j^n}{\Delta x},$$

$$\frac{\partial u}{\partial t} \rightarrow \frac{u_j^{n+1} - u_j^n}{\Delta t}$$

in the differential equation and obtain

$$u_j^{n+1} = u_j^n - \frac{\Delta t}{\Delta x} (u_{j+1}^n - u_j^n). \tag{10.11}$$

If the initial function $u_j^0 = u(x_j, 0)$ is known, we can compute u_j^1 for all j, then u_j^2 and so on.

Let us now try an experiment. An initial pulse is defined as in Sect. 9.2 with its center at $x = 0.2$ as

$$u(x, 0) = e^{-800(x - 0.2)^2}.$$

The solution at time t is

$$u(x, t) = e^{-800(x - t - 0.2)^2},$$

i.e., the pulse has moved a distance t to the right. We do the computation in the x-interval $[0, 1]$ with $N = 1/\Delta x$ grid points, and choose $\Delta t = 0.8 \Delta x$. Figure 10.7(a) shows the result at $t = 0.038$ for $N = 400$. It is centered properly at $x = 0.238$, but the peak is too high. It is reasonable to assume that a finer grid should give better results, since the difference scheme approximates the differential equation more closely. However, we are in for a surprise. With half the step size in both directions,

Fig. 10.6 Computational grid

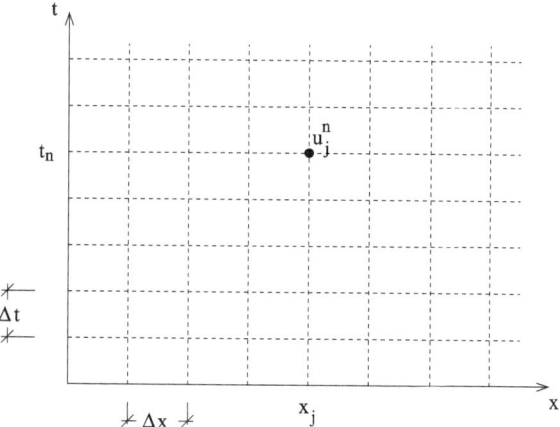

we get the result shown in Fig. 10.7(b). The solution goes completely wrong and shows strange oscillations. What has happened?

We have defined the derivative as a limit of a *forward difference* $(u(x + \Delta x, t) - u(x, t))/\Delta x$. As noted earlier it is of course also possible to define it as a limit of a *backward difference* $(u(x, t) - u(x - \Delta x, t))/\Delta x$. When using this as the basis for our difference approximation, we get

$$u_j^{n+1} = u_j^n - \frac{\Delta t}{\Delta x}(u_j^n - u_{j-1}^n). \tag{10.12}$$

It turns out that the numerical solution now behaves well, and we can compute it over long time. Figure 10.8 shows the result at $t = 0.4$ with the same data as above. For $N = 400$ the pulse is centered at the right position $x = 0.6$, but the top is too low. With half the step size we get a better result as shown in Fig. 10.8(b).

The first computation with forward differences in the approximation is an example of an *unstable* computation, while the second one with *backward differences* is a *stable* computation. Actually, there is a simple explanation for the bad behavior of the first one. At any given grid point (x_j, t_{n+1}) the approximation doesn't use any point to the left at the previous time level. Since the pulse is moving to the right, we must know what is coming in from the left. The second approximation takes this into account.

Actually, there is not plain sailing with the second approximation either. We do the same computation with $\Delta t = 1.2\Delta x$, and the result is shown in Fig. 10.9 at $t = 0.32$. Severe oscillations have occurred, even quite far from the pulse, and obviously the numerical solution is useless. Apparently, the time step must be chosen small enough in order to retain the stability. This is in accordance with the discussion about stability domains for ODE. For PDE the theoretical analysis is harder but, by using Fourier analysis to be presented in the next section, one can show that stability requires the condition $\Delta t \le \Delta x$.

Let us next compute the solution to the heat conduction problem (2.20). We construct a difference scheme by using the three point expression used in the definition of the second derivative $\partial^2 u/\partial x^2$ and a forward finite difference for $\partial u/\partial t$. With

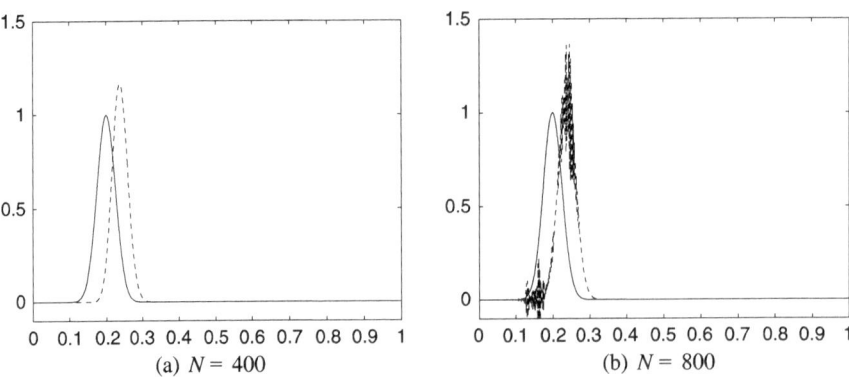

Fig. 10.7 Solution of (10.11), $t = 0$ (—) and $t = 0.038$ (−−)

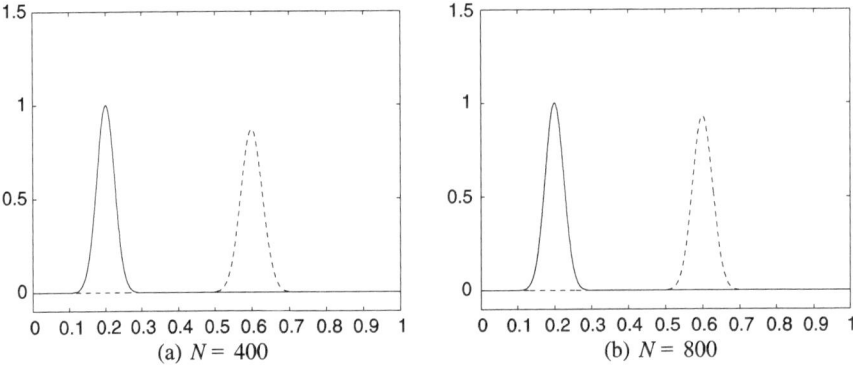

Fig. 10.8 Solution of (10.12) with $\Delta t = 0.8 \Delta x$, $t = 0$ (—) and $t = 0.4$ (−−)

Fig. 10.9 Solution of (10.12) with $\Delta t = 1.2 \Delta x$, $t = 0$ (—) and $t = 0.32$ (−−)

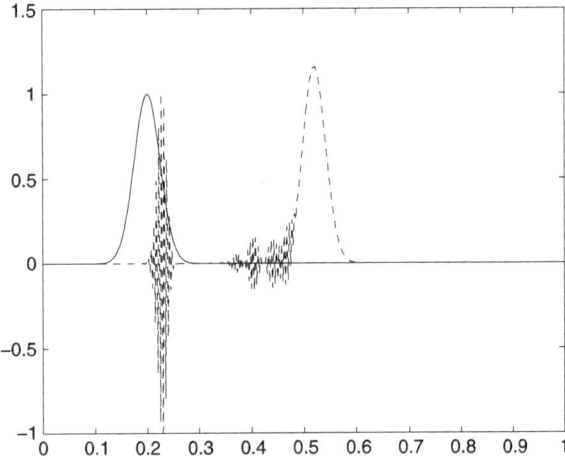

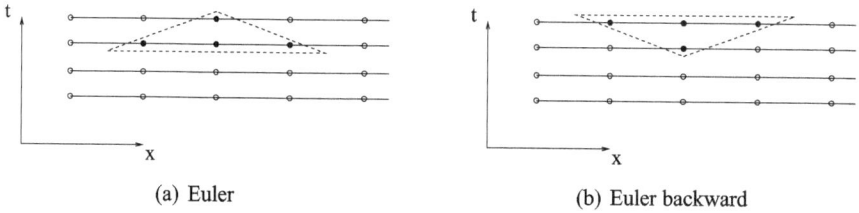

(a) Euler	(b) Euler backward

Fig. 10.10 Computational stencils for the heat equation

$N + 1$ grid points in the x-direction including the boundary points, the resulting scheme is

$$u_j^{n+1} = u_j^n + \frac{\Delta t}{\Delta x^2}(u_{j-1}^n - 2u_j^n + u_{j+1}^n),$$
$$j = 1, 2, \ldots, N - 1, \ n = 0, 1, \ldots,$$
$$u_0^{n+1} = 1,$$
$$u_N^{n+1} = 1,$$
$$u_j^0 = f_1(x_j),$$

as illustrated in Fig. 10.10(a). With u_j^0 known for all j, u_j^1 can be computed for all j, and so on until we reach the final time level.

Even if the solution is obtained by stepping forward in time, one could still use an approximation based on the Euler backward scheme, just as for ordinary differential equations discussed above. The forward difference in time is replaced by a backward difference, which gives

$$u_j^{n+1} = u_j^n + \frac{\Delta t}{\Delta x^2}(u_{j-1}^{n+1} - 2u_j^{n+1} + u_{j+1}^{n+1}),$$
$$j = 1, 2, \ldots, N - 1, \ n = 0, 1, \ldots,$$
$$u_0^{n+1} = 1,$$
$$u_N^{n+1} = 1,$$
$$u_j^0 = f_1(x_j), \tag{10.13}$$

see Fig. 10.10(b). This is an implicit scheme, and it requires more computation for each time step. Since all grid points at the new time level t_{n+1} are coupled to each other, this complication is more severe compared to ODE. We must solve a large system of equations for each time step.

For a well posed problem, there is still a possibility that the numerical scheme goes wrong as we saw for the equation $\partial u / \partial t + \partial u / \partial x = 0$ above. We run the first explicit approximation above with two different time steps. Figure 10.11 shows the result for the step size $\Delta t = 0.000310$ and $\Delta t = 0.000315$.

Apparently there is a critical limit somewhere in between these two values. Indeed, by using analytical tools the theoretical stability limit on Δt can be found. In the next section we shall describe how this can be done.

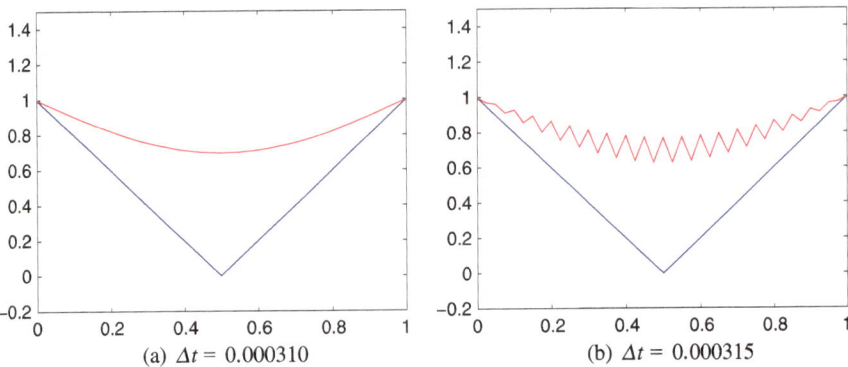

Fig. 10.11 The heat equation

The implicit scheme above does not have any stability restriction on the time step. It is *unconditionally stable*, which is typical for implicit schemes.

The consistency and convergence concepts can be defined in analogy with ordinary differential equations. A *consistent* difference scheme approaches formally the differential equation as $\Delta x \to 0$ and $\Delta t \to 0$. It is *convergent* if for any fixed $t = T$ the numerical solution converges to the true solution:

$$\|u_j^{T/\Delta t} - u(x_j, T)\| \to 0 \quad \text{as } \Delta x \to 0, \ \Delta t \to 0.$$

The norm is a measure of the discrete function at a given time level corresponding to vector norms.

The consistency is usually easy to verify by a direct application of Taylor expansions, but convergence is not. However, there is a fundamental theorem saying that a consistent scheme is convergent if it is stable. Therefore, stability analysis is the key to the construction of accurate difference schemes. In the next section we shall indicate how it can be carried out by using a Fourier technique.

Exercise 10.4 Show that the difference method (10.13) requires the solution of a tridiagonal system (see Sect. 3.3) for each time step. Write down the system in detail.

Exercise 10.5 Suggest a difference method for $\partial u/\partial t = \partial^2 u/\partial x^2$ that uses a combination of $(u_{j-1}^n - 2u_j^n + u_{j+1}^n)/\Delta x^2$ and $(u_{j-1}^{n+1} - 2u_j^{n+1} + u_{j+1}^{n+1})/\Delta x^2$ for approximation of $\partial^2 u/\partial x^2$.

10.3 Fourier Stability Analysis

In this section we shall describe how one can analyze a given difference scheme for a partial differential equation with respect to its stability properties. We shall limit ourselves to the simplest form of analysis, which is based on the Fourier transform.

Even if the solutions are nonperiodic, it turns out that the analysis of a modified problem with periodic solutions gives significant information about the stability. Let us discuss the solution of the heat equation above, and the difference scheme

$$u_j^{n+1} = u_j^n + \frac{\Delta t}{\Delta x^2}(u_{j-1}^n - 2u_j^n + u_{j+1}^n),$$

$$j = 0, 1, \ldots, N, \ n = 0, 1, \ldots, \qquad (10.14)$$

$$u_j^0 = f_j.$$

Here we have canceled the boundary conditions, and assume periodicity instead: $u_{j+N+1}^n = u_j^n$. For difference methods, only the grid values are accounted for. Therefore, the solution at any time level t_n can be represented as a discrete Fourier series as was demonstrated in Sect. 6.2. We write the series in the form

$$u_j^n = \sum_{k=-N/2}^{N/2} c_k^n e^{ikx_j}, \quad j = 0, 1, \ldots, N,$$

where the coefficients are defined by

$$c_k^n = \frac{1}{2\pi} \sum_{j=0}^{N} u_j^n e^{-ikx_j} \Delta x.$$

The coefficients c_k^n are now time dependent, and the idea is to investigate how these coefficients are behaving when time increases.

At a first glance, it seems like a complication to study these coefficients instead of the original grid values u_j^n. But there are two facts that show why it is a good idea:

1. The difference scheme takes a particularly simple form when it is formulated in terms of the Fourier coefficients.
2. The behavior of the Fourier coefficients is directly related to the behavior of the original grid values via the discrete Parseval's relation (6.10).

We introduce the Fourier series into the difference approximation of the second space derivative and obtain

$$u_{j-1}^n - 2u_j^n + u_{j+1}^n = \sum_{k=-N/2}^{N/2} c_k^n (e^{ikx_{j-1}} - 2e^{ikx_j} + e^{ikx_{j+1}})$$

$$= \sum_{k=-N/2}^{N/2} c_k^n e^{ikx_j} q(\xi),$$

where

$$q(\xi) = e^{-i\xi} - 2 + e^{i\xi}, \quad \xi = k\Delta x.$$

The whole difference scheme can now be written as

$$\sum_{k=-N/2}^{N/2} \left(c_k^{n+1} - \left(1 + \sigma q(\xi)\right)c_k^n\right)e^{ikx_j} = 0,$$

where $\sigma = \Delta t / \Delta x^2$. In Sect. 6.2 it was demonstrated that the $N + 1$ grid functions e^{ikx_j} (which can also be considered as vectors) are linearly independent. By definition this means that each one of the coefficients in the sum must be zero, i.e.,

$$c_k^{n+1} = \left(1 + \sigma q(\xi)\right)c_k^n, \quad k = 0, 1, \ldots, N, \quad |\xi| \leq \pi. \tag{10.15}$$

This is quite a simplification! The original difference scheme couples neighboring points in space to each other, and the whole set of variables must be treated together. On the contrary, there is no coupling between the Fourier coefficients for different k-values, and the development with time can be handled separately for each one of them. This is in exact analogy with the "continuous" Fourier transform and differential operators as discussed in Sect. 6.1. A differential operator is replaced by multiplication by a number by using the Fourier transform also in that case.

Knowledge about the behavior of the Fourier coefficients is transferred to the solution of the difference scheme by the discrete Parseval's relation. If we can make sure that the Fourier coefficients do not grow with time, i.e.,

$$|c_k^{n+1}| \leq |c_k^n|, \quad k = 0, 1, \ldots, N,$$

then

$$\sum_{j=0}^{N} |u_j^{n+1}|^2 \Delta x = 2\pi \sum_{k=-N/2}^{N/2} |c_k^{n+1}|^2 \leq 2\pi \sum_{k=-N/2}^{N/2} |c_k^n|^2$$

$$= \sum_{j=0}^{N} |u_j^n|^2 \Delta x \leq \cdots \leq \sum_{j=0}^{N} |u_j^0|^2 \Delta x.$$

If we order the grid values u_j^n in a vector $\mathbf{u}^n$, then the norm is defined by

$$\|\mathbf{u}^n\|^2 = \sum_{j=0}^{N} |u_j^n|^2 \Delta x,$$

and we have

$$\|\mathbf{u}^n\|^2 \leq \|\mathbf{u}^0\|^2.$$

This could be used as the definition of stability. However, a more reasonable definition is to allow a constant in the estimate:

A difference approximation is stable if the solution satisfies

$$\|\mathbf{u}^n\| \leq K\|\mathbf{u}^0\|,$$

where K is a constant independent of n and $\mathbf{u}^0$.

We are now in a very good position. By making sure that the Fourier coefficients satisfy the *von Neumann condition*

$$|c_k^{n+1}| \leq |c_k^n|,$$

we have a final stability estimate for the solution. But (10.15) shows that this condition is satisfied if

$$|1 + \sigma q(\xi)| \leq 1.$$

By using the trigonometric interpretation of $e^{i\xi}$, we get

$$q(\xi) = e^{-i\xi} - 2 + e^{i\xi} = 2\cos\xi - 2,$$

leading to the inequality

$$|1 - 2\sigma(1 - \cos\xi)| \le 1.$$

Since $\cos\xi$ never exceeds one, the critical point is $\cos\xi = -1$. This leads to the final condition

$$\sigma = \frac{\Delta t}{\Delta x^2} \le \frac{1}{2}.$$

The difference scheme is stable if the time step is chosen small enough:

$$\Delta t \le \frac{\Delta x^2}{2}.$$

It may seem that the periodicity assumption on the solutions is too restrictive, making the stability result of little value. But it is not. It is actually a necessary stability condition, and it is often sufficient as well. The heat conduction problem above with the temperature specified at both boundaries is such an example. Furthermore, if the heat conduction coefficient depends on x and t, so that the differential equation is

$$\frac{\partial u}{\partial t} = \frac{\partial}{\partial x}\left(a(x,t)\frac{\partial u}{\partial x}\right)$$

with $a(x,t) > 0$, then the corresponding generalized difference scheme has the stability limit

$$\Delta t \le \frac{\Delta x^2}{2\max_{x,t} a(x,t)}.$$

Let us take another look at the transformation procedure used above. If the grid functions are organized as vectors

$$\mathbf{u}^n = \begin{bmatrix} u_0^n \\ u_1^n \\ \vdots \\ u_N^n \end{bmatrix},$$

then we can consider the difference scheme as a relation between the two vectors $\mathbf{u}^n$ and $\mathbf{u}^{n+1}$ connected by a matrix Q:

$$\mathbf{u}^{n+1} = Q\mathbf{u}^n.$$

For our example with periodic solutions, the matrix is

$$Q = \begin{bmatrix} 1-2\sigma & \sigma & & & & & \sigma \\ \sigma & 1-2\sigma & \sigma & & & & \\ & \sigma & 1-2\sigma & \sigma & & & \\ & & \ddots & \ddots & \ddots & & \\ & & & \sigma & 1-2\sigma & \sigma \\ \sigma & & & & \sigma & 1-2\sigma \end{bmatrix}.$$

In Sect. 6.2 we introduced the matrix F for the Fourier transform, and we multiply with it from the left, getting

$$F\mathbf{u}^{n+1} = FQF^{-1}F\mathbf{u}^n.$$

With the new vector $\mathbf{v}^n = F\mathbf{u}^n$ and $\Lambda = FQF^{-1}$, we get

$$\mathbf{v}^{n+1} = \Lambda\mathbf{v}^n.$$

The vector $\mathbf{v}^n$ is the Fourier transform of $\mathbf{u}^n$, and it has the elements c_k^n. Consequently we have a system of equations for each one of the Fourier coefficients c_k^n, and by (10.15) we can see that $\Lambda = FQF^{-1}$ is a diagonal matrix:

$$\Lambda = \mathrm{diag}\left(1 - \sigma q\left(\left(-\frac{N}{2}\right)\Delta x\right), 1 - \sigma q\left(\left(-\frac{N}{2} + 1\right)\Delta x\right), \ldots,\right.$$
$$\left.1 - \sigma q\left(\frac{N}{2}\Delta x\right)\right).$$

By these arguments we have shown that the application of the Fourier transform is equivalent to diagonalizing the corresponding vector/matrix formulation of the scheme. It is then very easy to do the analysis, since we are now dealing with a set of scalar equations.

For more general initial-boundary value problems, a different kind of theory is required. However, the Fourier type stability analysis is still very powerful. Indeed it is often the only type of analysis that is done for many realistic application problems, and quite often it leads to the correct stability limit on Δt.

Exercise 10.6 Consider the PDE $\partial u/\partial t = \partial^2 u/\partial x^2$ with periodic boundary conditions. Prove that the ODE system that is obtained by discretizing in space by using the standard second order difference operator is stiff (see definition in Sect. 10.1).

Exercise 10.7 Write down the Euler backward difference scheme corresponding to (10.13), but now for the periodic case. Derive the exact form of the system of equations that must be solved for advancing this scheme one step. Compare the form to the nonperiodic case (Exercise 10.4).

Exercise 10.8 Use the Fourier method to prove that the Euler backward method in Exercise 10.7 is unconditionally stable.

Exercise 10.9 Consider the PDE $\partial u/\partial t = a\partial u/\partial x$ and the leap-frog difference scheme

$$u_j^{n+1} = u_j^{n-1} + a\frac{\Delta t}{\Delta x}(u_{j+1}^n - u_{j-1}^n).$$

Use Fourier analysis to derive the stability condition.

10.4 Several Space Dimensions

Problems in one space dimension are almost exclusively used as model problems for analysis and preliminary investigations. Real life problems have almost always at least two space dimensions, and we shall make a few comments on these.

The differential equation has the form

$$\frac{\partial u}{\partial t} = \mathcal{P}(\partial x, \partial y)u$$

with proper initial and boundary conditions. Finite difference methods are not well suited for problems where the computational domain is irregular, since both the construction of the computational grid and the analysis become more complicated. However, for regular geometries, we can use *structured grids*, and the easiest 2D-case is a rectangle $0 \le x \le a, 0 \le y \le b$. The grid is defined by

$$(x_{j_1}, y_{j_2}) = (j_1 \Delta x, j_2 \Delta y), \quad j_1 = 0, 1, \ldots, N_1, \ N_1 \Delta x = a,$$
$$j_2 = 0, 1, \ldots, N_2, \ N_2 \Delta y = b,$$

in the x, y-plane, see Fig. 10.12.

The solution $u(x_{j_1}, y_{j_2}, t_n)$ is approximated by $u^n_{j_1 j_2}$. The Fourier analysis is easily generalized from 1D. The grid function is transformed as

$$u^n_{j_1 j_2} = \sum_{k_1=-N_1/2}^{N_1/2} \sum_{k_2=-N_2/2}^{N_2/2} c^n_{k_1 k_2} e^{i(k_1 x_{j_1} + k_2 y_{j_2})},$$

$$j_1 = 0, 1, \ldots, N_1, \ j_2 = 0, 1, \ldots, N_2,$$

where the coefficients are defined by

$$c^n_{k_1 k_2} = \frac{1}{(2\pi)^2} \sum_{j_1=0}^{N_1} \sum_{j_2=0}^{N_2} u^n_{j_1 j_2} e^{-i(k_1 x_{j_1} + k_2 y_{j_2})} \Delta x \Delta y.$$

The wave numbers k_1 and k_2 correspond to the wave number k in 1D. After discretization in time and Fourier transformation of the difference scheme in space, we get a number of scalar relations of the type

$$v^{n+1} = q(\xi, \eta)v^n, \quad 0 \le |\xi|, |\eta| \le \pi,$$

where $\xi = k_1 \Delta x, \eta = k_2 \Delta y$. Also in the 2D-case we have obtained a number of simple algebraic equations instead of a difficult partial differential equation. We simply have to make sure that the amplification factor satisfies the inequality $|q(\xi, \eta)| \le 1$ for $0 \le \xi, \eta < 2\pi$, making the difference scheme stable.

Difference schemes can be used for other computational domains than rectangles. As long as we can map the domain to a rectangle we are in good shape. For example, if the boundaries are circular, we use the well known polar coordinates r and θ defined by

$$x = r \cos \theta, \quad y = r \sin \theta.$$

Fig. 10.12 Two-dimensional grid

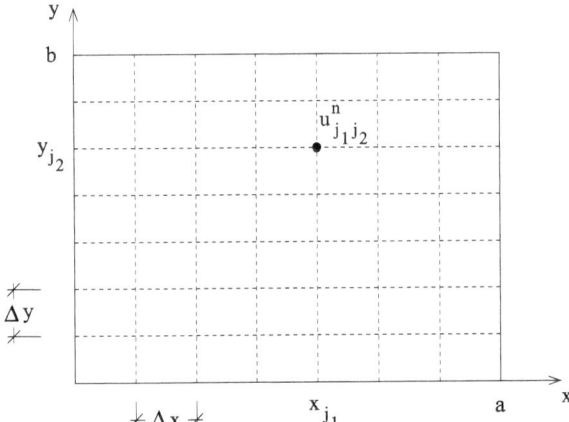

Fig. 10.13 Mapping by using polar coordinates

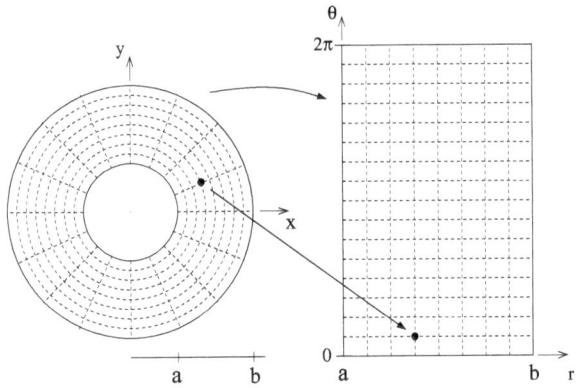

Figure 10.13 shows the mapping for a domain between two circles with radius a and b respectively (the scales are changed in the right figure). The computation is done on the computational grid in the r, θ-plane to the right.

When changing the coordinate system, we must also change the dependent variables and the differential equation. If $u(x, y)$ is a given function, then we get a new function by

$$u(x, y) \rightarrow u(r\cos\theta, r\sin\theta) \rightarrow v(r, \theta).$$

The new differential equation is obtained by using the relations

$$\frac{\partial v}{\partial r} = \frac{\partial u}{\partial x}\frac{\partial x}{\partial r} + \frac{\partial u}{\partial y}\frac{\partial y}{\partial r} = \cos\theta\frac{\partial u}{\partial x} + \sin\theta\frac{\partial u}{\partial y},$$

$$\frac{\partial v}{\partial \theta} = \frac{\partial u}{\partial x}\frac{\partial x}{\partial \theta} + \frac{\partial u}{\partial y}\frac{\partial y}{\partial \theta} = -r\sin\theta\frac{\partial u}{\partial x} + r\cos\theta\frac{\partial u}{\partial y},$$

leading to

$$\frac{\partial u}{\partial x} = \frac{1}{r}\left(r\cos\theta\frac{\partial v}{\partial r} - \sin\theta\frac{\partial v}{\partial \theta}\right).$$

$$\frac{\partial u}{\partial y} = \frac{1}{r}\left(r\sin\theta\frac{\partial v}{\partial r} + \cos\theta\frac{\partial v}{\partial \theta}\right).$$

These relations can be further differentiated to obtain higher order derivatives in the new coordinates.

In Sect. 2.3 we introduced the gradient of a function. We must be careful when transferring this concept to a new coordinate system. The direct translation $[v_r, v_\theta]^T$ doesn't work. The definition of the gradient is that it is the vector pointing in the direction where the function has the strongest growth, and with a magnitude that equals this growth rate. Then we must take into account the geometric properties of the new system, and for polar coordinates it turns out that the gradient is

$$\nabla v(r,\theta) = \begin{bmatrix} \frac{\partial v}{\partial r}(r,\theta) \\ \frac{1}{r}\frac{\partial v}{\partial \theta}(r,\theta) \end{bmatrix}.$$

We now go back to the heat equation

$$\frac{\partial u}{\partial t} = \frac{\partial^2 u}{\partial x^2} + \frac{\partial^2 u}{\partial y^2}, \tag{10.16}$$

and change variables such that

$$v(r,\theta,t) = u\big(x(r,\theta), y(r,\theta), t\big).$$

Then it can be shown that the equation takes the form

$$\frac{\partial v}{\partial t} = \frac{\partial^2}{\partial r^2} + \frac{1}{r}\frac{\partial}{\partial r} + \frac{1}{r^2}\frac{\partial^2}{\partial \theta^2}.$$

The boundaries would be difficult to represent by a rectangular grid in the original (x, y)-coordinates, and the mapping makes it possible to represent the boundaries exactly while still keeping a structured grid.

There is a technical difficulty with these coordinates if the computational domain contains the center point $r = 0$, since the coefficients $1/r$ and $1/r^2$ become infinite there. Since the physics doesn't know anything about coordinate systems, this singularity has to be an artificial effect caused by the choice of coordinates. One can avoid the problem by excluding the point $r = 0$ from the computational grid. However, one should be aware that these coordinates are no good anyway, since the coordinate lines in the original Cartesian system converge at the center point resulting in a very small step size $\Delta\theta$. We shall discuss this further in Sect. 17.4, where the same type of problem occurs at the poles of the globe when doing atmospheric simulations.

There are classic coordinate systems for many different types of geometry, and these should of course be used for computing. We saw above that the differential equation got a different and more complicated form. In Appendix A.2, the most

Fig. 10.14 Part of irregular
domain and computational
domain

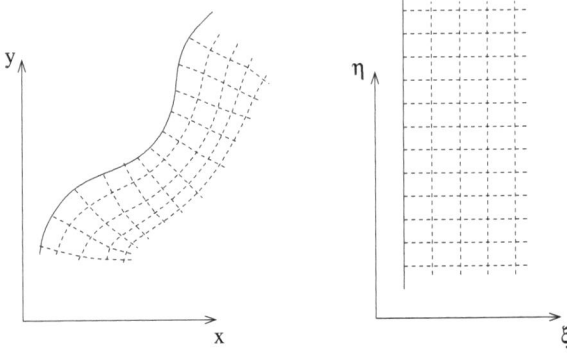

common differential operators are listed in polar, cylindrical and spherical coordinates.

It may be possible to use structured grids even for other domains where no obvious coordinate system is available. In 2D, a general transformation is

$$\xi = \xi(x, y), \qquad \eta = \eta(x, y),$$

with a unique inverse transformation

$$x = x(\xi, \eta), \qquad y = y(\xi, \eta).$$

The transformed differential equation contains derivatives of the variables x, y with respect to ξ, η and, if the transformation has been constructed by some numerical procedure, these are not known by any explicit expression. However, they can be approximated by finite differences, and in this way we can still keep a structured grid. If the final computation can be carried out on a uniform rectangular grid, we are in good shape, since the algebraic operations can be organized in an efficient way. This usually outweighs the extra complication caused by the fact that the differential equation contains more terms. There is also the advantage that the boundary can be represented exactly, with the exact form of boundary conditions. Figure 10.14 shows part of a domain with a curved boundary, and part of the computational domain.

In many applications, the solution varies on a much smaller scale in some parts of the domain than in others, which means that an efficient method should use a finer grid in those parts. A typical application where this occurs is fluid dynamics, where the solution may have very sharp gradients near solid walls. These are called *boundary layers*, and they require a fine grid. Figure 10.15 shows such a case with a simple geometry with the original coordinates to the left, and the computational coordinates to the right.

There is actually another way of handling irregular domains with curved boundaries. The problem is to construct a grid that is structured all over the domain. This difficulty is partially avoided by constructing a local grid near the boundary, and then couple it to one or more rectangular grids in the remaining part of the domain without requiring that the grid points match each other at the edges. This is called *overlapping grids*, and an example is shown in Fig. 10.16.

Fig. 10.15 Boundary layer grid

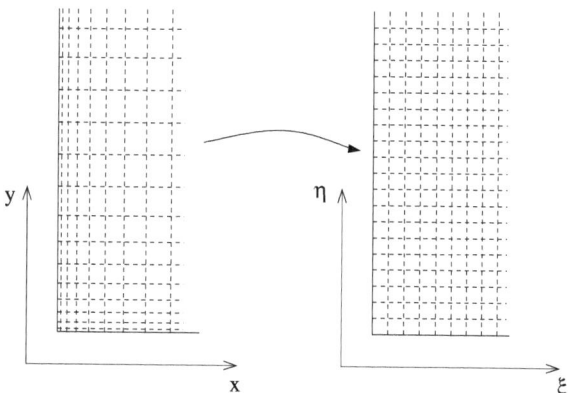

Fig. 10.16 Overlapping grids

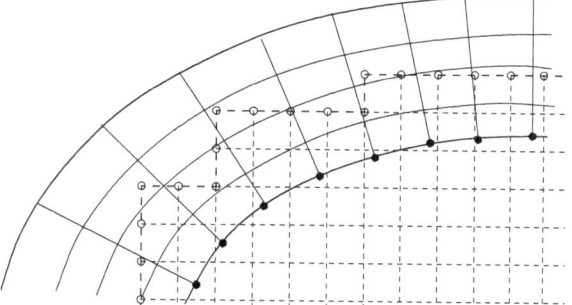

The grid nearest the boundary is called a *curvilinear grid*, and is constructed such that the boundary conditions are easy to approximate. But there is now a new problem. Each one of computational domains has a new boundary, and since they are located in the inner part of the domain, there are no boundary conditions to pick up from the original problem. This is of course as it should be, since each grid must be coupled to the other one, and that coupling is obtained through new boundary conditions at these artificial boundaries. The most straightforward method is to define the boundary values by interpolation from the other grid. A certain point at the inner edge of the curvilinear grid (filled circle) is given a value that is interpolated from the nearest surrounding points in the rectangular grid. In the same way, a point at the outer edge of the rectangular grid (open circle) is interpolated from the nearest surrounding points in the curvilinear grid. The number of points used for interpolation is determined by the accuracy of the main difference scheme. A higher order scheme requires higher order interpolation using more points, otherwise the accuracy goes down.

Exercise 10.10 Write down the explicit difference scheme corresponding to (10.14) but now for the two-dimensional equation (10.16). Derive the stability condition.

Chapter 11
Finite Element Methods

Finite element methods (FEM) were used very early for problems in structural mechanics. Such problems often have a natural discretization by partitioning the structure in a number of finite elements, and this gave the name to this class of methods.
This kind of mechanical approach merged with the more mathematical approach that
gained momentum in the 1960s. FEM are more flexible than finite difference methods in the sense that irregular domains can more easily be represented accurately.
In this chapter we shall discuss boundary and initial-boundary value problems, and
in the latter case assume that finite difference methods are used for discretization in
time.

In the final section we discuss discontinuous Galerkin methods (DG), which have
become very popular recently. They are closely connected to FEM, and are therefore
included in this chapter, even if they could be considered as a separate class of
methods by themselves.

11.1 Boundary Value Problems

In order to describe finite element methods, we begin by discussing boundary value
problems, and consider first the example

$$\frac{d^2u}{dx^2} + au = F, \quad 0 \le x \le 1,$$
$$u(0) = 0,$$
$$u(1) = 0,$$

(11.1)

where a is a constant, and $F = F(x)$ is a known function. (This problem does not
have a unique solution for all a, but we assume here that a is such that it has.)
There are several versions of the finite element method, but the most common one is
closely connected to the least square method. We choose a set of basis functions for
representation of the approximate solution $v(x)$, and determine the coefficients such
that the error becomes minimal. For the least square method described in Chap. 8,

B. Gustafsson, *Fundamentals of Scientific Computing*,
Texts in Computational Science and Engineering 8,
DOI 10.1007/978-3-642-19495-5_11, © Springer-Verlag Berlin Heidelberg 2011

the error is simply defined as $\|u - v\|$ measured in a certain norm, and then we require that the error is orthogonal to the basis functions. Unfortunately, this doesn't lead anywhere for the problem (11.1), since the solution u is not known. There are no data to feed into the least square algorithm.

Another approach would be to minimize the *error in the differential equation*, i.e., we look for an approximating function v such that $\|d^2v/dx^2 + av - F\|^2$ becomes as small as possible. This is a direct analogue of the least square method for overdetermined algebraic systems $A\mathbf{x} = \mathbf{b}$ as discussed at the end of Sect. 8.3. The solution is there given by the system $A^T A\mathbf{x}^* = A^T\mathbf{b}$, and we can use the same formal procedure for the differential equation. However, in this case the matrix $A^T A$ is replaced by a differential operator that is of order four. This is a severe complication, since the original problem (11.1) contains only a second order derivative.

To get a simpler algorithm, we require instead that the error in the differential equation $d^2v/dx^2 + av - F$, also called the *residual*, is orthogonal to all basis functions. But before deriving these equations, we shall reformulate the differential equation.

Assume that $\phi(x)$ is a certain function that will be specified later. We start by multiplying the differential equation by ϕ and then integrating it over the interval $[0, 1]$, obtaining

$$\int_0^1 \left(\frac{d^2u}{dx^2} + au\right)\phi \, dx = \int_0^1 F\phi \, dx. \tag{11.2}$$

If ϕ represents a basis function, this relation resembles the orthogonality condition. However, because we are aiming for approximation by piecewise polynomials, we encounter a problem here. Consider the roof functions discussed in Sect. 5.1.2 as basis functions. The approximating function $v(x)$ has the form

$$v(x) = \sum_{j=1}^{N} c_j\phi_j(x),$$

where the coefficients c_j are to be determined. Since each function $\phi_j(x)$ is a straight line between the nodes, the function $v(x)$ is piecewise linear as shown in Fig. 11.1(a).

We now want to replace $u(x)$ by $v(x)$ in (11.2), but here is where the new difficulty arises. Figure 11.1(b) shows the derivative dv/dx of the piecewise linear function v. It is a well defined step function, but it has a discontinuity at every node. Equation (11.2) contains the second derivative, i.e., the first derivative of dv/dx. It is well defined between the nodes, but what is it at the nodes? This question requires some abstract mathematics that is not well suited for practical computation, and we need some other technique to overcome this difficulty. The trick is to do so called *integration by parts*, see (2.14). Since the function v is zero at the end points of the interval $[0, 1]$, the identity

$$\int_0^1 \frac{du}{dx}(x)v(x)\,dx = -\int_0^1 u(x)\frac{dv}{dx}(x)\,dx \tag{11.3}$$

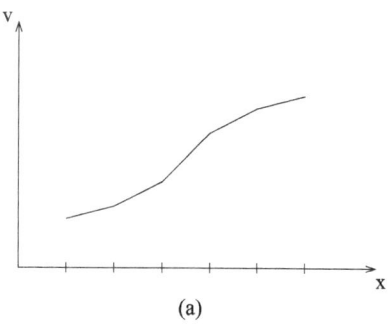
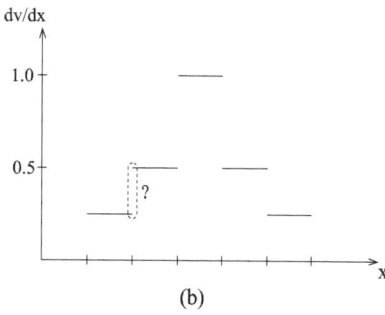

Fig. 11.1 Piecewise linear function (**a**) and its derivative (**b**)

follows, provided that the functions u and v are such that the integrals exist. The differentiation operator d/dx is transferred from one of the functions to the other one.

By applying integration by parts to the first term in (11.2), we obtain

$$\int_0^1 \left(-\frac{du}{dx}\frac{d\phi}{dx} + au\phi \right) dx = \int_0^1 F\phi \, dx. \tag{11.4}$$

This is called the *weak form* of the differential equation. With the scalar product defined by an integral (see for example Sect. 5.1.1), we can write the equation as

$$\left(-\frac{du}{dx}, \frac{d\phi}{dx} \right) + (au, \phi) = (F, \phi). \tag{11.5}$$

Solutions of the original problem (11.1) with bounded second derivatives are called *strong solutions*. Obviously they satisfy the weak form (11.5) as well, but there are weak solutions that do not satisfy the original strong form.

When choosing the basis functions ϕ_j for the solution u, we get a larger class to choose from if there are less restrictions on the differentiability. In mathematical terms, all possible functions of the form $v = \sum_j c_j \phi_j(x)$ constitute a *function space* $\mathcal{S}$. Formally we write it as

$$\mathcal{S} = \left\{ v(x) : \|v\|^2 + \left\| \frac{dv}{dx} \right\|^2 < \infty, \ v(0) = v(1) = 0 \right\}. \tag{11.6}$$

The approximation of the true solution u is to be chosen from this space, and the trick with integration by parts makes $\mathcal{S}$ larger than it would be otherwise. There are more functions in $\mathcal{S}$ compared to the original space $\mathcal{S}^*$, where it is natural to require that also the second derivatives are bounded:

$$\mathcal{S}^* = \left\{ v(x) : \|v\|^2 + \left\| \frac{dv}{dx} \right\|^2 + \left\| \frac{d^2v}{dx^2} \right\|^2 < \infty, \ v(0) = v(1) = 0 \right\}.$$

We use the notation $\mathcal{S}^* \subset \mathcal{S}$ to indicate that all functions in $\mathcal{S}^*$ are included in the larger space $\mathcal{S}$. With N basis functions $\{\phi_j\}$, the approximation problem has N *degrees of freedom*, i.e., there are N unknown coefficients $\{c_j\}$ to be determined.

Fig. 11.2 Function spaces
and FEM solution v
approximating u

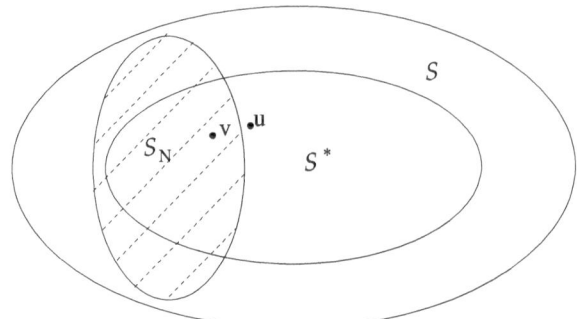

One can also say that the approximating function space $\mathscr{S}_N$ has dimension N, while
the true solution space $\mathscr{S}$ has infinite dimension. The boundary value problem has a
unique solution u in $\mathscr{S}$, but this one cannot be found. Instead we look for a solution
v in the smaller space $\mathscr{S}_N$. In abstract form, the original problem in weak form is
defined by:

Find the function $u \in \mathscr{S}$ such that (11.5) is satisfied for all $\phi \in \mathscr{S}$.

The approximative solution is defined by:

Find the function $v \in \mathscr{S}_N \subset \mathscr{S}$ such that (11.5) is satisfied for all $\phi \in \mathscr{S}_N$.

Figure 11.2 shows a schematic picture of the relation between the function
spaces.

Since the basis functions $\{\phi_j\}$ span the space $\mathscr{S}_N$, the last problem can be formulated as

$$\int_0^1 \left(-\frac{d}{dx}\left(\sum_{j=1}^N c_j \phi_j(x) \right) \frac{d\phi_k}{dx}(x) + a \sum_{j=1}^N c_j \phi_j(x) \phi_k(x) \right) dx$$

$$= \int_0^1 F(x)\phi_k(x)\,dx, \quad k = 1, 2, \ldots, N. \tag{11.7}$$

The only quantities that are unknown here are the coefficients c_k. We reformulate
the system as

$$\sum_{j=1}^N \left(-\int_0^1 \frac{d\phi_j}{dx}(x) \frac{d\phi_k}{dx}(x) + a \int_0^1 \phi_j(x)\phi_k(x)\,dx \right) c_j$$

$$= \int_0^1 F(x)\phi_k(x)\,dx, \quad k = 1, 2, \ldots, N. \tag{11.8}$$

This is a linear system with N equations for the N unknowns c_j. This method of
finding an approximation is called the *Galerkin method*.

With the least square method applied to the original differential equation, we
would minimize $\|d^2v/dx^2 + av - F\|^2$. In Chap. 8 the condition for a minimum
of the squared norm was derived for the standard function approximation problem.

If this condition is generalized to the differential equation, we would have products $d^2\phi_j/dx^2 \cdot d^2\phi_k/dx^2$ occurring in the system for the coefficients c_j without any possibility of bringing down the order of the derivatives. Any attempt of integration by parts would lower the order on one basis function, but increase the order on the other one, and the situation would become even worse.

With piecewise linear basis functions the product $\phi_j(x)\phi_k(x)$ is zero for most j, k. Only for $k = j - 1$, j, $j + 1$ is there any contribution to the integrals. The same conclusion holds for the product $d\phi_j/dx \cdot d\phi_k/dx$. The integrals on the left hand side of (11.8) are easily computed, but for the integrals on the right hand side we may be forced to use numerical methods.

Once the coefficients c_j have been computed, the values of the approximation at the nodes are particularly easy to compute. Since the piecewise linear basis functions $\phi_j(x)$ are zero at all nodes except at x_j, we have $v(x_j) = c_j$.

If the distribution of nodes is uniform with a step size Δx we get a particularly simple system where all the equations have the form

$$\left(1 + \frac{a\Delta x^2}{6}\right)c_{j-1} + \left(-2 + \frac{4a\Delta x^2}{6}\right)c_j + \left(1 + \frac{a\Delta x^2}{6}\right)c_{j+1}$$

$$= \Delta x^2 \int_0^1 F(x)\phi_j(x)\,dx, \quad j = 1, 2, \ldots, N. \tag{11.9}$$

This system involves the two extra coefficients c_0 and c_{N+1}. However, with the extended representation

$$v(x) = \sum_{j=0}^{N+1} c_j\phi_j(x),$$

the boundary conditions imply that $c_0 = c_{N+1} = 0$, and therefore we have as many equations as unknowns.

The system can be written in matrix/vector form. With

$$\mathbf{c} = \begin{bmatrix} c_1 \\ c_2 \\ \vdots \\ c_N \end{bmatrix}, \qquad K = \begin{bmatrix} -2 & 1 & & \\ 1 & -2 & 1 & \\ & \ddots & \ddots & \ddots \\ & & -2 & 1 \end{bmatrix},$$

$$M = \frac{a\Delta x^2}{6}\begin{bmatrix} 4 & 1 & & \\ 1 & 4 & 1 & \\ & \ddots & \ddots & \ddots \\ & & 4 & 1 \end{bmatrix},$$

$$\mathbf{F} = \Delta x^2 \begin{bmatrix} \int_0^1 F(x)\phi_1(x)\,dx \\ \int_0^1 F(x)\phi_2(x)\,dx \\ \vdots \\ \int_0^1 F(x)\phi_N(x)\,dx \end{bmatrix},$$

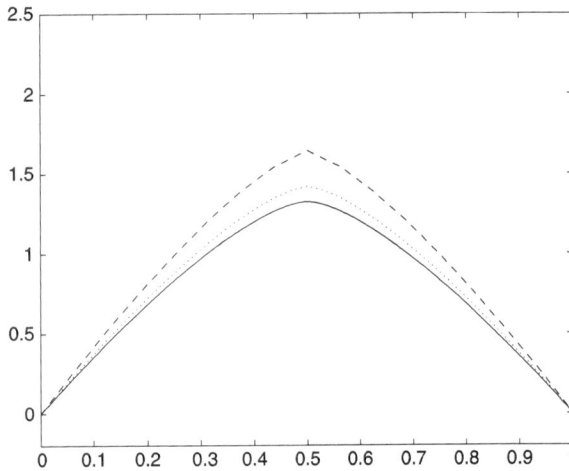

Fig. 11.3 Finite element solution of (11.1), $N = 15 (--)$, $N = 31 (\cdots)$, exact (—)

the system is

$$K\mathbf{c} + M\mathbf{c} = \mathbf{F}. \tag{11.10}$$

The matrix K is the *stiffness matrix*, and M is the *mass matrix*.

We note that the system is very similar to a difference scheme and, after dividing by Δx^2, we recognize the three point approximation of $d^2 u/dx^2$. However, instead of the more natural representation c_j of $v(x_j)$, the finite element uses a weighted average over three points with the weights $1/6$, $4/6$, $1/6$. Another difference is that it provides a solution everywhere, and not only at the grid points. With our choice of basis functions, the solution is piecewise linear, i.e., it is a straight line between the nodes. The solution is shown in Fig. 11.3 for $a = 1$ and $F(x) = -1/(0.02 + |x - 0.5|)$ for two different resolutions. The convergence towards the true solution for increasing N is clearly seen.

So far we have used approximations $v(x)$ that are straight lines between the nodes, i.e., $v(x)$ has the form $a_0 + a_1 x$ in each interval. A more accurate solution is obtained if we allow for higher order polynomials, $a_0 + a_1 x + \cdots + a_p x^p$, and in Sect. 5.1.2 quadratic and cubic polynomials were discussed. The nonzero part of the basis functions $\phi_j(x)$ is wider, but still spread over very few subintervals. The integrals $\int_0^1 \phi_j(x)\phi_{j+1}(x)\, dx$ etc. are still easy to compute.

Even if the grid is nonuniform, only neighboring basis functions ϕ_j give any contribution to the integrals. This is the reason for calling the procedure a *finite element* method. For linear functions this means that each equation in the system for the coefficients c_j has the form

$$a_{j-1}c_{j-1} + a_j c_j + a_{j+1}c_{j+1} = F_j,$$

and these equations can be written in the matrix/vector form (11.10) with different but still tridiagonal stiffness and mass matrices. For higher order polynomials, more neighboring coefficients become involved, but still very few compared to the total number. This means that we are dealing with *sparse* matrices, and we shall discuss later how these systems can be solved effectively.

The generalization to two space dimensions is mathematically straightforward, but of course computationally more demanding. Consider the extension of the problem (11.1) to a two-dimensional domain Ω with boundary Γ

$$\frac{\partial^2 u}{\partial x^2} + \frac{\partial^2 u}{\partial y^2} + au = F \quad \text{in } \Omega,$$

$$u = 0 \quad \text{at } \Gamma.$$

This is the well known Poisson equation. In order to bring the order of the derivatives down one step, we need the generalization of integration by parts, which is the Green's theorem. We first multiply the equation by a function $v(x, y)$ and take the scalar product:

$$\int\int_\Omega \left(\frac{\partial^2 u}{\partial x^2} + \frac{\partial^2 u}{\partial y^2} + au \right) v \, dx \, dy = \int\int_\Omega F v \, dx \, dy.$$

On replacing u by $v\partial u/\partial x$ and v by $v\partial u/\partial y$ in the divergence theorem (2.19), and taking into account that v vanishes at the boundary, we get

$$\int\int_\Omega \left(\frac{\partial^2 u}{\partial x^2} + \frac{\partial^2 u}{\partial y^2} \right) v \, dx \, dy = -\int\int_\Omega \left(\frac{\partial u}{\partial x} \frac{\partial v}{\partial x} + \frac{\partial u}{\partial y} \frac{\partial v}{\partial y} \right) dx \, dy.$$

This gives the weak Galerkin formulation

$$-\int\int_\Omega \left(\frac{\partial u}{\partial x} \frac{\partial \phi}{\partial x} + \frac{\partial u}{\partial y} \frac{\partial \phi}{\partial y} + au \right) v \, dx \, dy = \int\int_\Omega F v \, dx \, dy.$$

The scalar product and norm are defined by

$$(u, v) = \int\int_\Omega u(x, y)v(x, y) \, dx \, dy, \qquad \|u\|^2 = (u, u).$$

The regularity requirement on u and ϕ is now that the first derivatives should be bounded. The proper function space is

$$\mathscr{S} = \left\{ v(x)/ \|v\|^2 + \left\| \frac{dv}{dx} \right\|^2 + \left\| \frac{dv}{dy} \right\|^2 < \infty, \ v = 0 \text{ at the boundary } \Gamma \right\}.$$

For the finite element method we need a subspace $\mathscr{S}_N$ of $\mathscr{S}$. With the basis functions $\{\phi_j(x, y)\}$, the Galerkin method is formulated precisely as in the 1D-case. The remaining question is how to choose $\mathscr{S}_N$ and the basis functions.

Two-dimensional piecewise polynomials were discussed in Sect. 5.1.2, and it was shown that triangles are very natural as the geometrical basis not only for linear polynomials, but also for quadratic and cubic polynomials. The computational domain is therefore divided into triangles, as shown in Fig. 11.4. For piecewise linear functions, the solution is a plane in the 3-D space corresponding to each triangle. The shaded area shows where one basis function is nonzero.

Even if rectangles could be used as finite elements with other types of approximating function spaces, for example bilinear, triangles are better from another point of view. If the domain is irregular, it is much easier to approximate the boundary by triangle sides. If the sides of each triangle and each rectangle are of the order

Fig. 11.4 Triangular grid for finite elements

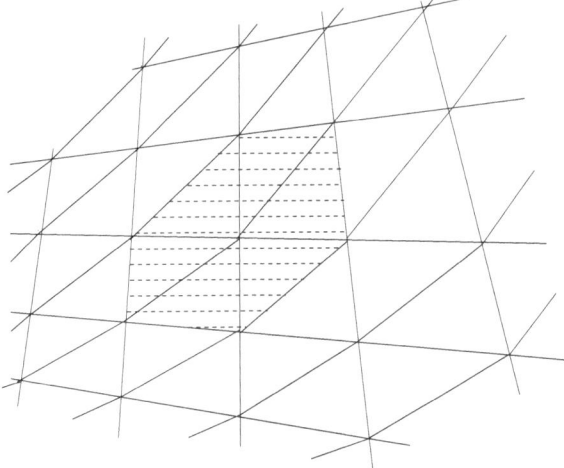

Δx, the distance between the true and the approximative boundary is $\mathcal{O}(\Delta x^2)$ with triangles and $\mathcal{O}(\Delta x)$ with rectangles, see Fig. 11.5.

As we showed in Sect. 5.1.2, one can use general quadrilaterals instead of rectangles. They give more flexibility, and in certain situations they are a good choice. For certain types of boundary conditions it is an advantage to have the edges perpendicular to the boundary, and Fig. 11.6 shows an example of a computational domain, where this is arranged easily by the use of quadrilaterals.

Once the grid and the type of polynomials have been chosen, the finite element method can be defined. With the ansatz

$$v(x, y) = \sum_{j=1}^{N} c_j \phi_j(x, y)$$

for the numerical solution the coefficients are found by solving the system

$$\sum_{j=1}^{N} \left(-\int\int_{\Omega} \left(\frac{\partial \phi_j(x, y)}{\partial x} \frac{\partial \phi_k(x, y)}{\partial x} + \frac{\partial \phi_j(x, y)}{\partial y} \frac{\partial \phi_k(x, y)}{\partial y} \right) dx \, dy \right.$$

$$\left. + a \int\int_{\Omega} \phi_j(x, y) \phi_k(x, y) \, dx \, dy \right) c_j$$

$$= \int\int_{\Omega} F(x, y) \phi_k(x, y) \, dx \, dy, \quad k = 1, 2, \ldots, N,$$

that corresponds to (11.8). This system of equations for the coefficients is much larger than for one-dimensional problems. However, since the basis functions are zero in all but a few of the subdomains, the system is still sparse, and allows for fast solution methods.

The basis functions are usually chosen to be zero at all nodes $\{x, y\}_j$ except one. This means that $v(\{x, y\}_j) = c_j$ at the nodes. By using the basis functions, the

Fig. 11.5 Boundary approximation with triangles and rectangles

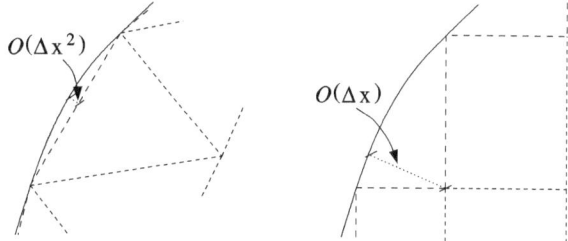

Fig. 11.6 Quadrilateral grid

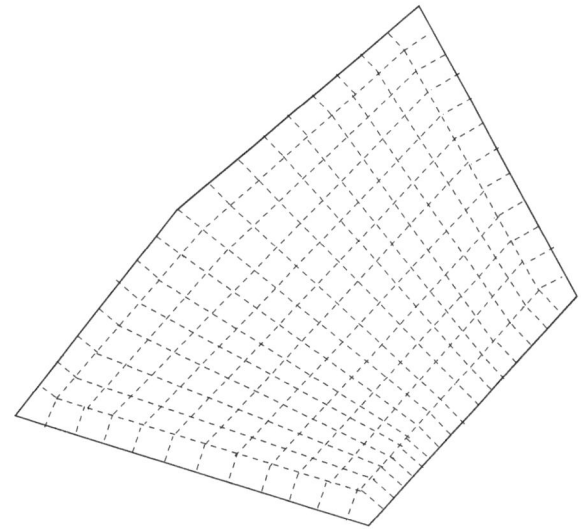

system for $\{c_j\}$ is easily derived. However, for practical computational purposes, the coefficient matrix for the system is not computed this way. This process is called the *assembly* of the matrix. It is done by going over all finite elements one by one, and computing their contribution to the matrix elements. There is a connection between the nodal values v_j and the coefficients of the local polynomial at each one of the finite elements. For example, in 1D with linear elements, we have for $x_j \leq x \leq x_{j+1}$

$$v(x) = v_j + \frac{x - x_j}{x_{j+1} - x_j}(v_{j+1} - v_j) = \left(1 - \frac{x - x_j}{x_{j+1} - x_j}\right)v_j + \frac{x - x_j}{x_{j+1} - x_j}v_{j+1},$$

and

$$\frac{dv}{dx} = \frac{v_{j+1} - v_j}{x_{j+1} - x_j},$$

where $v(x_j) = v_j$ and $v(x_{j+1}) = v_{j+1}$. In the same way the basis functions can be expressed in each subinterval in terms of their values at the nodes, which is one for one node and zero for all others. For example,

$$\phi_j(x) = 1 - \frac{x - x_j}{x_{j+1} - x_j},$$

and

$$\frac{d\phi_j}{dx} = -\frac{1}{x_{j+1} - x_j}.$$

The system (11.7) is rewritten in its original form as

$$-\int_0^1 \frac{dv}{dx}(x)\frac{d\phi_k}{dx}(x)\,dx + \int_0^1 av(x)\phi_k(x)\,dx = \int_0^1 F(x)\phi_k(x)\,dx,$$
$$k = 1, 2, \ldots, N.$$

Each integral is now partitioned into its pieces corresponding to the subintervals

$$\int_0^1 = \sum_j \int_{x_j}^{x_{j+1}},$$

and computed in terms of the unknowns v_j and v_{j+1}. For our example only two basis functions contribute for each subinterval. Furthermore, for a given ϕ_k, the corresponding equation contains only the three unknowns v_{k-1}, v_k, v_{k+1} and, by assembling all contributions, we get a tridiagonal system with the stiffness and mass matrices that were already derived above. For the integrals on the right hand side we may have to use a numerical method for the integration.

The same assembly principle is used with any type of piecewise polynomials and in any number of space dimensions. As an example, consider the triangular grid in Fig. 11.4. The integrals are computed over each triangle and added together. The basis function $\phi_k(x, y)$ associated with the node in the middle of the shaded area influences the integrals over the six triangles in the same area. This means for linear polynomials that, in equation number k, there will be 7 unknown nodal values $c_j = v_j$ involved. For quadratic polynomials there are nodes also in the middle of each triangle edge, and there will be 19 unknowns in each equation.

When comparing to finite difference methods, there is a similarity in the sense that the nodal values v_j are carried as unknowns. However, the brief discussion of Hermite polynomials in Sect. 5.1.2 shows that the derivative nodal values can be introduced as well. The assembly process is carried out in the same way, but now with a slightly more complicated connection between the nodal values and the integrals. However this computation is carried out once and for all, and the matrices are stored for use with any number of right hand sides.

When deriving error estimates for the FEM solutions, it can be shown that this problem can be referred back to a pure function approximation problem. How close to a given function in $\mathscr{S}$ can we come by using an approximation in $\mathscr{S}_N$? We assume that the distance between the nodes is of the order Δx (also in several dimensions), and that the given function is smooth. Then it can be shown that, with piecewise polynomials of degree p, the error is $\mathscr{O}(\Delta x^{p+1})$. This result requires some other conditions, in particular on the regularity of the finite elements. For example, the triangles must have a uniform lower bound on the smallest angle. Furthermore, the boundary and the boundary conditions must be approximated accurately enough. We saw above that triangles give an $\mathscr{O}(\Delta x^2)$ error when it comes to approximation of the boundary which corresponds to linear polynomials. But how is this change of boundary possible when it comes to the Galerkin formulation? The condition

Fig. 11.7 Modification of the computational domain

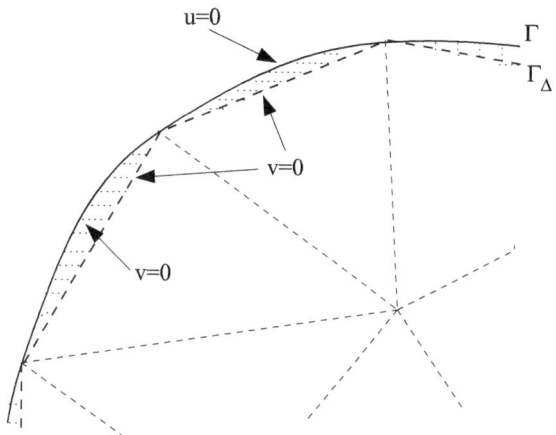

$\mathscr{S}_N \subset \mathscr{S}$ requires that every function in $\mathscr{S}_N$ must satisfy the boundary conditions exactly. But there is now a numerical boundary Γ_Δ made up of the outer triangle edges, that approximates the true boundary Γ. Assume that we substitute the true boundary condition

$$u = 0, \quad (x, y) \in \Gamma$$

with

$$v = 0, \quad (x, y) \in \Gamma_\Delta.$$

Since the FEM solution is not even defined on the true boundary, the condition $\mathscr{S}_N \subset \mathscr{S}$ is not fulfilled. However, there is a simple fix for this problem. The computational domain Ω_Δ is extended such that it agrees with Ω, and v is defined as zero in the whole strip between the two boundaries, see Fig. 11.7.

We still have a piecewise linear function space, but with a special form near Γ. This leads to a final error estimate of second order for the whole solution.

With quadratic or higher degree polynomials we must do better to keep the higher order accuracy, and use some sort of curved edge for the triangles nearest to the boundary.

Exercise 11.1 Derive the FEM system of equations obtained with piecewise linear polynomials on a nonuniform node distribution for the problem (11.1), i.e., derive the generalized form of (11.9).

Exercise 11.2 Consider the boundary value problem

$$\frac{d^4u}{dx^4} + au = F, \quad 0 \le x \le 1,$$

$$u(0) = \frac{\partial u}{\partial x}(0) = 0,$$

$$u(1) = \frac{\partial u}{\partial x}(1) = 0.$$

(a) Use integration by parts to derive the weak form of the problem.
(b) Define the function space $\mathscr{S}$ for this case (corresponding to (11.6) for the second order equation).
(c) Discuss the choice of piecewise polynomials for the finite element method for this problem. Why are cubic Hermite polynomials a proper choice, while quadratic polynomials are not (see Sect. 5.1.2)?

11.2 Initial-Boundary Value Problems

As an example of an initial-boundary value problem, we consider the problem

$$\frac{\partial u}{\partial t} = \frac{\partial^2 u}{\partial x^2}, \quad 0 \le x \le 1, \ 0 \le t,$$
$$u(0, t) = 0,$$
$$u(1, t) = 0,$$
$$u(x, 0) = f(x).$$
(11.11)

The finite element method for time dependent initial–boundary value problems is very similar to the procedure for boundary value problems. The basis functions $\phi_j(x)$ are also in this case independent of t, and they satisfy the boundary conditions. The integration by parts is carried out as above, and we obtain

$$\int_0^1 \frac{\partial u}{\partial t} \phi \, dx = - \int_0^1 \frac{\partial u}{\partial x} \frac{\partial \phi}{\partial x} \, dx.$$

The approximation $v(x, t)$ is

$$v(x, t) = \sum_{j=1}^N c_j(t) \phi_j(x),$$

where the coefficients $c_j(t)$ now are time dependent. The initial function is also approximated as a combination of the basis functions:

$$f(x) \approx \sum_{j=1}^N d_j \phi_j(x).$$

These coefficients d_j are obtained by the least square method resulting in the system

$$\sum_{j=1}^N d_j \int_0^1 \phi_j(x) \phi_k(x) \, dx = \int_0^1 f(x) \phi_k(x) \, dx, \quad k = 1, 2, \ldots, N.$$

This is a linear system of equations for the N coefficients d_j. It is sparse, and the particular choice of roof functions ϕ_j gives a tridiagonal system. The integrals on the right hand side must be computed by some numerical method in the general case.

With this representation of $f(x)$, the final system for computation of the coefficients $c_j(t)$ is

$$\sum_{j=1}^{N} \frac{dc_j}{dt} \int_0^1 \phi_j(x)\phi_k(x)\,dx$$

$$= -\sum_{j=1}^{N} c_j \int_0^1 \frac{d\phi_j(x)}{dx}\frac{d\phi_k(x)}{dx}\,dx, \quad k = 1, 2, \ldots, N, \qquad (11.12)$$

$$c_k(0) = d_k, \quad k = 1, 2, \ldots, N.$$

This system is a set of ordinary differential equations, and together with the initial conditions it constitutes an initial value problem for the N unknown functions $c_j(t)$. The finite element method has reduced the partial differential equation to a system of ordinary differential equation. This way of discretizing in space and leaving time continuous, is called the *method of lines*. A graph in the x, t plane would show a collection of vertical lines, one for each x-node.

It remains to solve the initial value problem, and again we have to use a numerical method. Usually a difference method is applied as described in Sect. 10.1, but other methods can be used as well, for example the finite element method. There is a special feature that we have to take into consideration here. If the coefficients c_j and d_j are ordered as vectors

$$\mathbf{c} = \begin{bmatrix} c_1 \\ c_2 \\ \vdots \\ c_N \end{bmatrix}, \qquad \mathbf{d} = \begin{bmatrix} d_1 \\ d_2 \\ \vdots \\ d_N \end{bmatrix},$$

the system (11.12) can be written in vector/matrix form

$$M\frac{d\mathbf{c}}{dt} = K\mathbf{c},$$

$$\mathbf{c}(0) = \mathbf{d}, \qquad (11.13)$$

where M is the mass matrix and K is the stiffness matrix. Even if an explicit difference method is used for the time integration, for example the Euler method

$$M(\mathbf{c}_{n+1} - \mathbf{c}_n) = \Delta t\, K \mathbf{c}_n,$$

$$\mathbf{c}_0 = \mathbf{d},$$

there is a system of equations to be solved for each time step. However, we recall that M is a sparse matrix, making the solution procedure simpler.

Error estimates are derived much in the same way as for boundary value problems, and we may expect an $\mathcal{O}(\Delta x^{p+1})$ error for p-degree polynomials. The stability concept was an important issue for time dependent difference approximations, and one may ask how it comes into the picture for FEM. The answer is that it is essential also here, but it follows automatically if the original problem is well posed in a certain sense. However, the condition $\mathcal{S}_N \subset \mathcal{S}$ in the formulation of the Galerkin method is essential. It means that the choice of approximating piecewise polynomials must be made very carefully for irregular domains. We saw above how this is done for linear polynomials on triangles in the case of curved boundaries. For higher degree approximations it requires some extra effort to retain the accuracy.

Exercise 11.3 Consider the initial-boundary value problem (11.11) and the finite element method with piecewise linear polynomials on a uniform node distribution.

(a) Derive the explicit form of the matrices M and K in (11.13).
(b) Assume periodic solutions in space and use Fourier analysis to derive the stability condition for the Euler forward and the Euler backward methods applied to (11.13).
(c) Estimate the number of arithmetic operations in each time step for the two methods.

11.3 Discontinuous Galerkin Methods

An essential property for finite element methods based on the Galerkin principle is that the approximating subspace $\mathscr{S}_N$ consists of globally defined functions. Even if the basis functions are nonzero only at a small number of elements, they are defined everywhere with at least continuity over the element boundaries. One effect of this is that there is a coupling between neighboring points leading to a coupling between all the nodal values for the full system. The mass matrix is a band matrix, and even if an explicit difference method is used for integration in time, there is a system of equations to be solved at each time step. This may be a significant disadvantage, and one would like to get rid of it. This can be achieved by introducing local basis functions, leading to *discontinuous Galerkin methods*.

We shall consider a *conservation law*

$$\frac{\partial u}{\partial t} + \frac{\partial f(u)}{\partial x} = 0,$$

where $f(u)$ is called the *flux function*, which in general is a nonlinear function of $u = u(x,t)$. Integration by parts over an interval $[x_1, x_2]$ gives

$$\frac{d}{dt} \int_{x_1}^{x_2} u\, dx = -\int_{x_1}^{x_2} \frac{\partial f(u)}{\partial x} u\, dx = f\big(u(x_1,t)\big) - f\big(u(x_2,t)\big). \qquad (11.14)$$

The label "conservation law" for this class of PDE is motivated by the fact that, if u vanishes outside an interval $[x_1, x_2]$, the integral of the solution is independent of time:

$$\frac{d}{dt} \int_{x_1}^{x_2} u\, dx = 0,$$

The form (11.14) is actually the basis for *finite volume methods*. The integral is considered as a "cell average", and the u-values at the end points x_1 and x_2 are approximated in terms of the neighboring cell averages. The advantage with such methods is that the conservation property is retained with a proper discretization of the integral. They will be further discussed in Sect. 17.3.

Discontinuous Galerkin methods are similar to finite volume methods. We multiply the original conservation form by a function $\phi(x)$ and integrate over an interval $I_j = [x_{j-1/2}, x_{j+1/2}]$:

$$\int_{I_j} \frac{\partial u}{\partial t} \phi \, dx + \int_{I_j} \frac{\partial f(u)}{\partial x} \phi \, dx = 0,$$

which after integration by parts becomes

$$\int_{I_j} \frac{\partial u}{\partial t} \phi \, dx + f\big(u(x_{j+1/2})\big) \phi(x_{j+1/2})$$

$$- f\big(u(x_{j-1/2})\big) \phi(x_{j-1/2}) - \int_{I_j} f(u) \frac{\partial \phi}{\partial x} \, dx = 0.$$

In contrast to the Galerkin method, the flux function is replaced by a numerical flux function $g = g(u^-, u^+)$ that depends on both the left and right limits

$$u^-(x) = \lim_{\delta \to 0} u(x - \delta), \quad \delta > 0,$$

$$u^+(x) = \lim_{\delta \to 0} u(x + \delta), \quad \delta > 0.$$

At all points where u is continuous, we require $g(u^-, u^+) = f(u)$, but here we allow also discontinuous functions u. In this case the numerical solution v satisfies

$$\int_{I_j} \frac{\partial v}{\partial t} \phi \, dx + g\big(v^-(x_{j+1/2}), v^+(x_{j+1/2})\big) \phi(x_{j+1/2})$$

$$- g\big(v^-(x_{j-1/2}), v^+(x_{j-1/2})\big) \phi(x_{j-1/2}) - \int_{I_j} f(v) \frac{\partial \phi}{\partial x} \, dx = 0.$$

The numerical solution is defined on each subinterval I_j as a linear function expressed in terms of the two basis functions

$$\phi_j = \frac{1}{\Delta x_j}(x_{j+1/2} - x), \qquad \psi_j = \frac{1}{\Delta x_j}(x - x_{j-1/2}), \quad x \in I_j,$$

where Δx_j is the length of I_j. Figure 11.8 shows what they look like at two neighboring subintervals.

In each subinterval, the numerical solution has the form

$$v(x, t) = a_j(t)\phi_j(x) + b_j(t)\psi_j(x), \quad x \in I_j,$$

where the coefficients a_j and b_j are to be determined. With N intervals I_j in the domain, the weak form of the problem is now obtained as

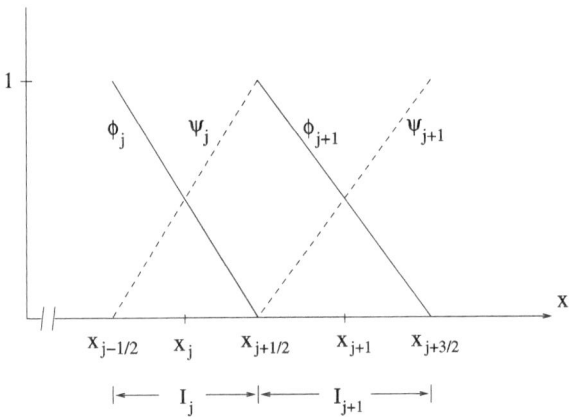

Fig. 11.8 Piecewise linear discontinuous basis functions

$$\int_{x_{j-1/2}}^{x_{j+1/2}} \left(\frac{da_j}{dt}\phi_j + \frac{db_j}{dt}\psi_j \right)\phi_j\,dx + g\big(v^-(x_{j+1/2}), v^+(x_{j+1/2})\big)\phi_j(x_{j+1/2})$$

$$- g\big(v^-(x_{j-1/2}), v^+(x_{j-1/2})\big)\phi_j(x_{j-1/2})$$

$$- \int_{x_{j-1/2}}^{x_{j+1/2}} f(a_j\phi_j + b_j\psi_j)\frac{\partial\phi_j}{\partial x}\,dx = 0,$$

$$\int_{x_{j-1/2}}^{x_{j+1/2}} \left(\frac{da_j}{dt}\phi_j + \frac{db_j}{dt}\psi_j \right)\psi_j\,dx + g\big(v^-(x_{j+1/2}), v^+(x_{j+1/2})\big)\psi_j(x_{j+1/2})$$

$$- g\big(v^-(x_{j-1/2}), v^+(x_{j-1/2})\big)\psi_j(x_{j-1/2})$$

$$- \int_{x_{j-1/2}}^{x_{j+1/2}} f(a_j\phi_j + b_j\psi_j)\frac{\partial\psi_j}{\partial x}\,dx = 0,$$

$$j = 1, 2, \ldots, N.$$

The coupling between the different subintervals is obtained solely by the choice of flux function g. In order to get a simple illustration we choose the equation

$$\frac{\partial u}{\partial t} + \frac{\partial u}{\partial x} = 0,$$

i.e., $f(u) = u$. Since information is flowing from left to right, it is natural to make the solution in I_j dependent on the action in I_{j-1}. The simplest choice is

$$g(u^-, u^+) = u^-,$$

and from the form of the basis functions we get

$$g\big(v^-(x_{j-1/2}), v^+(x_{j-1/2})\big) = v^-(x_{j-1/2}) = b_{j-1},$$

$$g\big(v^-(x_{j+1/2}), v^+(x_{j+1/2})\big) = v^-(x_{j+1/2}) = b_j.$$

The final approximation then becomes

$$\int_{x_{j-1/2}}^{x_{j+1/2}} \left(\frac{da_j}{dt}\phi_j + \frac{db_j}{dt}\psi_j \right) \phi_j \, dx + b_{j-1} - \int_{x_{j-1/2}}^{x_{j+1/2}} (a_j\phi_j + b_j\psi_j)\frac{\partial \phi_j}{\partial x} \, dx = 0,$$

$$\int_{x_{j-1/2}}^{x_{j+1/2}} \left(\frac{da_j}{dt}\phi_j + \frac{db_j}{dt}\psi_j \right) \psi_j \, dx + b_j - \int_{x_{j-1/2}}^{x_{j+1/2}} (a_j\phi_j + b_j\psi_j)\frac{\partial \psi_j}{\partial x} \, dx = 0,$$

$$j = 1, 2, \dots, N.$$

Since the basis functions are known and have a simple form, the integrals can be computed. For simplicity we assume a uniform node distribution separated by a distance Δx. For each subinterval we get when dropping the subscript

$$\int \phi^2 \, dx = \frac{h}{3}, \qquad \int \phi\psi \, dx = \frac{h}{6}, \qquad \int \psi^2 \, dx = \frac{h}{3},$$

$$\int \phi \frac{\partial \phi}{\partial x} \, dx = -\frac{1}{2}, \qquad \int \psi \frac{\partial \phi}{\partial x} \, dx = -\frac{1}{2},$$

$$\int \phi \frac{\partial \psi}{\partial x} \, dx = \frac{1}{2}, \qquad \int \psi \frac{\partial \psi}{\partial x} \, dx = \frac{1}{2},$$

leading to the ODE system

$$\begin{aligned}
\frac{\Delta x}{3}\frac{da_j}{dt} + \frac{\Delta x}{6}\frac{db_j}{dt} &= b_{j-1} - \frac{1}{2}b_j, \\
\frac{\Delta x}{6}\frac{da_j}{dt} + \frac{\Delta x}{3}\frac{db_j}{dt} &= \frac{1}{2}a_j - \frac{1}{2}b_j, \quad j = 1, 2, \dots, N.
\end{aligned} \tag{11.15}$$

Since information is flowing from left to right, the equation requires a boundary condition at the left boundary, i.e., b_0 is known.

The advantage compared to the Galerkin method is now obvious. There is no coupling of da_j/dt and db_j/dt to the corresponding coefficients at neighboring intervals, i.e., the mass matrix has a simple block diagonal form. Each 2×2 system is easily reformulated *before* the computation, and we get

$$\frac{da_j}{dt} = \frac{1}{\Delta x}(4b_{j-1} - 3a_j - b_j),$$

$$\frac{db_j}{dt} = \frac{1}{\Delta x}(-2b_{j-1} + 3a_j - b_j), \quad j = 1, 2, \dots, N.$$

This form is the semidiscrete approximation, and allows for a fast explicit ODE-solver in time, just as for finite difference methods.

When putting the pieces of the solution on each subinterval together, we realize that there is a dual representation of v at each node $x_{j+1/2}$. But this is of little practical trouble. When evaluating and/or plotting the solution, we simply use $v(x) = v^-(x)$ everywhere (or $v(x) = v^+(x)$).

In contrast to Galerkin methods, the stability does not follow automatically for the semidiscrete approximation. The necessary analysis techniques are outside the scope of this book, where we limit the stability analysis to the standard Fourier technique.

Fig. 11.9 Eigenvalues of
$\hat{Q}(\xi)$, $0 \leq \xi \leq 2\pi$

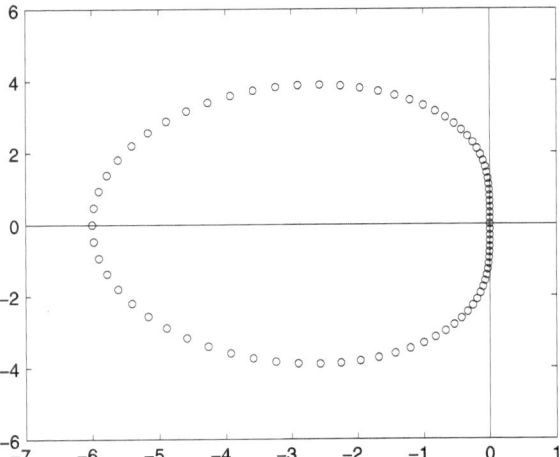

Fig. 11.10 $u(x, 6)$ $(-)$,
$v(x, 6)$ $(--)$, $r = 1$, $N = 20$

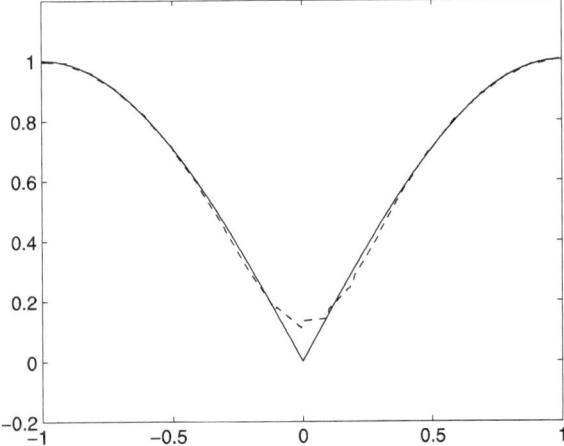

For our example, the original PDE problem with periodic solutions in space have a constant norm in time. The Fourier transformed system for the vectors **a** and **b** representing the discrete solution is

$$\frac{d}{dt}\begin{bmatrix}\hat{a}\\\hat{b}\end{bmatrix} = \frac{1}{\Delta x}\begin{bmatrix}-3 & 4e^{-i\xi} - 1\\3 & -2e^{-i\xi} - 1\end{bmatrix}\begin{bmatrix}\hat{a}\\\hat{b}\end{bmatrix}.$$

The eigenvalues of the coefficient matrix $\hat{Q}$ on the right hand side are shown in Fig. 11.9 for discrete ξ-values and $\Delta x = 1$. All of them are located in the left half plane, with the single eigenvalue zero at the imaginary axis. It is easily shown that $\hat{Q}$ can be transformed to diagonal form and, referring back to Sect. 10.1, we conclude that the solutions are bounded independent of Δx and t.

Fig. 11.11 $u(x, 6)$ $(-)$,
$v(x, 6)$ $(--)$, $r = 3$, $N = 20$

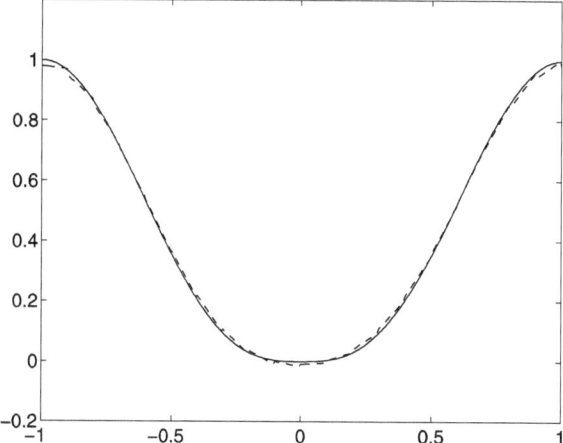

Fig. 11.12 $u(x, 6)$ $(-)$,
$v(x, 6)$ $(--)$, $r = 5$, $N = 20$

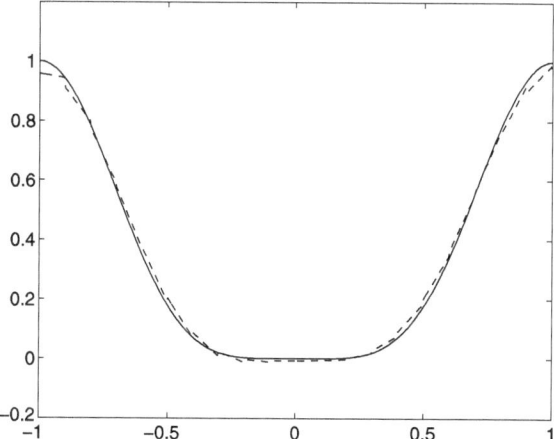

As a numerical test we solved the periodic problem for $-1 \le x \le 1$, $0 \le t \le 6$ with the initial data

$$u(x, 0) = \left| \sin \frac{\pi x}{2} \right|^r.$$

The nodes are defined such that $x_j = -1 + j\Delta x$. At $t = 6$ the wave has made 3 complete rotations, and is back to its original position.

Figures 11.10, 11.11, 11.12 show the solution

$$v = \sum_{j=1}^{N} (a_j \phi_j + b_j \psi_j)$$

for $r = 1, 3, 5$ and $N = 20$.

The case $r = 1$ is the most difficult one, since there is a sharp cusp at $x = 0$, but the results are quite good. On the scale used in the pictures, it is hard to see the

Fig. 11.13 $u(x, 6)$,
$r = 1\ (-),\ v(x, 6),\ N = 20$

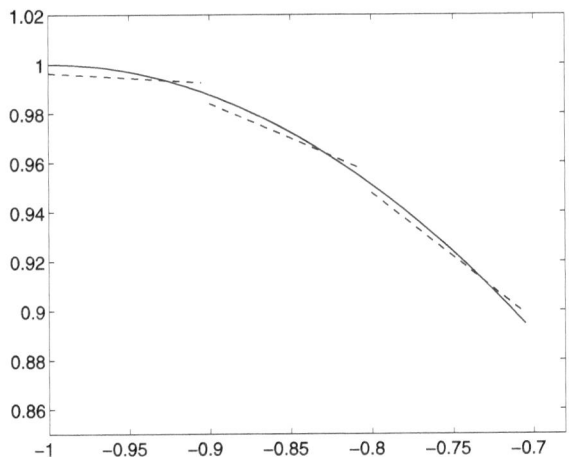

discontinuities at the nodes $x_{j+1/2}$. In Fig. 11.13, showing the solution near $x = -1$ at another scale, they are clearly visible.

The basis functions ϕ_j and ψ_j were chosen in order to illustrate the block diagonal structure of the mass matrix. In our case, we could as well have chosen them such that a diagonal mass matrix is obtained without solving the 2×2 systems. It is easily obtained by choosing the orthogonal linear functions $\phi_j = 1$ and $\psi_j = 2(x - x_j)/h_j$.

Higher order methods are obtained by simply raising the degree of the polynomials in each subinterval. There is still no coupling between the time derivatives in different subintervals, but the local systems become larger. However, as noted above, by choosing orthogonal polynomials, even this minor problem is eliminated.

The generalization to several space dimensions is done in analogy with the Galerkin methods. The polynomials are constructed locally in each element such as triangles in two space dimensions. The basis functions associated with a certain triangle T_j are zero at all other triangles, and the solution v becomes discontinuous at the edges of each triangle.

DG methods is currently a very active area of research, and the methods are now being used for an increasing number of applications. For a complete description of the current state of the art, we recommend the recent book by Hesthaven and Warburton [14].

Exercise 11.4 Carry out the analysis for the example above, but now using quadratic polynomials on each subinterval. Derive the semidiscrete system corresponding to (11.15).

Chapter 12
Spectral Methods

Earlier in this book we have demonstrated how functions can be represented by a series

$$u(x) = \sum_{k=1}^{\infty} a_k \phi_k(x), \qquad (12.1)$$

where $\{\phi_k(x)\}$ are certain basis functions. By choosing the ϕ_k properly, the co-efficients a_k can sometimes be found without too much effort, and hopefully the coefficients are such that the function is well represented by the first few terms.

When looking for the solution to a differential equation, we cannot expect that the finite sum can satisfy the differential equation everywhere. Here we shall use the collocation principle to relax the conditions on the approximate solution. We require only that the differential equation is satisfied at the collocation points, or grid points as described in Sect. 9.1.

The fundamental interpretation of spectral methods was originally that the differential equation is transformed such that we work all the time with the coefficients a_k as a representation of the function instead of the usual explicit function values. But this leads to severe limitations when it comes to differential equations with variable coefficients or nonlinear problems. Instead we shall use the spectral representation for only a part of the solution procedure, which leads to something called *pseudo-spectral methods*. We shall begin with an introductory example.

12.1 An Example with a Sine Expansion

We gave several examples of orthogonal basis functions in Chap. 7, for which the computation becomes particularly simple. Here we shall use still another set of orthogonal functions, namely $\phi_k(x) = \sin(k\pi x)$ on the interval $0 \le x \le 1$. The scalar product and norm are defined by

$$(f, g) = \int_0^1 f(x)g(x)\,dx, \qquad \|f\|^2 = (f, f),$$

B. Gustafsson, *Fundamentals of Scientific Computing*,
Texts in Computational Science and Engineering 8,
DOI 10.1007/978-3-642-19495-5_12, © Springer-Verlag Berlin Heidelberg 2011

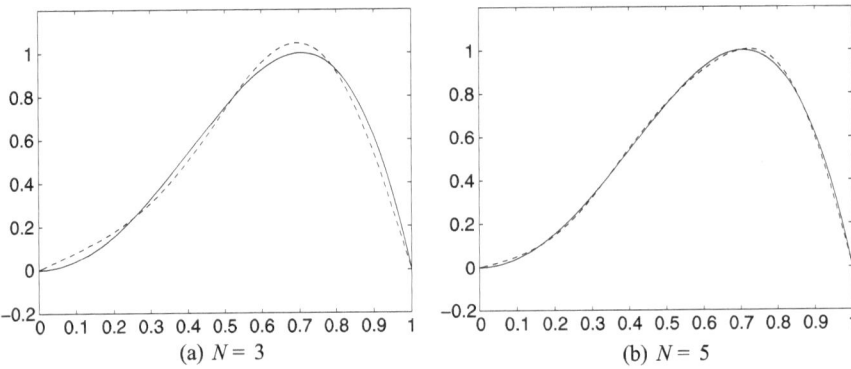

Fig. 12.1   Sine expansion $(--)$ of the function $x^2(1-x^2)$ $(—)$

and we have

$$\|\sin(k\pi x)\|^2 = \frac{1}{2}, \quad k = 1, 2, \dots.$$

Consider the function $u(x) = x^2(1-x^2)$. It is easy enough to handle as it is, but for illustration we try to write it as a sine series. The coefficients are found by integration and, by orthogonality,

$$a_k = 2\int_0^1 u(x)\sin(k\pi x)\,dx, \quad k = 1, 2, \dots.$$

These integrals are easily computed for our function $u(x)$, and we get the explicit formula

$$a_k = \left(\frac{1}{k\pi} + \left(\frac{12}{(k\pi)^2} + 1\right)\left(-\frac{1}{k\pi} + \frac{2}{(k\pi)^3}\right)\right)\cos k\pi - \left(\frac{12}{(k\pi)^2} + 1\right)\frac{16}{(k\pi)^3}.$$

It can be shown that the infinite sum *converges* to the function $u(x)$. Figure 12.1 shows the finite approximating sum

$$\sum_{k=1}^N a_k \sin(k\pi x)$$

for $N = 3$ and $N = 5$.

We consider again the simple heat conduction problem

$$\begin{aligned}
\frac{\partial u}{\partial t} &= \frac{\partial^2 u}{\partial x^2}, \quad 0 \le x \le 1,\ 0 \le t, \\
u(0, t) &= 0, \\
u(1, t) &= 0, \\
u(x, 0) &= f(x).
\end{aligned} \tag{12.2}$$

The boundary conditions with zero temperature are not very realistic, but we use them here in order to illustrate the basic numerical method. We now assume that there is a finite sum representation with time dependent coefficients a_k, giving the approximation

$$u(x,t) = \sum_{k=1}^{N} a_k(t) \sin(k\pi x). \qquad (12.3)$$

When plugging this into the differential equation, we get

$$\sum_{k=1}^{N} \frac{da_k(t)}{dt} \sin(k\pi x) = -\sum_{k=1}^{N} (k\pi)^2 a_k(t) \sin(k\pi x).$$

This is an example of *separation of variables*, which is often used as a mathematical tool. The x- and t-dependence are separated, with the x-dependence confined to $\sin(k\pi x)$ and the t-dependence to $a_k(t)$. The orthogonality of the functions $\sin(k\pi x)$ implies that this equation is satisfied for all x and t if and only if it holds for each term by itself:

$$\frac{da_k(t)}{dt} = -(k\pi)^2 a_k(t), \quad k = 1, 2, \ldots, N.$$

This is a system of ordinary differential equations, which is a significant reduction in complexity compared to a partial differential equation. Since each basis function $\sin(k\pi x)$ satisfies the boundary conditions, the whole sum does so as well. Hence, we need only an initial condition for the system. This condition is obtained by approximating the initial function by a sine expansion as well:

$$f(x) \approx \sum_{k=1}^{N} b_k \sin(k\pi x).$$

The coefficients b_k are found by the least square method as demonstrated earlier, and we get the initial value problem

$$\frac{da_k(t)}{dt} = -(k\pi)^2 a_k(t), \quad k = 1, 2, \ldots, N,$$
$$a_k(0) = b_k, \quad k = 1, 2, \ldots, N.$$

This system is usually solved by a finite difference method as described in Sect. 10.1. The final solution is then obtained by evaluating the sum (12.3) for any given t.

The whole procedure is an example of a *spectral method*. The name originates from the concept of the *spectrum* of an operator. In our example the operator is the differential operator d^2/dx^2 acting on functions $u(x)$ with $u(0) = u(1) = 0$. The relation

$$\frac{d^2}{dx^2} \sin(k\pi x) = -(k\pi)^2 \sin(k\pi x) \qquad (12.4)$$

is a remarkable one, since it converts a differential operator to a simple multiplication acting on the same function. Referring back to Sect. 3.4 we note the similarity with eigenvalues of a matrix. A complicated matrix is applied to a certain vector and returns the same vector multiplied by a number. For our case, the numbers

$$\lambda_k = -(k\pi)^2, \quad k = 1, 2, \ldots$$

are the eigenvalues of the *operator* d^2/dx^2, and the functions $\sin(k\pi x)$ are the *eigenfunctions*. The whole set of eigenvalues $\{\lambda_k\}$ is called the *spectrum*. We can also see the eigenfunctions and the eigenvalues as the possible solutions to the boundary value problem

$$\frac{d^2u}{dx^2} - \lambda u = 0, \quad 0 \le x \le 1,$$
$$u(0) = 0,$$
$$u(1) = 1.$$

Such a problem is called a *Sturm–Liouville* problem after the French mathematicians Charles-Francois Sturm (1803–1855) and Joseph Liouville (1809–1892). This type of problem is an example where for a given value of λ there is not a unique solution, and there are two possibilities. If λ does not belong to the set $\{\lambda_k\}$ as defined above, there is no solution at all (we don't include the trivial solution $u(x) \equiv 0$ here). If, on the other hand, $\lambda = -(k\pi)^2$ for some integer k, there are infinitely many solutions since $c \sin(k\pi x)$ is a solution for any constant c.

Unfortunately, the method described above does not work well for more general forms of the problem. If the heat equation is changed such that the heat conduction coefficient depends on x resulting in the differential equation

$$\frac{\partial u}{\partial t} = \kappa(x)\frac{\partial^2 u}{\partial x^2}, \tag{12.5}$$

we are in trouble. Even if there is an expansion

$$\kappa(x) = \sum_{k=1}^{N} \hat{\kappa}_k \sin(k\pi x),$$

the product $\kappa(x)u(x, t)$ cannot be directly represented as a sine expansion with N terms. Therefore, the spectral method in the form described above is not well suited for problems like this one. But there is an alternative.

We introduce a grid

$$x_j = j\Delta x, \quad j = 0, 1, \ldots, N+1, \quad (N+1)\Delta x = 1$$

in space just as for difference methods. Let us assume that a certain function $u(x)$ is known at the grid points x_j. We now define an interpolating function

$$v(x) = \sum_{k=1}^{N} a_k \sin(k\pi x) \tag{12.6}$$

such that

$$v(x_j) = u(x_j), \quad j = 1, 2, \ldots, N.$$

This is in complete analogy with the Fourier interpolation leading to the discrete Fourier transform as described in Sect. 6.2. The set of coefficients a_k is the *discrete sine transform* of the set $\{v(x_j)\}$. Even if the interval $[0, 1]$ is only half a period for the function $\sin \pi x$, the fast Fourier transform FFT can still be used for computation of the coefficients.

We now go back to our example. The interpolating function (12.6) can of course be used for the obvious purpose of representing the grid function for all x in between the grid points. But here we use it for another purpose. We want to construct an approximation of the derivative in space. For a grid function known only at the grid points, we are limited to finite differences. Here we are in a better position since $v(x)$ is known everywhere when the coefficients a_k are known. The derivative of the approximation function $v(x)$ is well defined, and the second derivative of $v(x)$ is easily obtained as

$$\frac{d^2v}{dx^2}(x) = w(x) = \sum_{k=1}^{N} b_k \sin(k\pi x) = -\sum_{k=1}^{N} (k\pi)^2 a_k \sin(k\pi x).$$

Here is where the importance of an accurate approximation $v(x)$ for *all* x becomes obvious. At the grid points the accuracy of the function values is by definition infinite for any interpolating function, since the approximation agrees exactly with the given values. However, if there are severe oscillations in between the grid points, the derivative is not going to be accurate at the grid points. Orthogonality of the basis functions is the key to good approximations, as we have shown in Chaps. 6 and 7, and here we benefit from this property of the sine functions.

With the new coefficients b_k known, the grid values $w(x_j)$ can be computed via the inverse sine transform, also this one in $\sim N \log N$ arithmetic operations.

We now have the foundation for a new type of solution method. Symbolically we form a vector $\mathbf{v}$ containing the grid values $v_j = v(x_j)$, and compute the coefficients a_k. This can be seen as a matrix/vector multiplication $\mathbf{a} = T\mathbf{v}$, where the matrix T is the discrete sine transform. The differentiation is represented by the diagonal matrix

$$D = -\pi^2 \text{diag}(1, \ 4, \ 9, \ldots, N^2), \tag{12.7}$$

and we get the new coefficients b_k by $\mathbf{b} = DT\mathbf{v}$. The new grid values w_j are finally obtained by $\mathbf{w} = T^{-1}DT\mathbf{v}$, where T^{-1} is the inverse sine transform.

In this way the partial differential equation $\partial u/\partial t = \partial^2 u/\partial x^2$ is transferred to an ODE system for the vector $\mathbf{v}$:

$$\frac{d\mathbf{v}}{dt} = T^{-1}DT\mathbf{v}. \tag{12.8}$$

This system requires a numerical method for integration in time, and usually the discretization is done by using a difference method. The simplest discretization is the one we have used before:

$$\mathbf{v}^{n+1} = \mathbf{v}^n + \Delta t \, T^{-1}DT\mathbf{v}^n.$$

The complete procedure is the *pseudo-spectral method*. The prefix *pseudo* is introduced because there is no complete transformation to spectral space where we would work solely with the coefficients a_k. In contrast to the spectral method we are working with a grid in space, switching back and forth between the original space and the spectral space for each time step.

For our simple example, the pseudo-spectral method actually gives the same result as the spectral method if the same discretization is used in time. The great advantage with the pseudo-spectral method is that it can easily be generalized to more complicated problems. For the problem (12.5) we form the diagonal matrix

$$
K = \begin{bmatrix} \kappa(x_1) & & & \\ & \kappa(x_2) & & \\ & & \ddots & \\ & & & \kappa(x_N) \end{bmatrix}.
$$

The system of ODE now becomes

$$
\frac{d\mathbf{v}}{dt} = KT^{-1}DT\mathbf{v},
$$

which is also solved by a difference method.

The pseudo-spectral method has very high order of accuracy, since it uses every grid point for approximating the x-derivative. This means that, even for very high accuracy requirements, the grid can be quite coarse, and it is very efficient for problems where the solution is smooth. On the other hand, we demonstrated in Sect. 6.1 that trigonometric functions don't handle discontinuous functions very well. Much research has been done to develop various types of filter to eliminate this disadvantage, but for problems in several space dimensions there is still work to do.

Exercise 12.1 Consider the problem

$$
\begin{aligned}
\frac{\partial u}{\partial t} &= \frac{\partial^2 u}{\partial x^2} + au, \quad 0 \le x \le 1, \ 0 \le t, \\
u(0, t) &= 0, \\
u(1, t) &= 0, \\
u(x, 0) &= f(x),
\end{aligned}
$$

i.e., (12.2) with an extra term au added to the differential equation.

(a) Prove that $\sin(k\pi x)$ are still the eigenfunctions of the new differential operator in space, but that the eigenvalues change to $\lambda_k = -(k\pi)^2 + a$.
(b) Prove that the spectral and the pseudo-spectral methods are equivalent for this problem.

12.2 Fourier Methods

In the previous section we used sine functions which are a special case of more general trigonometric functions. They are particularly convenient for second order differential operators, since we are getting the sine functions back after two differentiations. For general problems with p-periodic solutions $u(x + p) = u(x)$ for all x, we can use the complex Fourier grid functions and the FFT for fast computation as described in Sect. 6.2. Periodic solutions are quite frequent in applications, and the computation in such a case can be limited to one period. The size of the period can be modified by a proper scaling, and a convenient choice is $[0, 2\pi]$, with the basis functions

$$\phi_k(x) = e^{ikx}.$$

The differentiation of the interpolating function $v(x)$ yields the same basis functions for the new vector $\mathbf{w}$:

$$a_k = \frac{1}{2\pi} \sum_{j=0}^{N} v_j e^{-ikx_j} \Delta x,$$

$$v(x) = \sum_{k=-N/2}^{N/2} a_k e^{ikx},$$

$$w_j = \sum_{k=-N/2}^{N/2} ik a_k e^{ikx_j}.$$

For a first order PDE

$$\frac{\partial u}{\partial t} = c(x,t) \frac{\partial u}{\partial x},$$

we define the diagonal matrices

$$C(t) = \begin{bmatrix} c(x_0, t) & & & & \\ & c(x_1, t) & & & \\ & & \ddots & & \\ & & & c(x_N, t) \end{bmatrix},$$

$$D = i \begin{bmatrix} -\frac{N}{2} & & & \\ & -\frac{N}{2} + 1 & & \\ & & \ddots & \\ & & & \frac{N}{2} \end{bmatrix}.$$

The pseudo-spectral method with these basis functions (often called the *Fourier method*) gives the ODE system

$$\frac{d\mathbf{v}}{dt} = C(t) F^{-1} D F \mathbf{v},$$

where F is the matrix representing the FFT as explained in Sect. 6.2. For discretization in time we can use the leap-frog scheme

$$\mathbf{v}^{n+1} = \mathbf{v}^{n-1} + 2\Delta t C(t_n) F^{-1} D F \mathbf{v}^n.$$

Figure 12.2 shows the pseudo-spectral principle.

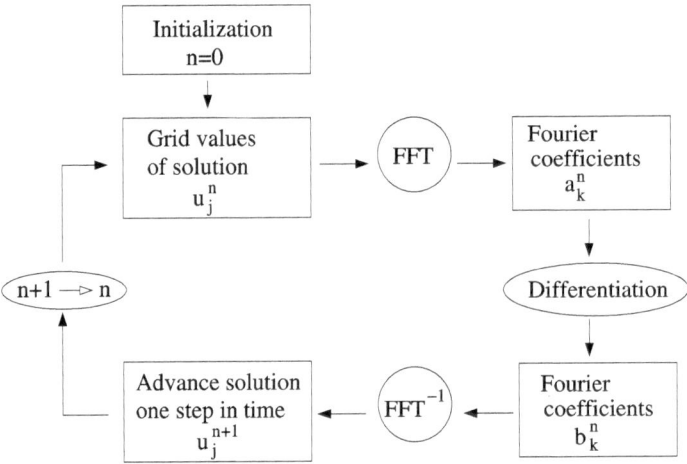

Fig. 12.2 The pseudo-spectral method

Let us now take a look at the stability of this scheme for the simplified case that $c(x,t) = c$ is a constant, i.e., $C = cI$. We showed how the Fourier transform is used for the stability analysis for difference schemes. For the Fourier method it is even more natural to use the same method of analysis, since the Fourier transformation is already involved. By multiplying the scheme from the left by F, and introducing the new vector $\mathbf{a}^n = F\mathbf{v}^n$, which contains the Fourier coefficients, we get

$$\mathbf{a}^{n+1} = \mathbf{a}^{n-1} + 2c\Delta t\, D\mathbf{a}^n.$$

But this is a set of scalar equations, and the maximal magnitude of the elements of D is $N/2$. By the results in Sect. 10.1, we get the stability limit

$$c\Delta t \frac{N}{2} \le 1,$$

and, since $(N+1)\Delta x = 2\pi$, this is equivalent to

$$c\Delta t \left(\frac{\pi}{\Delta x} - \frac{1}{2} \right) \le 1.$$

But Δx is arbitrarily small, so the final condition is

$$\frac{c\Delta t}{\Delta x} \le \frac{1}{\pi}.$$

Exercise 12.2 Apply the Fourier method to the problem $\partial u/\partial t = \partial^2 u/\partial x^2$ with periodic solutions.

(a) What is the stability condition for the Euler forward time discretization?
(b) Find the stability condition for the same time discretization of the sine expansion method applied to the problem (12.2) (with nonperiodic solutions). Compare the two conditions.

12.3 Polynomial Methods

In the previous sections we have used trigonometric expansions with the discrete Fourier transform as the basic tool. For nonperiodic solutions, there are many other types of functions that can be used. The class of orthogonal polynomials $\{P_k(x)\}$ that we have described in Chap. 7 is very common, in particular the Chebychev and Legendre polynomials. However, the differentiation is a little more complicated in these cases. The interpolating Chebychev polynomial is

$$v(x) = \sum_{n=0}^{N} a_n T_n(x),$$

which equals the given function values at the special grid points

$$x_j = \cos \frac{j\pi}{N}, \quad j = 0, 1, \ldots, N.$$

The coefficients a_n are obtained by the cosine transform. But what is the derivative $w(x)$ at the grid points? There is actually a recursive formula also for the derivative, and there is a connection to the polynomials themselves:

$$T'_{n+1}(x) = 2(n+1)T_n(x) + \frac{n+1}{n-1}T'_{n-1}(x),$$

$$(1-x^2)T'_n(x) = \frac{n}{2}\left(T_{n-1}(x) - T_{n+1}(x)\right).$$

With the derivative evaluated at the grid points, the inverse discrete Chebychev transform is used to get back to the original space represented by u_j.

Even if the cosine transform is fast, the direct discrete relation between the given values u_j and the computed derivatives $w_j = w(x_j)$ is sometimes used. This relation is represented by a matrix S corresponding to $F^{-1}DF$ in the Fourier case. After a little work we find that the elements s_{jn} of S are defined by

$$s_{jn} = \begin{cases} -(2N^2+1)/6, & j = n = 0, \\ -x_j/(2(1-x_j^2)), & j = n = 1, 2, \ldots, N-1, \\ (2N^2+1)/6, & j = n = N, \\ c_j(-1)^{j+n}/(c_n(x_j - x_n)), & j \neq n, \end{cases}$$

where

$$c_j = \begin{cases} 1, & j = 1, 2, \ldots, N-1, \\ 2, & j = 0, N. \end{cases}$$

The distribution of the grid points is such that they cluster near the boundaries. This may be an advantage if the true solution has boundary layers as we have seen above. On the other hand, it is a disadvantage if the solution has no special features near the boundaries. We are wasting grid points where they are not needed. However, the clustering is there, and nothing can be done about it as long as we want the interpolation and differentiation to be correct.

There is also another disadvantage that follows from this clustering. The step size Δx near the boundaries is of the order $1/N^2$ and, if an explicit time discretization is used, the step size Δt in time must be kept of the order $1/N^2$ as well.

In several space dimensions and special geometries, there may be special types of suitable functions. In Sect. 17.4 we discuss this for the case when the computational domain is a sphere.

Part IV
Numerical Methods for Algebraic Equations

Chapter 13
Numerical Solution of Nonlinear Equations

Numerical methods for differential equations naturally lead to equations or systems of equations that require numerical solution methods. In fact, these equations are most often the central part of the whole solution procedure, and they must therefore be solved as effectively as possible. In this chapter, we shall describe the basic principles for the construction and analysis of such methods.

13.1 Scalar Nonlinear Equations

In Sect. 10.1 it was shown how certain types of numerical methods for ordinary differential equations lead to algebraic equations that must be solved in each time step. Linear scalar equations with only one unknown are trivial to solve, but nonlinear equations are not. Only in very special cases can they be solved by direct methods but, in general, iterative methods are required. We start by guessing a solution, and then use a step by step algorithm to compute new and better approximations at each step. We want to solve the equation $f(x) = 0$ for the unknown x, where we assume that $f(x)$ has a known algebraic form, and can be computed for any given x-value. One difficulty is that in general there are many solutions. For example, we know that the equation $x^2 - 1 = 0$ has the two solutions $x = 1$ and $x = -1$. Often we are aiming for just one solution, and we may have some rough idea about its location. We choose an initial guess as accurately as we can, and then we hope that the algorithm takes us to the right solution.

The most commonly used method is the classic *Newton–Raphson method*. The essence of the method was first developed by Isaac Newton (1643–1727) and was later described in a different way by Joseph Raphson (1678–1715). It can be derived geometrically as demonstrated in Fig. 13.1.

We want to find the point x^* where the curve $f(x)$ passes through the x-axis. Assume that we have an approximate solution x_j and that we can compute $f(x_j)$. The figure shows the tangent to the curve at that point, and we recall that the slope of the tangent is $f'(x_j)$. The tangent doesn't deviate much from the curve in the immediate neighborhood of x_j and, if we are sufficiently close to the true solution,

B. Gustafsson, *Fundamentals of Scientific Computing*,
Texts in Computational Science and Engineering 8,
DOI 10.1007/978-3-642-19495-5_13, © Springer-Verlag Berlin Heidelberg 2011

Fig. 13.1
Newton–Raphson's method

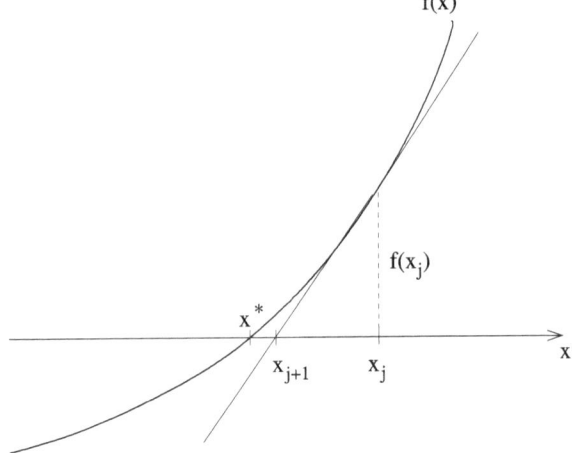

the intersection with the x-axis should be close to x^*. The intersection can be computed, and this point is chosen as our next approximation x_{j+1}. The slope is found geometrically as $f'(x_j) = f(x_j)/(x_j - x_{j+1})$, which we rewrite as

$$x_{j+1} = x_j - \frac{f(x_j)}{f'(x_j)}. \tag{13.1}$$

Another way of deriving this formula is to use the Taylor expansion around a given approximation x_j. We have

$$0 = f(x^*) = f(x_j) + (x^* - x_j)f'(x_j) + \mathcal{O}\!\left(|x^* - x_j|^2\right).$$

If the last term is disregarded, we get

$$x^* \approx x_j - \frac{f(x_j)}{f'(x_j)}$$

and, by substituting x^* by x_{j+1}, we have the formula (13.1).

The repeated use of this formula with the initial value x_0 is the *Newton–Raphson method*. If we have a good idea about the location of the solution, the method is very efficient. As an example we take the equation $x^5 - 3x + 1 = 0$, which was solved by iteration in Sect. 1.1. Newton–Raphson's method is

$$x_{j+1} = x_j - \frac{x_j^5 - 3x_j + 1}{5x^4 - 3},$$

and with $x_0 = 1$ it produces the values

$$x_0 = 1.0000,$$
$$x_1 = 1.5000,$$
$$x_2 = 1.3165,$$
$$x_3 = 1.2329,$$

$$x_4 = 1.2154,$$
$$x_5 = 1.2146,$$
$$x_6 = 1.2146,$$

i.e., we obtain the solution with 4 accurate decimals in five iterations. But wait a minute! Looking back to the iteration in Sect. 1.1 we find that it produced the solution $x = 0.3347$, and that with the same initial value $x_0 = 1$. This is of course no contradiction. Both values are correct solutions to the equation, after all there are five different roots to choose from. We shall discuss this further below.

However, there is not always plain sailing with the Newton–Raphson method. The function $f(x)$ may not be as smooth as in our example and, furthermore, we may not have a very good initial guess x_0. In both cases the tangent in our figure may not intersect the x-axis very close to the solution. Indeed, the new value x_{j+1} may actually be further away than the previous value x_j, and the method does not converge at all. Many software packages use a strategy which picks a large number of different initial values x_0 with the hope that at least one of them is in the right neighborhood, such that the iteration produces the right solution. There are also other iterative methods that do not use the derivative of the function, which sometimes is not known, or is difficult to compute.

The convergence analysis for iterative solvers is quite difficult for the general case. The reason is that, if the initial guess is far from the true solution (or one of the true solutions), the restrictions on the function $f(x)$ are severe in order to get a good idea about the behavior of the iterations. A nonlinear function has too much freedom to change in the domain. But if we know that we are in the neighborhood of the solution, it is possible to verify convergence. We assume that the iterative formula is

$$x_{j+1} = g(x_j), \quad j = 0, 1, \dots. \tag{13.2}$$

Each new approximation is obtained by using only one previous approximation. If the original equation is $f(x) = 0$, there are of course many ways of rewriting it as $x = g(x)$, but it can always be done. A simple way is

$$x = x - f(x),$$

which obviously has the same solution. But there is no guarantee that the iteration converges with this choice of $g(x)$. The most general convergence result is the following:

Suppose that $|g'(x)| \leq \alpha < 1$ everywhere in the interval $I = \{x : |x - x^*| \leq d\}$, where $x^* = g(x^*)$. If $x_0 \in I$, then the points x_j generated by the iteration formula (13.2) have the properties

- $x_j \in I, j = 0, 1, \dots.$
- $x_j \to x^*$ as $j \to \infty.$
- There is no other solution in I.

The iteration process can be interpreted geometrically. The function $y = g(x)$ and the straight line $y = x$ are drawn, and the intersection of the two (or one of the

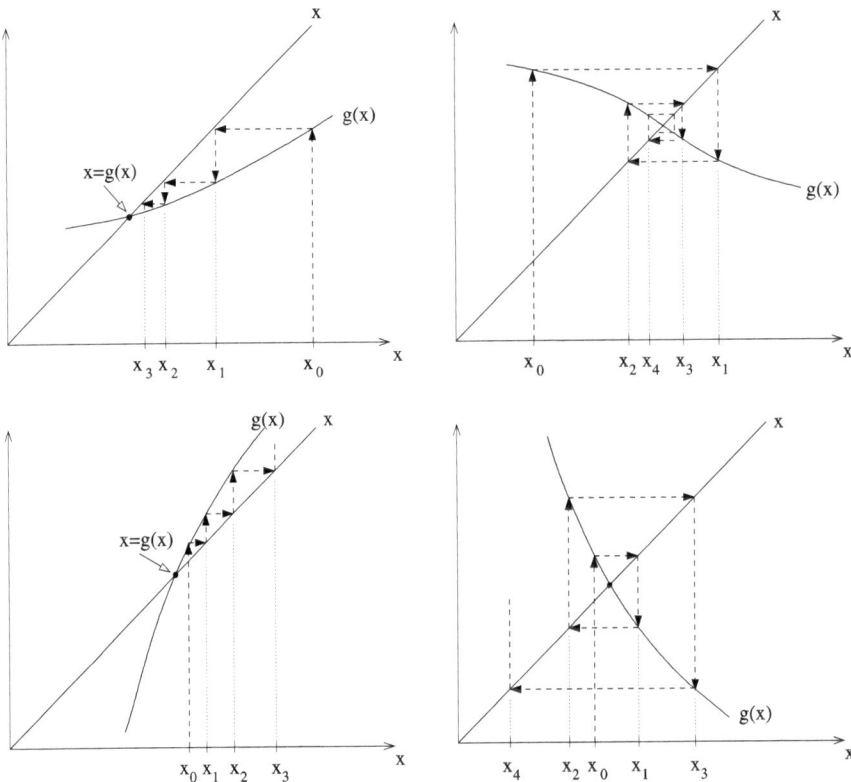

Fig. 13.2 Converging and diverging solutions of $x = g(x)$

intersections) is the solution. The initial guess is x_0 and the point $(x_0, g(x_0))$ is found. A horizontal line through that point intersects the line $y = x$. This gives the new point $x_1 = g(x_0)$, and the process is then repeated, showing either convergence or divergence. Figure 13.2 shows the iteration process for two cases with $|g'| < 1$ and two cases with $|g'| > 1$.

When solving time dependent differential equations, Newton–Raphson's method is perfect in the sense that a good initial value is known in almost all cases. When solving for the solution u^{n+1} at the new time level, we simply use u^n as the initial guess.

Let us finally take a closer look at the example above, where we arrived at different solutions with two different methods. Figure 13.3 shows the first iteration with the Newton–Raphson method to the left, and the method

$$x_{j+1} = g(x_j),$$

$$g(x) = \frac{x^5}{3} + \frac{1}{3}$$

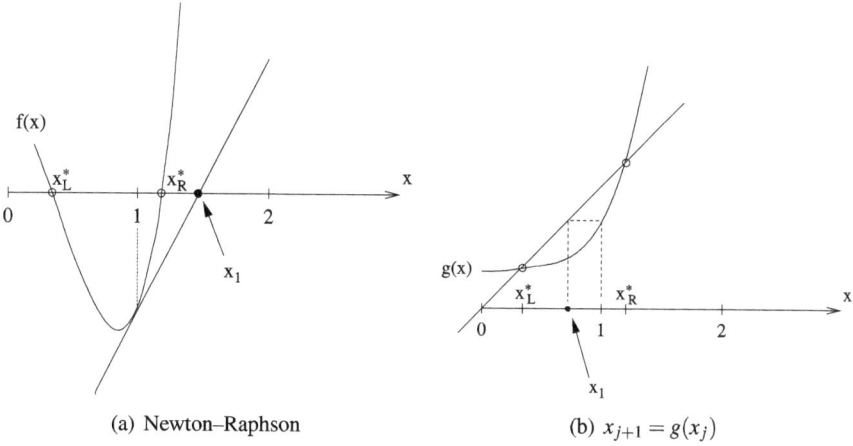

(a) Newton–Raphson   (b) $x_{j+1} = g(x_j)$

Fig. 13.3 Two iteration methods for $f(x) = x^5 - 3x + 1 = 0$

to the right. There are 3 real roots, and the figure shows the two larger ones x_L^* and x_R^*. In the first case the tangent at $x = x_0 = 1$ cuts the x-axis to the right of x_0 causing the subsequent iterations to converge to x_R^*, while in the second case the iteration turns to the left and converges to the left alternative x_L^*.

Exercise 13.1 Write a computer program for the iteration method (13.2) and for the Newton–Raphson method. Use different initial values x_0, and try to get the roots of the equation $x^5 - 3x + 1 = 0$ by both methods. Use also complex initial values to find the complex roots.

Exercise 13.2 The Newton–Raphson method is a special case of the general iteration formula (13.2). Assume that $f'(x^*)$ and $f''(x^*)$ are nonzero, where x^* is the solution. Prove that the convergence condition on $g(x)$ is satisfied if the approximation x_j is sufficiently close to x^*.

Exercise 13.3 The Newton–Raphson method has *quadratic convergence*, which means that $|x_{j+1} - x_j| \le c|x_j - x_{j-1}|^2$, where c is a constant. Prove this property by using the Taylor expansion of $f(x_j)$ around x_{j-1} and then using the equality $x_j - x_{j-1} = -f(x_{j-1})/f'(x_{j-1})$.

13.2 Vector Equations

So far we have treated scalar equations, i.e., $f(x)$ is a single number for any given number x. In most applications, we are dealing with systems of equations. An implicit difference method for a system of ODE gives rise to a system of equations to be solved for each time step. For example, the system (10.3) is obtained for the backward Euler method. In order to simplify the notation, we introduce the vector $\mathbf{v}$ by

$$\mathbf{v} = \begin{bmatrix} v_1 \\ v_2 \\ \vdots \\ v_N \end{bmatrix},$$

and write the system in the standard form corresponding to the scalar case

$$\mathbf{f}(\mathbf{v}) = \mathbf{0},$$

where the vector $\mathbf{f}$ also has N elements. When trying to write down the Newton–Raphson method, we encounter a problem. We get something that looks like $\mathbf{f}'(\mathbf{v})$ in the denominator, and what is that? Not only do we have many independent variables v_j, but also many elements f_j in the vector $\mathbf{f}$.

We go back to the basis for the Newton–Raphson' method, which is the Taylor expansion. Assuming that $\mathbf{v}^*$ is the true solution and that $\mathbf{v}$ is an approximate solution that we want to improve, we have according to Sect. 4.2 for each element of $\mathbf{f}$

$$0 = f_j(\mathbf{v}^*) = f_j(\mathbf{v}) + \sum_{k=1}^{N} \frac{\partial f_j}{\partial v_k}(\mathbf{x})(v_k^* - v_k) + \mathcal{O}(\|\mathbf{v}^* - \mathbf{v}\|^2), \quad j = 1, 2, \ldots, N.$$

Here $\|\mathbf{v}^* - \mathbf{v}\|$ is any reasonable norm, for example the usual Euclidean vector norm. We write the expansion as a vector equation

$$0 = \mathbf{f}(\mathbf{v}^*) = \mathbf{f}(\mathbf{v}) + \frac{\partial \mathbf{f}}{\partial \mathbf{v}}(\mathbf{v})(\mathbf{v}^* - \mathbf{v}) + \mathcal{O}(\|\mathbf{v}^* - \mathbf{v}\|^2),$$

where $\partial \mathbf{f}/\partial \mathbf{v}$ is an $N \times N$ matrix. It is called the *Jacobian* of the vector $\mathbf{f}$, and we give it the notation $\mathbf{f}'$:

$$\mathbf{f}'(\mathbf{v}) = \begin{bmatrix} \frac{\partial f_1}{\partial v_1} & \frac{\partial f_1}{\partial v_2} & \cdots & \frac{\partial f_1}{\partial v_N} \\ \frac{\partial f_2}{\partial v_1} & \frac{\partial f_2}{\partial v_2} & \cdots & \frac{\partial f_2}{\partial v_N} \\ \vdots & & \ddots & \vdots \\ \frac{\partial f_N}{\partial v_1} & \frac{\partial f_N}{\partial v_2} & \cdots & \frac{\partial f_N}{\partial v_N} \end{bmatrix}.$$

When disregarding the squared terms, we get a new approximation $\tilde{\mathbf{v}}$ of $\mathbf{v}^*$ by

$$0 = \mathbf{f}(\mathbf{v}) + \frac{\partial \mathbf{f}}{\partial \mathbf{v}}(\mathbf{v})(\tilde{\mathbf{v}} - \mathbf{v}).$$

In order to avoid confusion with the notation $\mathbf{v}_j^n$ representing a vector at a grid point (x_j, t_n), we shall use the notation $\mathbf{v}^{(n)}$ for the approximation at iteration number n in an iterative scheme. The new approximation is obtained by

$$\mathbf{v}^{(n+1)} = \mathbf{v}^{(n)} - (\mathbf{f}')^{-1}(\mathbf{v}^{(n)})\mathbf{f}(\mathbf{v}^{(n)}).$$

This is the Newton–Raphson formula for nonlinear vector equations. For implicit methods for differential equations, the vector $\mathbf{v}^*$ represents the numerical solution $\mathbf{u}^{n+1}$ at the fixed time level t_{n+1}.

The inverse $(\mathbf{f}')^{-1}$ is almost always a dense matrix and, even if it could be computed, it should not be stored. Instead we solve the system of equations

$$\mathbf{f}'(\mathbf{v}^{(n)})(\mathbf{v}^{(n+1)} - \mathbf{v}^{(n)}) = -\mathbf{f}(\mathbf{v}^{(n)})$$

for the difference between the new and the old approximation. The Jacobian $\mathbf{f}'$ may be difficult or impossible to derive, but just as for the scalar case there are many other methods that do not use the Jacobian. The most general iteration formula is

$$\mathbf{v}^{(n+1)} = \mathbf{g}(\mathbf{v}^{(n)}),$$

where $\mathbf{g}$ has N components. For the scalar case we stated a sufficient condition for convergence and, as is often the case in numerical analysis, the role of the scalar derivative is taken over by the eigenvalues of the corresponding matrix, which is the Jacobian. Under certain conditions, it can be shown that a sufficient condition for convergence is

$$\rho\big(\mathbf{g}'(\mathbf{v})\big) < 1$$

near the solution $\mathbf{v}^*$, where $\rho(\mathbf{g}')$ is the spectral radius of the matrix. However, one should be aware that, just as it may be difficult to derive the Jacobian $\mathbf{f}'$, it may be difficult to estimate $\mathbf{g}'$.

For any type of iterative method, the systems arising for PDE applications are large but sparse and linear. Again we have broken down a nonlinear problem to a series of linear ones, which is so typical for computational mathematics. In the next chapter we shall discuss methods for linear systems.

Exercise 13.4 Write down the Newton–Raphson method for the system

$$u + d(u^2 - v) = 0,$$
$$v + d(v^3 - u) = 0,$$

(13.3)

where d is a constant.

Exercise 13.5 Write a computer program for the Newton–Raphson method applied to general 2×2-systems $\mathbf{f}(\mathbf{v}) = \mathbf{0}$, where the vector $\mathbf{f}$ and the Jacobian $\partial \mathbf{f}/\partial \mathbf{v}$ are given in separate functions. Use the program to solve (13.3).

Exercise 13.6 Consider the Euler backward difference method (10.13) for the heat equation $\partial u/\partial t = \partial^2 u/\partial x^2$. For the nonlinear equation $\partial u/\partial t = u\partial^2 u/\partial x^2$ the corresponding approximation is

$$u_j^{n+1} = u_j^n + \frac{\Delta t}{\Delta x^2} u_j^{n+1}(u_{j-1}^{n+1} - 2u_j^{n+1} + u_{j+1}^{n+1}).$$

Write down the system to be solved for the unknowns u_j^{n+1}, $j = 1, 2, \ldots, N - 1$, in each time step, and derive the Newton–Raphson method for solving it.

Chapter 14
Linear Systems of Equations

Implicit methods for initial value problems for linear systems of ODE or PDE lead
to linear algebraic systems in each time step. For linear boundary value problems we
arrive at one single system to solve. If the differential equations are nonlinear we saw
in the previous chapter, that the iterative Newton–Raphson method leads to linear
systems in each iteration. Apparently, the numerical solution of large linear systems
of equations is a key issue in computational mathematics. We shall demonstrate how
to solve them.

There are two different types of solution methods. Direct methods are based on
a scheme that computes the solution in a fixed number of steps that depends on
the number of unknowns. Iterative methods keep iterating with an initial guess as
starting point, and hopefully provide accurate solutions within a reasonable number
of iterations.

However, independent of the solution method, the *condition* of the system plays
an important role, and we shall first discuss how it effects the quality of the computed solution.

14.1 The Condition of a System

Not all systems of equations have a unique solution as we saw in Sect. 3.3, where
the system

$$x + 2y = a,$$
$$2x + 4y = 6$$

was discussed. The coefficient matrix is singular, and there is a solution only if
$a = 3$. But even in that case we are in trouble, since there are infinitely many solutions. In order to make sure that there is a unique solution, we must change the matrix such that it becomes nonsingular. However, as usual, when dealing with numerical computations, the situation is not as clearcut. We change the system slightly to

$$x + 2.01y = 3,$$
$$2x + 4y = 6.$$

B. Gustafsson, *Fundamentals of Scientific Computing*,
Texts in Computational Science and Engineering 8,
DOI 10.1007/978-3-642-19495-5_14, © Springer-Verlag Berlin Heidelberg 2011

The coefficient matrix is no longer singular, and we can write down the unique solution $x = 3$, $y = 0$. We now perturb the right hand side a little as well:

$$x + 2.01y = 3.01,$$
$$2x + 4y = 6.$$

Perhaps somewhat surprised we now find the unique solution $x = 1$, $y = 1$, i.e., it has changed completely. The change is quite large compared to the very small change in the right hand side. The reason is that coefficient matrix

$$A = [\mathbf{a}_1 \ \mathbf{a}_2] = \begin{bmatrix} 1 & 2.01 \\ 2 & 4 \end{bmatrix} \tag{14.1}$$

is *almost singular* in the sense that

$$\mathbf{a}_1 - 0.5\mathbf{a}_2 = \begin{bmatrix} 0.01 \\ 0 \end{bmatrix}.$$

Despite the fact that the coefficients on the left hand side are not small, the combination of the column vectors is almost the zero vector. This means that the determinant is small; we have $\mathrm{Det}(A) = -0.02$. We encounter here another case, where the problem is not well posed, sometimes also called an *ill conditioned problem*. A small perturbation in the given data gives rise to a large perturbation in the solution. When the problem is such already before any numerical methods are applied, we must be extra careful when designing the algorithms.

In Chap. 8 a more realistic ill conditioned example was mentioned. If standard polynomials are used for least square approximation, then the matrix that arises for finding the coefficients in the polynomial becomes almost singular, and the situation becomes worse when the degree of the polynomials increases. One way of avoiding this difficulty is to use orthogonal polynomials as basis functions.

The conclusion from the example above is that any linear system of equations should first be analyzed and, if it turns out that the system is ill conditioned, we must expect trouble no matter what numerical method we come up with.

Let us formally derive an estimate of the condition. Given the system $A\mathbf{x} = \mathbf{b}$, we perturb the right hand side by a small vector $\delta\mathbf{b}$ such that we get a perturbation $\delta\mathbf{x}$ in the solution:

$$A(\mathbf{x} + \delta\mathbf{x}) = \mathbf{b} + \delta\mathbf{b}.$$

From this we get

$$\delta\mathbf{x} = A^{-1}\delta\mathbf{b},$$

which leads to

$$\|\delta\mathbf{x}\| = \|A^{-1}\delta\mathbf{b}\| \le \|A^{-1}\| \ \|\delta\mathbf{b}\|.$$

This is an estimate of the *absolute error*. Since the size depends on the scaling of the quantities, for example measuring in meters instead of millimeters, it is better to estimate the relative error $\|\delta\mathbf{x}\|/\|\delta\mathbf{x}\|$. We have

$$\|\mathbf{b}\| = \|A\mathbf{x}\| \le \|A\| \ \|\mathbf{x}\|,$$

or equivalently

$$\frac{1}{\|\mathbf{x}\|} \leq \frac{\|A\|}{\|\mathbf{b}\|}.$$

Finally we get

$$\frac{\|\delta\mathbf{x}\|}{\|\mathbf{x}\|} \leq \|A\| \, \|A^{-1}\| \frac{\|\delta\mathbf{b}\|}{\|\mathbf{b}\|}. \tag{14.2}$$

We have here a relation that connects the error of the solution to the error in the data. The constant involved is the condition number

$$\mathrm{cond}(A) = \|A\| \, \|A^{-1}\|$$

encountered earlier when dealing with differential equations. If A is an Hermitian matrix (symmetric in the real case), it was shown in Sect. 3.4 that the norm can be replaced by the maximal eigenvalue. The condition number is in this case

$$\mathrm{cond}(A) = \frac{\max_j |\lambda_j|}{\min_j |\lambda_j|}. \tag{14.3}$$

Exercise 14.1 Compute the condition number $\mathrm{cond}(A)$ for the matrix (14.1).

Exercise 14.2 Verify that the perturbation $\delta\mathbf{x}$ in the solution of the example above satisfies the estimate (14.2).

14.2 Direct Methods

Assuming that we have a well conditioned system, we shall now see how it can be solved. Here, as well as for many other problems, we find numerical algorithms that are based on a technique invented by some well known mathematician who lived and worked a long time ago. In the case of linear systems of equations it is Carl Friedrich Gauss (1777–1855). In order to illustrate his general technique, let us start with a simple example from school mathematics. We want to find the solution x, y to the system

$$\begin{aligned} x + y &= 2, \\ 2x - y &= -5. \end{aligned} \tag{14.4}$$

We first multiply the second equation by 0.5, and then subtract it from the first one. In this way we get a new equation where x is no longer present. We keep the first equation and get the new system

$$\begin{aligned} x + y &= 2, \\ 1.5y &= 4.5. \end{aligned}$$

The second equation is easily solved, and we substitute the solution $y = 3$ into the first equation, which is solved for x. The final solution is obtained as

$$\begin{aligned} x &= -1, \\ y &= 3. \end{aligned}$$

The *Gauss elimination* is just a generalization of this technique to larger systems. We call the unknowns x_1, x_2, ..., x_N, and illustrate the algorithm for the case $N = 8$. The system is written in the symbolic form

$$
\begin{bmatrix}
\times & \times & \times & \times & \times & \times & \times & \times \\
\times & \times & \times & \times & \times & \times & \times & \times \\
\times & \times & \times & \times & \times & \times & \times & \times \\
\times & \times & \times & \times & \times & \times & \times & \times \\
\times & \times & \times & \times & \times & \times & \times & \times \\
\times & \times & \times & \times & \times & \times & \times & \times \\
\times & \times & \times & \times & \times & \times & \times & \times \\
\times & \times & \times & \times & \times & \times & \times & \times
\end{bmatrix}
\begin{bmatrix}
\times \\ \times \\ \times \\ \times \\ \times \\ \times \\ \times \\ \times
\end{bmatrix}
=
\begin{bmatrix}
\times \\ \times \\ \times \\ \times \\ \times \\ \times \\ \times \\ \times
\end{bmatrix},
$$

where each $\times$ represents one term in the system. The first two equations are treated as in the example above. We multiply the second equation with a proper number such that the coefficient of the first unknown x_1 is the same in the first two equations. After subtraction of the second equation, we get a new equation where x_1 is eliminated, and it replaces the original second equation. Next we carry out exactly the same procedure for the third equation together with the first one, and obtain a new equation, where x_1 is eliminated. By continuing to the end, we get a new system that has the form

$$
\begin{bmatrix}
\times & \times & \times & \times & \times & \times & \times & \times \\
 & \times & \times & \times & \times & \times & \times & \times \\
 & \times & \times & \times & \times & \times & \times & \times \\
 & \times & \times & \times & \times & \times & \times & \times \\
 & \times & \times & \times & \times & \times & \times & \times \\
 & \times & \times & \times & \times & \times & \times & \times \\
 & \times & \times & \times & \times & \times & \times & \times \\
 & \times & \times & \times & \times & \times & \times & \times
\end{bmatrix}
\begin{bmatrix}
\times \\ \times \\ \times \\ \times \\ \times \\ \times \\ \times \\ \times
\end{bmatrix}
=
\begin{bmatrix}
\times \\ \times \\ \times \\ \times \\ \times \\ \times \\ \times \\ \times
\end{bmatrix}. \tag{14.5}
$$

We now leave the first equation as it is, and start over with the remaining system with $N - 1$ equations. The first two equations in that system are used to eliminate x_2, and the procedure continues as above. The result is a system of the form

$$
\begin{bmatrix}
\times & \times & \times & \times & \times & \times & \times & \times \\
 & \times & \times & \times & \times & \times & \times & \times \\
 & & \times & \times & \times & \times & \times & \times \\
 & & \times & \times & \times & \times & \times & \times \\
 & & \times & \times & \times & \times & \times & \times \\
 & & \times & \times & \times & \times & \times & \times \\
 & & \times & \times & \times & \times & \times & \times \\
 & & & \times & \times & \times & \times & \times
\end{bmatrix}
\begin{bmatrix}
\times \\ \times \\ \times \\ \times \\ \times \\ \times \\ \times \\ \times
\end{bmatrix}
=
\begin{bmatrix}
\times \\ \times \\ \times \\ \times \\ \times \\ \times \\ \times \\ \times
\end{bmatrix}.
$$

We now continue with the next subsystem with $N - 2$ equations, and continue until we have a triangular system of the form

$$
\begin{bmatrix}
\times \times \times \times \times \times \times \times \\
\times \times \times \times \times \times \times \\
\times \times \times \times \times \times \\
\times \times \times \times \times \\
\times \times \times \times \\
\times \times \times \\
\times \times \\
\times
\end{bmatrix}
\begin{bmatrix}
\times \\ \times \\ \times \\ \times \\ \times \\ \times \\ \times \\ \times
\end{bmatrix}
=
\begin{bmatrix}
\times \\ \times \\ \times \\ \times \\ \times \\ \times \\ \times \\ \times
\end{bmatrix}.
$$

The last equation is now solved for x_N, and this value is substituted into the second equation from the bottom, which gives the solution x_{N-1}. We continue like this until we have obtained the values for all the unknowns. This procedure of solving the triangular system is called *back substitution*.

The Gauss elimination for solution of a system $A\mathbf{x} = \mathbf{b}$ can be seen as a factorization of the coefficient matrix $A = LR$, where L is lower triangular and R is upper triangular. The reduction to a triangular system is described by the process

$$LR\mathbf{x} = \mathbf{b} \rightarrow R\mathbf{x} = L^{-1}\mathbf{b}.$$

With $\mathbf{y} = L^{-1}\mathbf{b}$, the back substitution is the process

$$R\mathbf{x} = \mathbf{y} \rightarrow \mathbf{x} = R^{-1}\mathbf{y}.$$

There are possible obstacles in Gauss elimination. For example, x_1 may not be present in the first equation, and then the procedure cannot even be started. This can of course also happen later in the process, such that the first unknown is missing in the first equation of a subsystem, causing the elimination procedure to break down. The remedy for this difficulty is to reorder the equations, such that we always have the necessary variable present. Actually, it can be shown that the best numerical results are obtained if we pick the equation with the largest coefficient of the first unknown as the upper one. This is called *row pivoting*. One can also use *column pivoting*, where the largest coefficient in the horizontal direction is chosen. This requires a reordering of the unknowns, not of the equations. *Complete pivoting* as opposed to *partial pivoting* means that we look for the largest coefficient in the whole submatrix that remains for elimination of unknowns. This requires reordering of equations and unknowns alike.

The systems of equations arising from discretization of partial differential equations are very large. With N unknowns, one can easily count the number of arithmetic operations W for a complete solution, which is approximately $N^3/3$. A realistic problem may have 10 million unknowns, and in that case we have $W \approx 3.3 \cdot 10^{20}$, which is an enormous number. The maximal speed of a high performance computer of today is of the order 10^{14} flops, i.e., it can carry out 10^{14} floating point arithmetic operations per second (the fastest supercomputer is about 20 times faster, but is terribly expensive, see Sect. 18.3). This means that the solution of one such system may take 38 days if the computer can work all the time at maximal speed, and this time is of course unacceptable. Furthermore, the high performance computers have a massively parallel architecture, which means that all the processors must work simultaneously for maximum speed. This is not possible for the Gauss elimination procedure, since it has a serial structure which is hard to parallelize.

A special type of direct method is to compute the inverse A^{-1} and then use a matrix/vector multiplication to obtain $\mathbf{x} = A^{-1}\mathbf{b}$. However, the number of arithmetic operations required for inversion of a matrix is proportional to N^3, i.e., it is of the same order as Gauss elimination, and with a larger constant multiplying N^3. Only if the system is to be solved for a large number of different right hand sides $\mathbf{b}$ could the explicit computation of the inverse be worthwhile.

Even if direct methods are expensive for very large problems, they are still useful when going down in size. Furthermore, they have been made very effective over the years in the sense that the algorithms are almost optimal within the theoretical bounds for direct methods. One example is the general LAPACK system that is particularly effective on parallel computers. It can not only solve linear systems, but also do various other matrix/vector operations, for example eigenvalue computation. For more information, see http://www.netlib.org/lapack/.

Exercise 14.3 Verify the operation count $\approx N^3/3$ for Gauss elimination.

Exercise 14.4 For a large system $A\mathbf{x} = \mathbf{b}$ to be solved for M different right hand side vectors $\mathbf{b}$, it may be worthwhile to store the triangular matrices L and R, and then solve the two systems $L\mathbf{y} = \mathbf{b}$, $R\mathbf{x} = \mathbf{y}$ for each vector b. Estimate the computational gain when using this procedure instead of doing the full Gauss elimination for each $\mathbf{b}$.

14.3 Iterative Methods

One conclusion to be made from the previous section describing direct methods, is that we need some other type of numerical method, and we turn to *iterative methods*. We have already discussed the Newton–Raphson iterative method for the solution of nonlinear equations, and the basic idea is the same here. We start from an initial guess, and then compute a better value by some algorithm that doesn't require too much work. The development of fast iterative methods has been an intensive research area during the last decades, and it is still going on. We shall first illustrate the simplest iterative method that was developed by the German mathematician Carl Jacobi (1804–1851).

We assume that all the diagonal elements in the coefficient matrix A are nonzero, and leave the diagonal terms on the left hand side of the system. All the other terms are moved to the right hand side:

$$
\begin{bmatrix} \times & & & & & & & \\ & \times & & & & & & \\ & & \times & & & & & \\ & & & \times & & & & \\ & & & & \times & & & \\ & & & & & \times & & \\ & & & & & & \times & \\ & & & & & & & \times \end{bmatrix} \begin{bmatrix} \times \\ \times \\ \times \\ \times \\ \times \\ \times \\ \times \\ \times \end{bmatrix} = \begin{bmatrix} \times & \times & \times & \times & \times & \times & \times \\ \times & & \times & \times & \times & \times & \times \\ \times & \times & & \times & \times & \times & \times \\ \times & \times & \times & & \times & \times & \times \\ \times & \times & \times & \times & & \times & \times \\ \times & \times & \times & \times & \times & & \times \\ \times & \times & \times & \times & \times & \times & \\ \times & \times & \times & \times & \times & \times & \times \end{bmatrix} \begin{bmatrix} \times \\ \times \\ \times \\ \times \\ \times \\ \times \\ \times \\ \times \end{bmatrix} + \begin{bmatrix} \times \\ \times \\ \times \\ \times \\ \times \\ \times \\ \times \\ \times \end{bmatrix} .
$$

Let us now define the method in mathematical terms. If D is the diagonal matrix containing the diagonal of A and zeros elsewhere, we can write the matrix as $A = D - B$. With an initial vector $\mathbf{x}^{(0)}$ given for the general case, the *Jacobi method* is then defined by

$$\mathbf{x}^{(n+1)} = D^{-1} B\mathbf{x}^{(n)} + D^{-1}\mathbf{b}, \quad n = 0, 1, \ldots . \tag{14.6}$$

This is a very simple algorithm, since the computation on the right hand side contains only additions and multiplications (the matrix $D^{-1}B$ and the vector $D^{-1}\mathbf{b}$ are stored once and for all). If the matrix B is *dense*, i.e., most of the coefficients are nonzero, then each iteration requires approximately $2N^2$ arithmetic operations. If the procedure converges within a small number of iterations, then we have a much faster algorithm compared to Gauss elimination. However, discretizations of partial differential equations lead to *sparse* matrices. If for example, there are 5 nonzero coefficients in each original equation, each iteration requires $8N$ arithmetic operations and, if the iterations converge rapidly to the right solution, we have a fast method even for very large systems.

The Jacobi method is a special form of the general iterative formula

$$\mathbf{x}^{(n+1)} = M\mathbf{x}^{(n)} + \mathbf{f}, \quad n = 0, 1, \ldots, \tag{14.7}$$

where $\mathbf{f}$ is a given vector. If the iterations converge, the solution is $\mathbf{x} = (I - M)^{-1}\mathbf{f}$. But how do we make sure that they converge?

Let us first assume that the matrix M has a full set of linearly independent eigenvectors that form the columns of the matrix T. By the arguments put forward in Sect. 3.3, we know that the inverse of T exists, and we multiply the iteration formula from the left by T^{-1}:

$$T^{-1}\mathbf{x}^{(n+1)} = T^{-1}MTT^{-1}\mathbf{x}^{(n)} + T^{-1}\mathbf{f}, \quad n = 0, 1, \ldots .$$

Here we have squeezed in the identity matrix TT^{-1} after M. By introducing the new vectors $\mathbf{y} = T^{-1}\mathbf{x}$ and $\mathbf{g} = T^{-1}\mathbf{f}$, and recalling the diagonalization procedure from Sect. 3.4, we get the new formula

$$\mathbf{y}^{(n+1)} = \Lambda\mathbf{y}^{(n)} + \mathbf{g}, \quad n = 0, 1, \ldots,$$

where Λ is a diagonal matrix containing the eigenvalues of M. After m iterations, we have the formula

$$\mathbf{y}^{(m)} = \Lambda^m\mathbf{y}^{(0)} + \sum_{n=0}^{m-1} \Lambda^n\mathbf{g}. \tag{14.8}$$

We note that

$$(I - \Lambda) \sum_{n=0}^{m-1} \Lambda^n = I - \Lambda^m,$$

i.e.,

$$\sum_{n=0}^{m-1} \Lambda^n = (I - \Lambda)^{-1}(I - \Lambda^m).$$

We assume that the eigenvalues λ_j satisfy

$$|\lambda_j| < 1, \quad j = 1, 2, \ldots, N,$$

i.e., $\rho(M) < 1$. The eigenvalues of $I - M$ are $1 - \lambda_j \neq 0$, which means that $I - M$ is nonsingular, and consequently the matrix $(I - M)^{-1}$ exists.

It is now time to let m tend to infinity. The iteration formula (14.8) is written as

$$y^{(m)} = \Lambda^{(m)} y^{(0)} + (I - \Lambda)^{-1}(I - \Lambda^m)\mathbf{g}.$$

The matrices in this formula are all diagonal, and in the limit we get

$$\mathbf{y}^{(\infty)} = \mathbf{0} + (I - \Lambda)^{-1}(I - 0)\mathbf{g} = (I - \Lambda)^{-1}\mathbf{g}.$$

After multiplication from the left by $T(I - \Lambda)$ we get

$$T\mathbf{y}^{(\infty)} - T\Lambda T^{-1}T\mathbf{y}^{(\infty)} = T\mathbf{g},$$

i.e.,

$$\mathbf{x}^{(\infty)} - M\mathbf{x}^{(\infty)} = \mathbf{f},$$

which gives the correct solution

$$\mathbf{x} = (I - M)^{-1}\mathbf{f}.$$

If the matrix M doesn't have a full set of eigenvectors, one can still show that the eigenvalue condition is sufficient for convergence, but it takes a little extra effort to prove it. Indeed one can show that the condition is also necessary for convergence.

The magnitude of the eigenvalues is often difficult to estimate, and other sufficient convergence conditions are sometimes applied. For the Jacobi method there is a very simple criterion: it converges if the coefficient matrix is *diagonally dominant*. By this we mean that the elements satisfy

$$|a_{jj}| > \sum_{k \neq j} |a_{jk}|, \quad j = 1, 2, \ldots, N.$$

The more the diagonal elements dominate the other elements in the same row, the faster is the convergence.

For the general iteration formula, a direct measure of the convergence rate is obtained by comparing the error

$$\|\mathbf{e}^{(n+1)}\| = \|\mathbf{x}^{(n+1)} - \mathbf{x}^{(\infty)}\|$$

with the previous one

$$\|\mathbf{e}^{(n)}\| = \|\mathbf{x}^{(n)} - \mathbf{x}^{(\infty)}\|.$$

By subtracting the iteration formula from the corresponding equation for $x^{(\infty)}$ we get rid of the given right hand side vector, and we have the homogeneous system

$$\mathbf{e}^{(n+1)} = M\mathbf{e}^{(n)}, \quad n = 0, 1, \ldots.$$

Without restriction we can assume that the diagonalization has already been done. The convergence rate is determined by the component in $\mathbf{e}$ that corresponds to the largest eigenvalue of M, which means that $\rho(M)$ should be small.

Let us go back to the Jacobi method and see how it works for the boundary value problem

$$\frac{\partial^2 u}{\partial x^2} = F(x), \quad 0 \le x \le 1,$$
$$u(0) = 0,$$
$$u(1) = 0,$$

with the standard difference approximation

$$u_{j-1} - 2u_j + u_{j+1} = \Delta x^2 F(x_j), \quad j = 1, 2, \ldots, N,$$
$$u_0 = 0,$$
$$u_{N+1} = 0,$$

where the step size is $\Delta x = 1/(N+1)$. The boundary values can be eliminated, and with

$$\mathbf{u} = \begin{bmatrix} u_1 \\ u_2 \\ \vdots \\ u_N \end{bmatrix}, \qquad \mathbf{b} = \Delta x^2 \begin{bmatrix} F_1 \\ F_2 \\ \vdots \\ F_N \end{bmatrix}, \qquad A = \begin{bmatrix} -2 & 1 & & & & \\ 1 & -2 & 1 & & & \\ & 1 & -2 & 1 & & \\ & & \ddots & \ddots & \ddots & \\ & & & & -2 & 1 \\ & & & & 1 & -2 \end{bmatrix},$$

the system is

$$A\mathbf{u} = \mathbf{b}.$$

With $A = D - B$, where D is the diagonal of A, the Jacobi method

$$\mathbf{u}^{(n+1)} = D^{-1}B\mathbf{u}^{(n)} + D^{-1}\mathbf{b}, \quad n = 0, 1, \ldots$$

can be written in the form

$$\mathbf{u}^{(n+1)} = \mathbf{u}^{(n)} - D^{-1}(A\mathbf{u}^{(n)} - \mathbf{b}), \quad n = 0, 1, \ldots.$$

The convergence rate is determined by the eigenvalues of the iteration matrix $M = I - D^{-1}A$. By using the usual trigonometric formulas and the ansatz

$$v_{kj} = \sin(k\pi x_j), \quad k = 1, 2, \ldots, N,$$

for the elements v_{kj} of the eigenvectors $\mathbf{v}_k$ of A, it is easy to show that the eigenvalues are

$$\mu_k = -4\sin^2\frac{k\pi\,\Delta x}{2}, \quad k = 1, 2, \ldots, N.$$

This means that the iteration matrix M has the eigenvalues

$$\lambda_k = 1 - 2\sin^2\frac{k\pi\,\Delta x}{2}, \quad k = 1, 2, \ldots, N.$$

The spectral radius never exceeds one, but there are two critical eigenvalues where it almost does, and that is for $k = 1$ and $k = N$. For $k = 1$ we have

$$\lambda_1 = 1 - 2\sin^2\frac{\pi\,\Delta x}{2} = 1 - \frac{(\pi\,\Delta x)^2}{2} + \mathcal{O}(\Delta x^4) \approx 1 - \frac{\pi^2}{2}\Delta x^2,$$

which shows that for a fine grid the eigenvalue becomes very close to one. A similar analysis for $k = N$ shows the same thing for the eigenvalue λ_N, and the conclusion is that the Jacobi method converges very slowly for this example.

There is a very natural modification of the Jacobi method, when looking for faster convergence. When the first vector element $x_1^{(n+1)}$ has been computed, why not use it in the computation of next element $x_2^{(n+1)}$? And then continue using both of the first elements in the next computation, and so on. Starting from the formula (14.7), we partition the matrix M into its lower triangular and upper triangular parts:

$$M = L + U = \begin{bmatrix} & & & & & & & \\ \times & & & & & & & \\ \times & \times & & & & & & \\ \times & \times & \times & & & & & \\ \times & \times & \times & \times & & & & \\ \times & \times & \times & \times & \times & & & \\ \times & \times & \times & \times & \times & \times & & \\ \times & \times & \times & \times & \times & \times & \times & \end{bmatrix}\begin{bmatrix} \times \\ \times \\ \times \\ \times \\ \times \\ \times \\ \times \\ \times \end{bmatrix} + \begin{bmatrix} \times & \times & \times & \times & \times & \times & \times \\ & \times & \times & \times & \times & \times & \times \\ & & \times & \times & \times & \times & \times \\ & & & \times & \times & \times & \times \\ & & & & \times & \times & \times \\ & & & & & \times & \times \\ & & & & & & \times \\ & & & & & & \end{bmatrix}\begin{bmatrix} \times \\ \times \\ \times \\ \times \\ \times \\ \times \\ \times \\ \times \end{bmatrix}.$$

(Recall that the diagonal of M is zero.) The iteration formula then is

$$(I - L)\mathbf{x}^{(n+1)} = U\mathbf{x}^{(n)} + \mathbf{f}, \quad n = 0, 1, \ldots. \tag{14.9}$$

This method is called the Gauss–Seidel method after Gauss and the German mathematician and astronomer Philipp Ludwig von Seidel (1821–1896).

After the introduction of modern computers, the development of fast iterative methods got a restart (like many other areas of numerical analysis). The most commonly used methods today are based on the so-called *conjugate gradient method*, and we shall describe the underlying principle here.

The original conjugate gradient method was presented in a famous paper by Magnus Hestenes and Aduard Stiefel in 1952. The system is $A\mathbf{x} = \mathbf{b}$, where A is a symmetric positive definite matrix. We consider first the *quadratic form*

$$f(\mathbf{x}) = \frac{1}{2}\mathbf{x}^T A\mathbf{x} - \mathbf{b}^T\mathbf{x}$$

for all vectors $\mathbf{x}$. For $N = 2$ we have

$$f(\mathbf{x}) = \frac{1}{2}\left(x_1(a_{11}x_1 + a_{12}x_2) + x_2(a_{21}x_1 + a_{22}x_2)\right) - (b_1x_1 + b_2x_2).$$

Looking for the minimum value of $f(\mathbf{x})$, we put the partial derivatives equal to zero (**grad** $f = \mathbf{0}$). Recalling the symmetry condition $a_{21} = a_{12}$, we get

$$\frac{\partial f}{\partial x_1} = a_{11}x_1 + a_{12}x_2 - b_1 = 0,$$

$$\frac{\partial f}{\partial x_2} = a_{12}x_1 + a_{22}x_2 - b_2 = 0.$$

But this can be written in the form

$$A\mathbf{x} - \mathbf{b} = \mathbf{0},$$

and this equation holds also in the N-dimensional space. It can be shown that there is no other local minimum of $f(\mathbf{x})$. (In 1D this is easily seen since $f(x) = Ax^2/2 - bx$.) Accordingly, finding the solution of $A\mathbf{x} = \mathbf{b}$ is equivalent to minimizing $f(\mathbf{x})$. There are many minimization methods, and the conjugate gradient method is one of them.

Two vectors $\mathbf{u}$ and $\mathbf{v}$ are orthogonal if $(\mathbf{u}, \mathbf{v}) = \mathbf{u}^T\mathbf{v} = 0$. If the matrix A is symmetric and positive definite, one can generalize the orthogonality concept and require $\mathbf{u}^T A\mathbf{u} = 0$. We say that the vectors are orthogonal with respect to this scalar product, or that they are conjugate. The conditions on A make sure that the generalized squared norm $\mathbf{u}^T A\mathbf{u} = \|\mathbf{u}\|_A^2$ is positive.

The minimization will be constructed as a search procedure, and we shall search along certain directions that are mutually conjugate. Having an initial guess $\mathbf{x}^{(0)}$, the question is now how to choose these conjugate directions. When choosing the first one, there is no other direction to relate to, and this gives us a certain freedom in the choice. There is a minimization method called the method of *steepest descent*, which means that, at any step in the iterative procedure, we choose the direction which points along the steepest downhill slope, i.e., in the direction of $-$**grad** $f(\mathbf{x}^{(0)})$ (see Sect. 2.3.1). But this is exactly the *residual*

$$\mathbf{r}^{(0)} = \mathbf{b} - A\mathbf{x}^{(0)},$$

and we choose the initial direction $\mathbf{d}^{(0)}$ as

$$\mathbf{d}^{(0)} = \mathbf{r}^{(0)}.$$

Let us now assume that at a certain point in the stepwise procedure we have an approximate solution point $\mathbf{x}^{(n)}$, a residual $\mathbf{r}^{(n)}$ and a direction $\mathbf{d}^{(n)}$. The question is how to choose the next point

$$\mathbf{x}^{(n+1)} = \mathbf{x}^{(n)} + \alpha\mathbf{d}^{(n)}.$$

It is quite natural to choose the parameter α such that $f(\mathbf{x}^{(n+1)})$ becomes as small as possible. This point is obtained by differentiating

$$f(\mathbf{x}^{(n)} + \alpha\mathbf{d}^{(n)}) = \frac{1}{2}(\mathbf{x}^{(n)} + \alpha\mathbf{d}^{(n)})^T A(\mathbf{x}^{(n)} + \alpha\mathbf{d}^{(n)}) - \mathbf{b}^T(\mathbf{x}^{(n)} + \alpha\mathbf{d}^{(n)})$$

with respect to α, and equating it to zero:

$$(\mathbf{d}^{(n)})^T A(\mathbf{x}^{(n)} + \alpha \mathbf{d}^{(n)}) - \mathbf{b}^T \mathbf{d}^{(n)} = 0.$$

This gives

$$\alpha = \alpha^{(n)} = \frac{(\mathbf{d}^{(n)})^T \mathbf{r}^{(n)}}{(\mathbf{d}^{(n)})^T A \mathbf{d}^{(n)}}.$$

We now have a new approximative solution point $\mathbf{x}^{(n+1)}$ and a new residual

$$\mathbf{r}^{(n+1)} = \mathbf{b} - A\mathbf{x}^{(n+1)}.$$

For $n = 0$, the procedure is identical to the steepest descent method, but the choice of a new direction $\mathbf{d}^{(n+1)}$ makes it different. We require that the new direction is conjugate to the previous ones. Picking the latest direction $d^{(n)}$ and using the ansatz

$$\mathbf{d}^{(n+1)} = \mathbf{r}^{(n+1)} + \beta \mathbf{d}^{(n)}$$

gives the condition

$$(\mathbf{d}^{(n)})^T A(\mathbf{r}^{(n+1)} + \beta \mathbf{d}^{(n)}) = 0,$$

resulting in the choice

$$\beta = \beta^{(n)} = -\frac{(\mathbf{d}^{(n)})^T A \mathbf{r}^{(n+1)}}{(\mathbf{d}^{(n)})^T A \mathbf{d}^{(n)}}.$$

It can be shown that all directions $\mathbf{d}^{(j)}$, $j = 0, 1, \ldots, n+1$, become mutually conjugate by this choice.

The full iteration step is now complete. We have a new solution point $\mathbf{x}^{(n+1)}$, a new solution residual $\mathbf{r}^{(n+1)}$ and a new direction $\mathbf{d}^{(n+1)}$:

$$\mathbf{x}^{(n+1)} = \mathbf{x}^{(n)} - \frac{(\mathbf{d}^{(n)})^T \mathbf{r}^{(n)}}{(\mathbf{d}^{(n)})^T A \mathbf{d}^{(n)}} \mathbf{d}^{(n)},$$

$$\mathbf{r}^{(n+1)} = \mathbf{b} - A\mathbf{x}^{(n+1)},$$

$$\mathbf{d}^{(n+1)} = \mathbf{r}^{(n+1)} - \frac{(\mathbf{d}^{(n)})^T A \mathbf{r}^{(n+1)}}{(\mathbf{d}^{(n)})^T A \mathbf{d}^{(n)}} \mathbf{d}^{(n)}.$$

This is called the *conjugate gradient method*.

We consider next the example

$$\begin{bmatrix} 2 & -1 & 0 \\ -1 & 2 & -1 \\ 0 & -1 & 2 \end{bmatrix} \begin{bmatrix} x_1 \\ x_2 \\ x_3 \end{bmatrix} = \begin{bmatrix} 1 \\ 1 \\ 1 \end{bmatrix},$$

and apply the conjugate gradient method. It turns out that with $\mathbf{x}^{(0)} = [1\ 1\ 1]^T$, the exact solution $\mathbf{x} = [1.5\ 2.0\ 1.5]^T$ is obtained after only 3 steps. Furthermore, no matter what initial solution $\mathbf{x}^{(0)}$ we choose, the final result is the same: the exact solution is obtained after three steps (sometimes even fewer). This can hardly be a coincidence, and we shall see why it is not.

Fig. 14.1 The conjugate
gradient method for $N = 2$

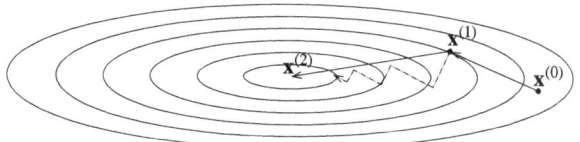

One can show that not only are all directions mutually conjugate, but also that all residuals are mutually orthogonal:

$$(\mathbf{r}^{(j)})^T \mathbf{r}^{(k)} = 0, \quad j \neq k.$$

This leads to a very interesting observation. In an N-dimensional space, there can be only N mutually orthogonal vectors. Therefore there can be only N nonzero residuals that are generated by the conjugate gradient procedure, and the conclusion is that after at most N steps we have a solution $\mathbf{x}^{(n)}$ that solves the system $A\mathbf{x}^{(n)} = \mathbf{b}$ exactly. Figure 14.1 shows how the procedure ends after two steps when $N = 2$. The dashed trail shows how the basic steepest descent method would handle the iterative procedure for the same initial point $\mathbf{x}^{(0)}$.

So what have we achieved with the conjugate gradient algorithm? Just another direct method for solving a system of linear equations? We saw that Gauss elimination requires $\mathcal{O}(N^3)$ arithmetic operations. For the conjugate direction method, the matrix/vector multiplication carried out for each step requires $\mathcal{O}(N^2)$ operations, and we take N steps. This means that here we also need $\mathcal{O}(N^3)$ operations. For sparse matrices, the count goes down but, compared to Gauss elimination, we still don't gain much, if anything. However, something interesting happens here.

At the n-th iteration we would like to know how small the error $\|\mathbf{x} - \mathbf{x}^{(n)}\|$ is compared to the initial error $\|\mathbf{x} - \mathbf{x}^{(0)}\|$. By a quite straightforward analysis we find that the space spanned by the first n directions $\mathbf{d}^{(n)}$ is equivalent to the so called *Krylov space* $\mathcal{K}_n$ spanned by the vectors

$$\mathcal{K}_n = \{\mathbf{r}^{(0)}, A\mathbf{r}^{(0)}, A^2\mathbf{r}^{(0)}, \ldots, A^{n-1}\mathbf{r}^{(0)}\}.$$

Furthermore, if we consider the least square problem of minimizing $\|\mathbf{x} - \mathbf{x}^{(n)}\|$ over the space $\mathcal{K}_n$, the solution is $\mathbf{x}^{(n)} - \mathbf{x}^{(0)}$. This leads to the conclusion that the initial error is reduced by a factor ε in

$$n \geq \frac{1}{2}\sqrt{\text{cond}(A)} \log \frac{2}{\varepsilon}$$

iterations. We recall from Sect. 3.4 that for symmetric matrices, the condition number is the ratio between the maximal and minimal eigenvalues. For PDE discretizations we typically have $\text{cond}(A) = \mathcal{O}(N)$, where N is very large, and the gain from $\mathcal{O}(N)$ iterations to $\mathcal{O}(\sqrt{N})$ operations is significant.

As a simple example we consider the boundary value problem

$$-\frac{d^2u}{dx^2} + au = F(x), \quad 0 \leq x \leq 1,$$
$$u(0) = 0,$$
$$u(1) = 0,$$

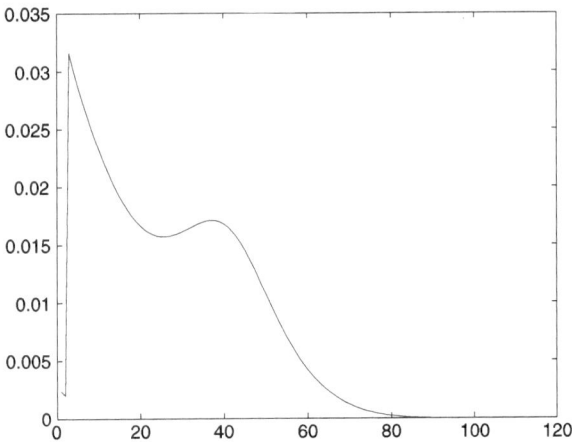

Fig. 14.2 The residual as a function of the number of iterations

and the standard three point difference approximation

$$-u_{j-1} + 2u_j - u_{j+1} + a\Delta x^2 u_j = \Delta x^2 F(x_j), \quad j = 1, 2, \ldots, N,$$
$$u_0 = 0, \tag{14.10}$$
$$u_{N+1} = 0$$

with $(N+1)\Delta x = 1$. Figure 14.2 shows the norm of the residual $\|\mathbf{b} - A\mathbf{x}^{(n)}\|$ as a function of n for the conjugate gradient method and $N = 160$. Obviously, the iterations can be stopped well before the theoretical end; how early depends on the required accuracy.

Considering the conjugate gradient method as an iterative method was a real breakthrough when it was invented. However, another surprise soon came up. It turned out that computations with the new method often went completely wrong, and the true solution was not obtained, not even after the full cycle with N iterations. The theory was not wrong, but it assumed exact arithmetic. On a computer we have to deal with rounding errors in each operation, and the method showed great sensitivity to these. In fact it is unstable in the sense that small perturbations cause increasingly large errors later in the computation.

The method was soon modified such that the operations are carried out in a different order leading to stable algorithms. Gene Golub (1932–2007) was one of the main players here, and there is today a large number of different versions of the conjugate gradient methods, also for other systems where the matrix A is not symmetric and positive definite. For a survey, see [7].

Exercise 14.5 Consider the system

$$2x + y = 2,$$
$$x - 3y = -2.5.$$

(a) Derive the matrix M in the iteration formula (14.7) for the Jacobi method, and compute the spectral radius $\rho(M)$.

(b) The same problem as (a), but now for the Gauss–Seidel method. Compare with $\rho(M)$ for the Jacobi method.

14.4 Preconditioning

The efficiency of most numerical methods depends on the type of problem. We have seen how the number of iterations for a given method depends on certain properties of the equations. A key concept is then various types of transformations. Before applying the numerical method, we try to transform the problem so that it fits the optimal conditions. For linear systems of equations, we saw above that the convergence rate for the conjugate gradient method is determined by the condition number of the coefficient matrix. Therefore, if it is possible, it makes sense to modify the system such that the condition number goes down. This is called *preconditioning*.

Let M be a nonsingular matrix with inverse M^{-1}. The original system

$$Ax = b$$

is premultiplied by M^{-1}, giving the new system

$$M^{-1}Ax = M^{-1}b.$$

The solution of this system is obviously the same as for the original one. The idea is now to choose M such that the new matrix $M^{-1}A$ has better properties when it comes to the convergence rate of the iterative method. The choice $M = A$ is optimal in the sense that the solution is obtained without any iteration at all. But finding M^{-1} is of course equivalent to solving the original system, and we have achieved nothing. We need a matrix M such that any system $My = c$ is easy to solve, but at the same time changes the properties of A in a favorable way from a convergence point of view. This is a balancing act, since the two criteria usually work against each other. The survey article [2] gives a good overview of various techniques.

14.5 Multigrid Methods

In the seventies a new type of solution method for linear systems emerged as an alternative to traditional methods. It was originally constructed for systems that arise from finite difference methods for boundary value problems, but has later been generalized to other types of system. It is today one of the most frequently used methods, and under certain conditions it is the fastest method available.

We illustrate it here for the standard model problem (14.10) which we write as

$$Au = b,$$

where $\mathbf{u}$ is the vector with elements $u_1, u_2, \ldots, u_N$ and $b_j = \Delta x^2 F(x_j)$. Here $\Delta x = 1/(N + 1)$, and for convenience we shall switch notation: $\Delta x \to h$. Since we are going to use different grids, we also introduce the notation $\mathbf{u}^h$ for the solution vector on the original grid, and A^h for the corresponding $N \times N$ matrix.

Let us first make an observation. If $\mathbf{v}$ is an approximate solution, $\mathbf{r} = \mathbf{b} - A\mathbf{v}$ is the residual. Since $\mathbf{v} = A^{-1}(\mathbf{b} - \mathbf{r})$, the true solution can formally be obtained in two steps:

$$A\mathbf{w} = \mathbf{r},$$
$$\mathbf{u} = \mathbf{v} + \mathbf{w}.$$

The fundamental multigrid idea is to compute $\mathbf{v}$ by one or more iterations on the original fine grid G_h, and then solve the first of the equations above on a coarser grid. At a first glance it looks like a bad idea to complicate things in this way, but it turns out that the convergence rate may be improved dramatically. We shall explain why.

The multigrid method needs an iterative method. In Sect. 14.3 the Jacobi iterative method was described. Here we shall use a slightly generalized version by introducing a parameter θ, and change the formula to

$$\mathbf{u}^{(n+1)} = \mathbf{u}^{(n)} - \theta D^{-1}(A\mathbf{u}^{(n)} - \mathbf{b}), \quad n = 0, 1, \ldots.$$

The method is called the *damped Jacobi method*, and it allows for a different weight of the correction term. The eigenvalues of the iteration matrix $M = I - \theta D^{-1}A$ are obtained in exactly the same way as for the original method, and they are for our example

$$\lambda_k = 1 - 2\theta \sin^2 \frac{k\pi h}{2}, \quad k = 1, 2, \ldots, N.$$

Convergence requires that $0 \le \theta \le 1$. The convergence rate on a fixed grid is very poor for all these values of θ. In fact the fastest convergence based on the largest eigenvalue magnitude is obtained with the original choice $\theta = 1$. It seems that the Jacobi method is a bad choice, but we shall see that it works much better in the multigrid setting.

We take a closer look at the convergence properties as a function of $\xi = k\pi h$ for $0 < \xi < \pi$. Each eigenvalue λ_k corresponds to an eigenvector with elements $\sin k\pi hj$, $j = 1, 2, \ldots, N$. For increasing k, the eigenvector can be seen as a grid function $v_j(k)$ that takes a more oscillatory behavior, and for $k = N$ it is essentially the function $(-1)^j$. It is then interesting to see what the convergence properties are for different wave numbers k. Figure 14.3 shows the eigenvalues $\lambda = \lambda(\xi)$.

No matter how the parameter θ is chosen, the convergence rate for small ξ is poor. However, if we look at the upper half of the ξ-interval, the choice $\theta = 0.5$ gives a very good convergence rate, since

$$\max_{\pi/2 \le \xi \le \pi} |\lambda(\xi)| = 0.5.$$

This means that the oscillatory part of the error will be eliminated quickly, while the smooth part will essentially stay untouched.

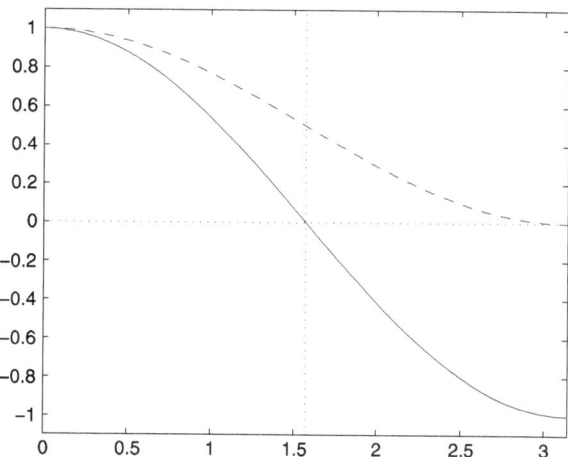

Fig. 14.3 Eigenvalues $\lambda = \lambda(\xi)$ of the damped Jacobi iteration, $\theta = 1$ (—), $\theta = 0.5$ (− −)

The trick is now to represent the remaining smooth part on a coarser grid by doubling the step size. In one dimension as we have here, the system becomes half as big, and a solution can be obtained with much less work. In several dimensions this effect becomes much more pronounced. In 2D, we gain a factor 4 in system size, in 3D a factor 8. Furthermore there is another effect. The highest possible wave number on the fine grid is $k \approx 1/h$, while it is $k \approx 1/(2h)$ on the twice as coarse grid. This means that the good convergence properties of the iterative method are converted to lower wave numbers on the coarse grid. This means in turn that the iterative method can be expected to converge faster.

Obviously, we need to handle transfers between nonidentical grids. It is most convenient to double the grid size such that the coarse grid is G_{2h}. Therefore we choose $N + 1$ as an even number. There are many ways of representing a certain vector $\mathbf{u}^h$ on G_{2h}. The most straightforward one is to simply pick the corresponding grid value such that u_{2j}^h becomes u_j^{2h} for $j = 0, 1, \ldots, (N+1)/2$. But for reasons that we shall see later, it is better to use

$$u_j^{2h} = \frac{1}{4}(u_{2j-1}^h + 2u_{2j}^h + u_{2j+1}^h), \quad j = 0, 1, \ldots, (N+1)/2.$$

This transfer is called a *restriction*, and it can be represented by the matrix

$$R = \frac{1}{4}\begin{bmatrix} 1 & 2 & 1 & & & \\ & 1 & 2 & 1 & & \\ & & & \ddots & & \\ & & & 1 & 2 & 1 \end{bmatrix}$$

for the inner points. We also need the opposite transfer, called a *prolongation*, by defining a fine grid vector from a coarse grid vector. It turns out that a good procedure is to use a similar principle, i.e., if a point in G_h coincides with a point in G_{2h}, it gets the full corresponding value, while the neighboring points gets an average of

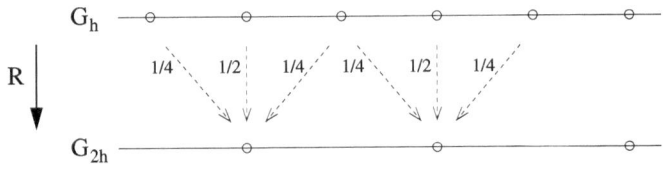

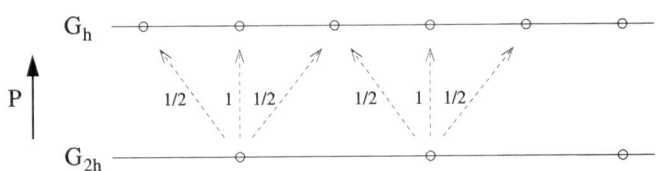

Fig. 14.4 Restriction to a coarser grid and prolongation to a finer grid

the two nearest points. The corresponding matrix P is exactly the transpose of R, except for a scaling factor 2:

$$
P = R^T = \frac{1}{2}
\begin{bmatrix}
1 & & & & & \\
2 & & & & & \\
1 & 1 & & & & \\
 & 2 & & & & \\
 & 1 & & & & \\
 & & \ddots & & & \\
 & & & 1 & \\
 & & & 2 & \\
 & & & 1 &
\end{bmatrix}.
$$

Figure 14.4 shows the transfer procedures.

We are now ready to define the whole *two grid method*. The damped Jacobi method is denoted by $\mathbf{u}^{(n+1)} = S\mathbf{u}^{(n)} + T\mathbf{b}$. With a given initial vector $\mathbf{u}^{(0)h}$, one iteration of the two grid method is:

$$
\begin{aligned}
\mathbf{v}^h &= S^h \mathbf{u}^{(0)h} + T^h \mathbf{b}^h, \\
\mathbf{r}^h &= \mathbf{b}^h - A^h \mathbf{v}^h, \\
\mathbf{r}^{2h} &= R\mathbf{r}^h, \\
A^{2h} \mathbf{w}^{2h} &= \mathbf{r}^{2h}, \\
\mathbf{w}^h &= P\mathbf{w}^{2h}, \\
\mathbf{u}^{(1)h} &= \mathbf{v}^h + \mathbf{w}^h.
\end{aligned}
$$

For a numerical test case we use

$$
F(x) = -\frac{1}{128}\left(\sin^2 \frac{\pi}{32} \sin(\pi x) + \sin^2 \frac{15\pi}{32} \sin(15\pi x) \right),
$$

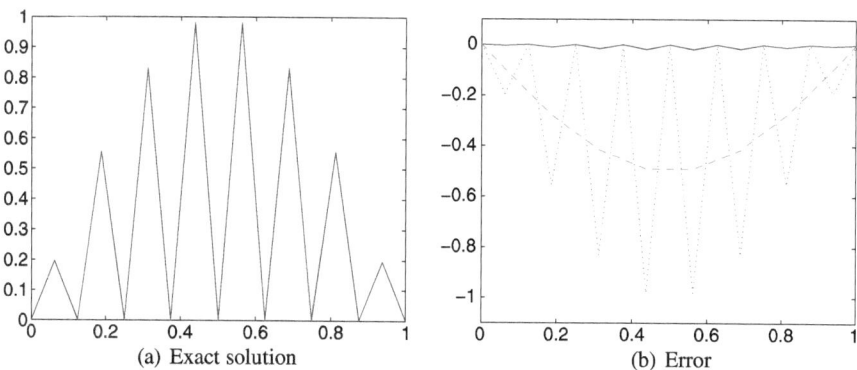

Fig. 14.5 Exact solution and error with the two grid method

which has the discrete solution

$$u_j = \frac{1}{2}\big(\sin(\pi x_j) + \sin(15\pi x_j)\big), \quad j = 0, 1, \ldots, N+1,$$

for $\Delta x = 1/16$. The solution has two distinct wave numbers as shown in Fig. 14.5(a), where the points u_j are connected by straight lines.

Figure 14.5(b) shows the error $u_j - u(x_j)$ at three different stages of the two grid algorithm. The dotted line is the initial error for the zero initial guess. The dashed line is after one iteration with the damped Jacobi method on the fine grid. The high oscillatory part of the error is almost completely eliminated. The solid line shows the final error after the coarse grid correction, which here has been done by an exact solution of the system $A^{2h}\mathbf{w}^{2h} = \mathbf{r}^{2h}$.

There are two obvious ways of modifying the algorithm. The first one is to apply the Jacobi iteration several times on each grid. The second one is to go to even coarser grids $G_{4h}, G_{8h}, \ldots$. This is possible if we choose $N + 1 = 2^p$, where p is large enough. Then we have the true *multigrid method*.

The implementation of the multigrid method is of course more troublesome for multi-dimensional problems, in particular for finite element methods on unstructured grids. However, for model problems, there is currently no other iteration method that can beat the multigrid method when it comes to convergence rate. But there are problems where the method is not robust enough, and it takes a certain amount of tuning to get it to work properly.

Part V
Applications

Chapter 15
Wave Propagation

There are many types of wave propagation, and there is no precise definition of it. The somewhat vague definition is that some sort of feature is transported through a medium with a certain typical speed. It can be a wave crest or some other feature that is recognizable at different times at different positions in space. If it is located at x_0 at time $t = t_0$, and has moved to another point x_1 at $t = t_1$, and the distance is $x_1 - x_0 = c(t_1 - t_0)$, then c is the *wave speed*. The classic areas for wave propagation are acoustics, electromagnetism and elasticity.

When simulating wave propagation, we are interested in the time evolution of the solution. A typical feature is that the energy is conserved for all time, at least in the ideal situation where there is no significant damping in the medium. In mathematics one needs strict definitions of the models, and for wave propagation these models are *hyperbolic* PDE. We shall begin by discussing these, and define their typical properties.

15.1 Hyperbolic PDE

In Sect. 9.1 conditions for hyperbolicity were given for second order PDE. However, for wave propagation the models are often formulated as first order systems, and we shall first discuss these.

The simplest differential equation for wave propagation is the 1D-transport equation that we have used several times for illustration of various things in this book:

$$\frac{\partial u}{\partial t} + c\frac{\partial u}{\partial x} = 0. \tag{15.1}$$

Any feature is transported in the positive x-direction by the velocity c. If there are no boundaries, the solution is simply

$$u(x,t) = f(x - ct),$$

where $f(x)$ is the initial function at $t = 0$. Every line $x - ct = a$, where a is a constant, is called a *characteristic* of the system. The initial value $f(a)$ is carried along

B. Gustafsson, *Fundamentals of Scientific Computing*,
Texts in Computational Science and Engineering 8,
DOI 10.1007/978-3-642-19495-5_15, © Springer-Verlag Berlin Heidelberg 2011

Fig. 15.1 Characteristics for
(15.1), $c = 1.5$

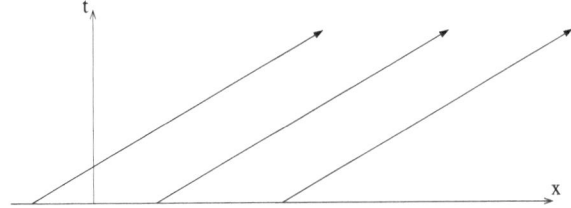

these characteristics when time increases. Figure 15.1 shows three characteristics in
the (x, t) plane for $c = 1.5$.

The transport equation is not a complete model for any of the three applications
mentioned above. However, since it has the typical properties for wave propagation,
it is one of the most studied PDE when it comes to analysis of numerical methods.
Even if there is a simple explicit formula for the solution, we shall use Fourier
analysis for investigation of the properties. In this way we can generalize it to more
general differential equations.

It was shown in Sect. 6.1 how a PDE can be Fourier transformed yielding

$$\frac{\partial \hat{u}}{\partial t} + ikc\hat{u} = 0,$$

where k is the wave number. The solution $\hat{u}(k, t)$ is

$$\hat{u}(k, t) = e^{-ikct}\hat{u}(k, 0),$$

i.e., the magnitude $|\hat{u}(k, t)|$ of the amplitude is independent of time for all wave
numbers. This is typical for wave propagation.

The transport equation is a special case of a *hyperbolic system*:

$$\frac{\partial \mathbf{u}}{\partial t} + A\frac{\partial \mathbf{u}}{\partial x} = 0,$$

where the matrix A can be diagonalized and has real eigenvalues λ_j. If $\Lambda = T^{-1}AT$
is a diagonal matrix containing the eigenvalues, we make the transformation $\mathbf{v} = T^{-1}\mathbf{u}$, and obtain

$$\frac{\partial \mathbf{v}}{\partial t} + \Lambda\frac{\partial \mathbf{v}}{\partial x} = 0.$$

This is a set of uncoupled equations, and we have

$$\frac{\partial v_j}{\partial t} + \lambda_j\frac{\partial v_j}{\partial x} = 0, \quad j = 1, 2, \ldots, m,$$

with exactly the same properties as the scalar equation above with m different wave
speeds. In accordance with the concept of characteristics, the variables v_j are called
the *characteristic variables*.

If boundaries are involved, for example at $x = 0$ and $x = 1$, we need boundary
conditions if the solution is nonperiodic. If a certain eigenvalue λ_j is positive, the
waves are moving from left to right, and if it is negative they are moving from right
to left. Consequently, the corresponding variable v_j must be given a prescribed value

at $x = 0$ in the first case, and at $x = 1$ in the second case. Note that a value can be prescribed only at one boundary for each variable. At the opposite boundary, the solution for that particular variable is uniquely determined by the process inside the domain.

Actually, the boundary conditions may have a more general form than the specification of the variables corresponding to the "ingoing" variables v_j, and we shall take a look at this. The transformation can always be constructed such that the eigenvalues appear in an a priori prescribed order in the diagonal. We assume that

$$\lambda_j > 0, \quad j = 1, 2, \ldots, r,$$
$$\lambda_j < 0, \quad j = r + 1, r + 2, \ldots, m,$$

and define the vectors

$$\mathbf{v}^I = \begin{bmatrix} v_1 \\ v_2 \\ \vdots \\ v_r \end{bmatrix}, \qquad \mathbf{v}^{II} = \begin{bmatrix} v_{r+1} \\ v_{r+2} \\ \vdots \\ v_m \end{bmatrix}.$$

There are $m - r$ sets of characteristics carrying the solution $\mathbf{v}^{II}$ towards the left boundary. At any given point in time, these values are available at the left boundary, and the ingoing solution $\mathbf{v}^I$ can be coupled to these by a boundary condition. Similarly, at the right boundary, the solution $\mathbf{v}^{II}$ transported into the domain can be coupled to the outgoing solution $\mathbf{v}^I$. Consequently, the complete set of boundary conditions has the general form

$$\mathbf{v}^I(0, t) = B^I \mathbf{v}^{II}(0, t) + \mathbf{g}^I(t),$$
$$\mathbf{v}^{II}(1, t) = B^{II} \mathbf{v}^I(1, t) + \mathbf{g}^{II}(t), \tag{15.2}$$

where $\mathbf{g}^I(t)$ and $\mathbf{g}^{II}(t)$ are given vector functions of t. The situation is illustrated in Fig. 15.2 for the case with one element in $\mathbf{v}^I$ and two elements in $\mathbf{v}^{II}$.

Sometimes there are zero eigenvalues of the matrix. This means that the characteristics are parallel to the t-axis, and the corresponding characteristic variable v_j is well defined by its initial value through the differential equation $\partial v_j / \partial t = 0$. In such a case, these variables may be included on the right hand side of the boundary conditions, and we get the generalized form

$$\mathbf{v}^I(0, t) = B^I \mathbf{v}^{II}(0, t) + C^I \mathbf{v}^{III}(0, t) + \mathbf{g}^I(t),$$
$$\mathbf{v}^{II}(1, t) = B^{II} \mathbf{v}^I(1, t) + C^{II} \mathbf{v}^{III}(1, t) + \mathbf{g}^{II}(t), \tag{15.3}$$

where the zero characteristic variables are collected in the vector $\mathbf{v}^{III}$.

The solution $\mathbf{u}$ is computed with the original system as a basis, and the transformation to diagonal form is made for the purpose of analysis only. Therefore, the true boundary conditions used for the computation have the form

$$L_0 \mathbf{u}(0, t) = \mathbf{g}_0(t),$$
$$L_1 \mathbf{u}(1, t) = \mathbf{g}_1(t).$$

Fig. 15.2 Boundary conditions for a hyperbolic system

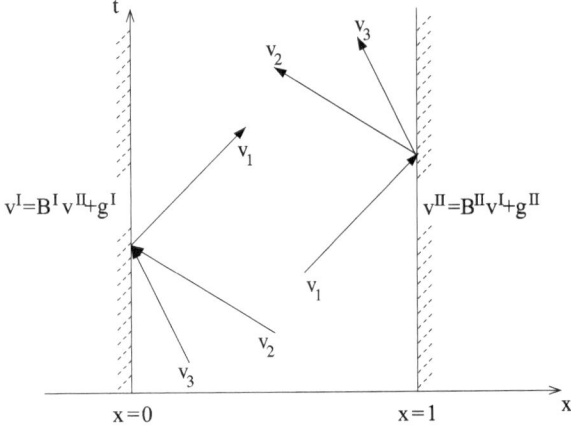

Here L_0 is an $r \times m$ matrix, L_1 is an $(m - r) \times m$ matrix, and $\mathbf{g}_0$, $\mathbf{g}_1$ are vectors with r and $m - r$ components respectively as shown below for $r = 1$ and $m = 3$ corresponding to Fig. 15.2:

$$x = 0: \quad \left[\times \times \times \right] \begin{bmatrix} \times \\ \times \\ \times \end{bmatrix} = \left[\times \right],$$

$$x = 1: \quad \begin{bmatrix} \times & \times & \times \\ \times & \times & \times \end{bmatrix} \begin{bmatrix} \times \\ \times \\ \times \end{bmatrix} = \begin{bmatrix} \times \\ \times \end{bmatrix}.$$

In order to be sure that these boundary conditions lead to a well posed problem, the transformation $\mathbf{u} = T\mathbf{v}$ is introduced, and it is checked that the correct form can be obtained. This means that, after transformation, the vector $\mathbf{v}^I$ can be solved for in terms of $\mathbf{v}^{II}$ and $\mathbf{v}^{III}$ at the left boundary, and that $\mathbf{v}^{II}$ can be solved for in terms of $\mathbf{v}^I$ and $\mathbf{v}^{III}$ at the right boundary.

Turning to multidimensional problems, we begin with the transport equation in 3D:

$$\frac{\partial u}{\partial t} + c_x \frac{\partial u}{\partial x} + c_y \frac{\partial u}{\partial y} + c_z \frac{\partial u}{\partial z} = 0.$$

The Fourier transform is

$$\frac{\partial \hat{u}}{\partial t} + i(c_x k_x + c_y k_y + c_z k_z)\hat{u} = 0,$$

where there are now three wave numbers corresponding to the Fourier mode

$$e^{i(k_x x + k_y y + k_z z)} = e^{ik_x x} e^{ik_y y} e^{ik_z z}.$$

Since the Fourier transform of the differential operator is purely imaginary, the conclusion about energy conservation holds precisely as for the 1D-case.

A hyperbolic system in 3D has the form

$$\frac{\partial \mathbf{u}}{\partial t} + A_x \frac{\partial \mathbf{u}}{\partial x} + A_y \frac{\partial \mathbf{u}}{\partial y} + A_z \frac{\partial \mathbf{u}}{\partial z} = 0.$$

Here it is not enough to require that each matrix has real eigenvalues. The condition on the matrices for hyperbolicity is stated in Fourier space. The differential operator

$$A_x \frac{\partial}{\partial x} + A_y \frac{\partial}{\partial y} + A_z \frac{\partial}{\partial z}$$

takes the form

$$i\hat{A} = i(k_x A_x + k_y A_y + k_z A_z), \tag{15.4}$$

and the requirement is that $\hat{A}$ can be diagonalized and has real eigenvalues for all k_x, k_y, k_z. This implies that each one of the matrices can be diagonalized, since any pair of wave numbers can be put equal to zero. On the other hand, it does *not* imply that the three matrices can be diagonalized by one and the same transformation. The requirement that there is a matrix T such that all matrices $T^{-1}A_x T$, $T^{-1}A_y T$, $T^{-1}A_z T$ are diagonal is a very restrictive condition, and it is seldom fulfilled in real applications. On the other hand, it is quite common that hyperbolic systems arising in various applications are *symmetrizable* (sometimes called *symmetric*), i.e., there is a matrix T such that the transformed matrices above are symmetric. These systems are much easier to analyze, and have a nice time dependent behavior.

Regarding the boundary conditions, we can generalize the results from the 1D-case. If part of the boundary is a plane $x = 0$, then we simply consider the case that the y-derivatives and the z-derivatives are zero, and then apply the diagonalization procedure giving the form (15.2) for the boundary conditions. The same procedure is then applied for the parts of the boundary satisfying $y = 0$ and $z = 0$.

But what should we do when the computational domain is not a parallelepiped? The solution is to consider the PDE locally at every boundary point (x_0, y_0, z_0). The normal $\mathbf{n}$ to the surface is identified, and then a local coordinate system (ξ, η, ζ) is defined with the ξ-axis pointing along the normal $\mathbf{n}$ as shown in Fig. 15.3. The system is rewritten in terms of these coordinates:

$$\frac{\partial \mathbf{w}}{\partial t} + A_\xi \frac{\partial \mathbf{w}}{\partial \xi} + A_\eta \frac{\partial \mathbf{w}}{\partial \eta} + A_\zeta \frac{\partial \mathbf{w}}{\partial \zeta} = 0.$$

The analysis then follows the lines above. It will be further discussed in Sects. 15.2 and 17.1.2.

It should be stressed that this way of designing the boundary conditions based on the simplified 1D-case, is necessary for wellposedness, but not sufficient. The theory for the multidimensional case is available, but is somewhat complicated to apply for a given system. By experience, the 1D-analysis used in practice often leads to boundary conditions that work very well for many applications in several space dimensions.

Sometimes first order systems of PDE are rewritten as second order systems (or scalar equations), where the order of both the time and space derivatives goes up one step. These are classified as hyperbolic as well, satisfying the conditions given in Sect. 9.1 for scalar equations in two space dimensions. We shall come back further to these second order formulations in the next application sections.

We shall now briefly discuss some of the main types of wave propagation.

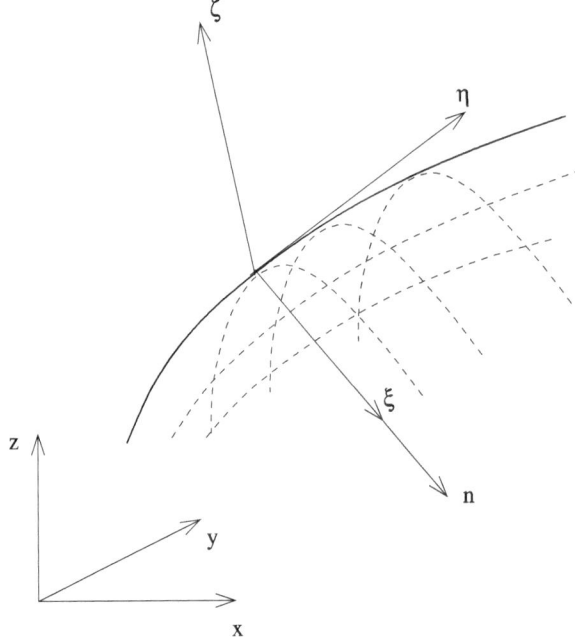

Fig. 15.3 Local coordinate system at the boundary

15.2 Acoustics

Acoustics is about propagation of sound, and it is a very important topic in the modern society. For example, when designing a new highway or a new airport, the generated noise in the immediate surroundings must be kept under control. Sound in air is a variation p of the pressure on a fast scale compared to the slowly varying atmospheric pressure p_0. The total pressure $p_0 + p$ is always positive, but p may be negative. A sound source is a combination of different frequencies ω that describes the number of pressure oscillations that take place during one second, and it is measured in the unit Hz $= 1/\text{sec}$ after the German physicist Heinrich Hertz (1857–1894). The sound corresponding to one particular frequency can be represented by a sine wave $\sin(2\pi\omega t)$, where there are ω full periods during each second.

The differential equation can be derived from the nonlinear Euler equations in fluid dynamics (see Sect. 17.1.2) by introducing a small perturbation to a given state. By disregarding quadratic and higher powers of the perturbation, we arrive at a system of linear differential equations that contain the unknown pressure p and the particle velocity vector $\mathbf{u}$ containing the components u, v, w in the three coordinate directions. In 1D, the system is

$$\frac{\partial}{\partial t}\begin{bmatrix} p \\ u \end{bmatrix} + \begin{bmatrix} 0 & c^2 \\ 1 & 0 \end{bmatrix}\frac{\partial}{\partial x}\begin{bmatrix} p \\ u \end{bmatrix} = 0.$$

Fig. 15.4 Solution to the
wave equation at
$t = 0 (—), 0.4 (--), 1.2 (-\cdot)$

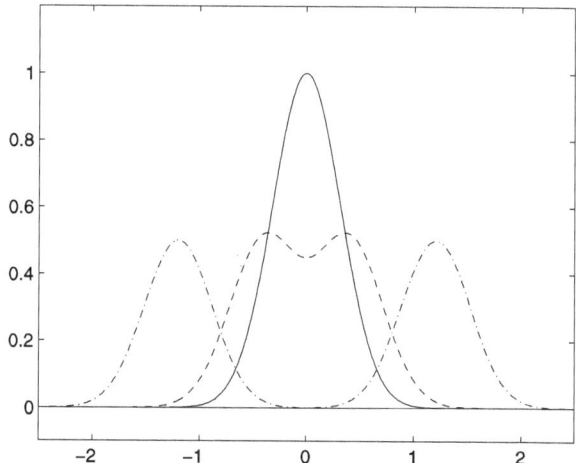

The eigenvalues of the coefficient matrix are $\pm c$, which represent the two sound waves, one propagating to the right and one propagating to the left.

A common situation is that the speed of sound $c = c(x)$ varies in space, but not in time. For example, if the salinity in ocean water varies, then the speed of sound does that as well. However, for a simulation of a process that lasts a few seconds over a domain with length of a few kilometers, the time variation of the salinity in time can be neglected, while the space variation cannot. A typical application of this type is submarine warfare, where the problem is to identify sonic information in an exact way such that the position of the enemy submarine can be determined. As is often the case in science, the military is in the forefront when it comes to acoustics research, but the refined methods are nowadays used for sophisticated non-military sonar equipment as well.

By differentiating the first equation with respect to t and using the second equation, the variable u is eliminated, and we arrive at the second order *wave equation*

$$\frac{\partial^2 p}{\partial t^2} = c(x)^2 \frac{\partial^2 p}{\partial x^2}.$$

If c is constant, the solution has the form

$$p(x, t) = f_1(x + ct) + f_2(x - ct),$$

representing the two waves identified above for the first order system.

Let us now consider a certain feature in p at $t = 0$ and $x = x_0$. When time progresses, one part of the solution stays constant along the line $x = x_0 - ct$ in the (x, t)-plane. We can identify this as wave propagation with velocity $dx/dt = -c$. For the other part of the solution the wave propagation has velocity c. Figure 15.4 shows how a pulse at $t = 0$ separates into two parts that move in opposite directions with velocity $c = \pm 1$.

The wave equation requires initial and boundary conditions. Since there is a second derivative in time, not only p must be prescribed at $t = 0$, but also $\partial p / \partial t$.

The Fourier transform of the equation is

$$\frac{d^2\hat{p}}{dt^2} = -c^2 k^2 \hat{p},$$

which has the solution

$$\hat{p}(k,t) = a_1 e^{ickt} + a_2 e^{-ickt}$$

corresponding to the two waves. Clearly, each component of the solution has a constant amplitude which is determined by the initial conditions. In Fourier space these conditions are

$$\hat{p}(k,0) = \hat{f}(k),$$
$$\frac{d\hat{p}}{dt}(k,0) = \hat{g}(k),$$

and we get the system

$$a_1 + a_2 = \hat{f}(k),$$
$$ikca_1 - ikca_2 = \hat{g}(k).$$

This system is easily solved, and the original solution p is obtained by an inverse Fourier transform.

By differentiating the second equation in the first order system, and then eliminating p, we get

$$\frac{\partial^2 u}{\partial t^2} = \frac{\partial}{\partial x}\left(c(x)^2 \frac{\partial u}{\partial x}\right),$$

which for $c(x) = $ const. is identical to the wave equation for p. Perturbations in either one of the variables p or u behaves in exactly the same way in that case.

Let us now take a look at the full 3D case. Assuming a constant speed of sound c, the system is

$$\frac{\partial}{\partial t}\begin{bmatrix} p \\ u \\ v \\ w \end{bmatrix} + \begin{bmatrix} 0 & c^2 & 0 & 0 \\ 1 & 0 & 0 & 0 \\ 0 & 0 & 0 & 0 \\ 0 & 0 & 0 & 0 \end{bmatrix}\frac{\partial}{\partial x}\begin{bmatrix} p \\ u \\ v \\ w \end{bmatrix} + \begin{bmatrix} 0 & 0 & c^2 & 0 \\ 0 & 0 & 0 & 0 \\ 1 & 0 & 0 & 0 \\ 0 & 0 & 0 & 0 \end{bmatrix}\frac{\partial}{\partial y}\begin{bmatrix} p \\ u \\ v \\ w \end{bmatrix}$$

$$+ \begin{bmatrix} 0 & 0 & 0 & c^2 \\ 0 & 0 & 0 & 0 \\ 0 & 0 & 0 & 0 \\ 1 & 0 & 0 & 0 \end{bmatrix}\frac{\partial}{\partial z}\begin{bmatrix} p \\ u \\ v \\ w \end{bmatrix} = 0.$$

To simplify the analysis, we look for plane waves that are propagating along the x-axis, i.e., all y- and z-derivatives are zero. The reduced system then becomes

$$\frac{\partial}{\partial t}\begin{bmatrix} p \\ u \\ v \\ w \end{bmatrix} + \begin{bmatrix} 0 & c^2 & 0 & 0 \\ 1 & 0 & 0 & 0 \\ 0 & 0 & 0 & 0 \\ 0 & 0 & 0 & 0 \end{bmatrix}\frac{\partial}{\partial x}\begin{bmatrix} p \\ u \\ v \\ w \end{bmatrix} = 0.$$

The only nonzero eigenvalues are $\pm c$, which shows that there is only one wave speed (with two possible signs) in the system, and it corresponds to particles moving in the same direction as the wave is propagating. These are the sound waves in the full 3D-space.

In analogy with the 1D case above, we can differentiate the system and obtain the full wave equation

$$\frac{\partial^2 p}{\partial t^2} = c^2 \left(\frac{\partial^2 p}{\partial x^2} + \frac{\partial^2 p}{\partial y^2} + \frac{\partial^2 p}{\partial z^2} \right).$$

Exactly the same equation is obtained with p replaced by u, v or w respectively. If there is no coupling at the boundaries, we can deal with scalar equations if we use the second order formulation. If, for example, we solve for the pressure this way, the velocities can then be obtained by a simple integration of the first order differential relations.

The boundary conditions depend on what type of domain in space we are dealing with. For a solid wall, the type of boundary condition follows from the first order formulation, and the 1D-version tells it all. The particle normal velocity u is obviously zero for all time at the wall, and it follows that $\partial u / \partial t = 0$ as well. The differential equation $\partial u / \partial t = -\partial p / \partial x$ then implies that $\partial p / \partial x = 0$, and this is the boundary condition for the second order wave equation for p. For the 3D-case there are three velocity components u, v, w, and it is the component that is pointing in the direction of the normal $\mathbf{n}$ that is zero. According to the discussion in the previous section, the PDE is rewritten in local coordinates with ξ pointing in the normal direction. With the new velocity components lined up with the new coordinates, the PDE is obviously the same, since the physics of acoustics doesn't know anything about coordinate directions. Accordingly, the boundary condition is $\partial p / \partial n = 0$, i.e., the normal derivative is zero.

Another case is where there is a known sound source at the boundary, in which case the pressure is prescribed. If the boundary surface is given by

$$x = x(r, s),$$
$$y = y(r, s),$$
$$z = z(r, s),$$

where the two parameters r and s vary within certain intervals, the boundary condition is

$$p\big(x(r, s), y(r, s), z(r, s)\big) = g(r, s).$$

In acoustic problems the domain is often open without any bounds in some directions. For example, the noise from a starting aircraft is spreading in all directions with no boundary except the ground. However, a computer simulation requires a finite domain, and we must introduce some artificial boundaries that are not present in the physical space. The problem is then that the differential equation requires some type of condition also at these boundaries. Construction of such conditions is not an easy task. Everything that is going on outside the computational domain is actually governed by the wave equation, and it is not at all certain that this can be described

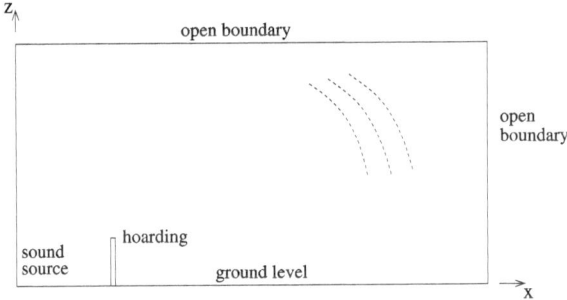

Fig. 15.5 Computational domain for noise simulation

by a local boundary condition in open space. But the situation is not hopeless. If we know that there is only one source, and furthermore, that the direction of the wave is approximately known, we can use conditions that simply let the waves out through the boundary. The concept *absorbing boundary conditions* is often used, pointing out the fact that no reflections should occur. Many different techniques exist, but it is still an active research area.

Let us now discuss an example closely related to real life. The city council in a certain city is planning to build a new street with heavy traffic through a residential area. People living there are coming in with various objections to the project, the main one being the high noise level that is expected. The city council proposes a hoarding along the street, but the inhabitants doubt if it will be effective. This is a typical situation where computer simulation comes in handy.

For preliminary simulations one tries to simplify the problem as much as possible. We assume that the street is directed along the y-axis, and the worst case traffic is homogeneous creating a continuous sound source along the street. The hoarding is also straight and parallel to the street. The computational domain is shown in Fig. 15.5.

The problem is solved by using the Acoustics module of the Comsol Multiphysics system. The lower boundary including the hoarding is modeled as a solid wall. At the left boundary, the source is modeled as a sine function with frequency 100 Hz which is decreasing in magnitude in the z-direction. The upper and the right boundaries are open with an absorbing boundary condition.

The geometry of the computational domain makes it possible to use a rectangular grid, but here we have been using a triangular grid, which is the most common type for FEM in two space dimensions. The grid is automatically generated by the system, and Fig. 15.6 shows what it looks like.

The interesting quantity for the simulation is the sound pressure level defined as

$$L_p = 10 \log \frac{|p|^2}{|p_0|^2}$$

in the unit dB. The reference value p_0 is taken as $p_0 = 2 \cdot 10^{-5}$ Pa, which corresponds to the lowest sound pressure level $L_p = 0$ that the human ear is able to hear.

The sound source at the left boundary is not tuned very accurately here. The purpose with the example is simply to demonstrate the effect of a change in the

Fig. 15.6 Finite element grid

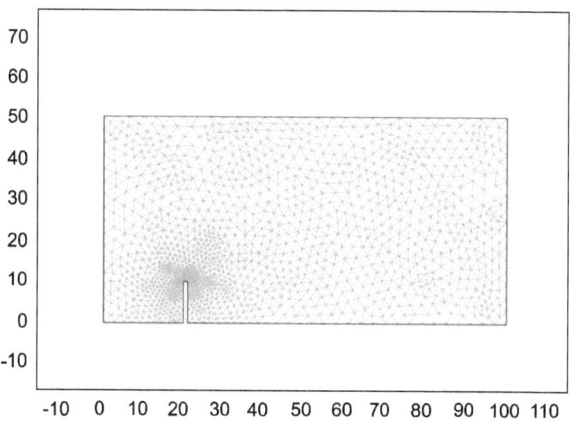

geometry. In this problem one might wonder if it is worthwhile to build a very high hoarding in order to get the sound pressure level sufficiently low. Figure 15.7 shows the result for a 5 meter high hoarding, while Fig. 15.8 shows the effect of increasing the height to 10 meters. There is a small effect in the noise level, but taking the cost and esthetic values into effect, it may be sufficient to build the lower alternative.

In this book we have used the Fourier transform in space for analysis of the differential equations, but in acoustics it is more natural to use it for transformation in time. Sound is a compound of different frequencies ω, and each one is represented by a Fourier mode $e^{i\omega t}$, where we again use the complex (but actually simpler) version. The solution has the form

$$p(x, y, z, t) = \sum_{\omega=-\infty}^{\infty} \hat{p}(x, y, z, \omega) e^{i\omega t},$$

and this is the basis for the Fourier transform of the PDE in time:

$$\frac{\partial^2 \hat{p}}{\partial x^2} + \frac{\partial^2 \hat{p}}{\partial y^2} + \frac{\partial^2 \hat{p}}{\partial z^2} + \frac{\omega^2}{c^2} \hat{p} = 0, \quad \omega = 0, \pm 1, \pm 2, \dots.$$

This is the *Helmholtz equation*, named after the German physicist Hermann von Helmholtz (1821–1894). It has a close resemblance to the Laplace equation, but the extra zero order term causes certain trouble. We have here a case where the principal part of the differential operator is elliptic and could be solved as a boundary value problem, see Sect. 9.1. However, the lower order term makes a fundamental difference in the properties. The complete problem including boundary conditions may not have a unique solution for certain values of ω/c. Mathematically speaking, the differential operator $\partial^2/\partial x^2 + \partial^2/\partial y^2 + \partial^2/\partial z^2$ may have an eigenvalue that equals $-\omega^2/c^2$, and this corresponds to the physical resonance effect. However, if we avoid these eigenvalues, it is possible to solve the equation for given frequencies, and in 3D an iterative solver is required. Much effort has been given to this problem, and there is still progress being made towards so-called fast Helmholtz solvers.

In practice we have to limit the computation to a finite and reasonable number of frequencies ω. Once the solution $\hat{p}$ has been found, the solution p in physical space is obtained via the inverse discrete Fourier transform.

Fig. 15.7 Pressure sound level, hoarding height 5 m

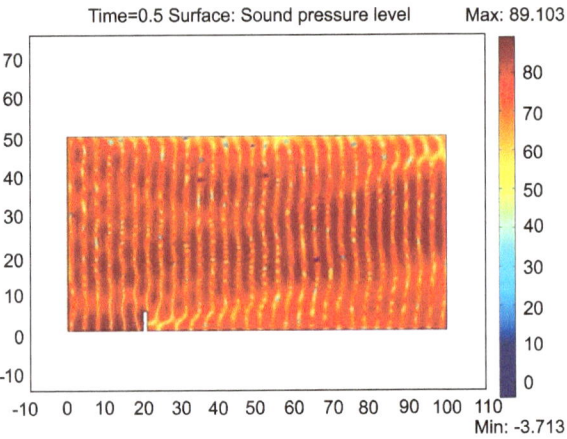

Fig. 15.8 Pressure sound level, hoarding height 10 m

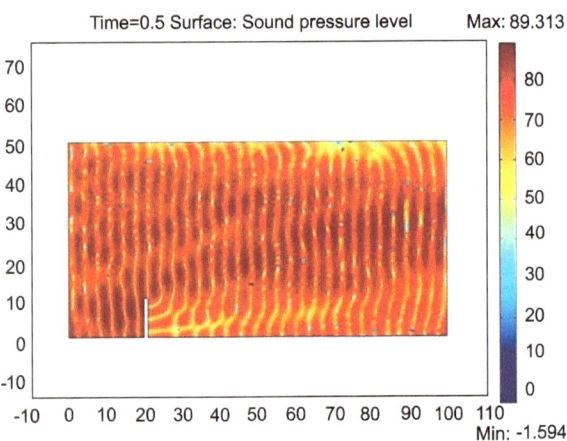

15.3 Electromagnetics

The Maxwell equations express the relation between the electric field E and magnetic field H. Before stating the equations, we need to define the parameters

ε_0 permittivity of vacuum
μ_0 permeability of vacuum
ε_r relative permittivity
μ_r relative permeability.

With $\varepsilon = \varepsilon_0 \varepsilon_r$ and $\mu = \mu_0 \mu_r$, the equations are in their simplest form (linear nonconductive materials)

$$\frac{\partial}{\partial t}\begin{bmatrix} E \\ H \end{bmatrix} + \begin{bmatrix} 0 & 1/\varepsilon \\ 1/\mu & 0 \end{bmatrix} \frac{\partial}{\partial x}\begin{bmatrix} E \\ H \end{bmatrix} = \mathbf{0}.$$

Note that this system of PDE has exactly the same structure as the system for acoustics. The coefficient matrix has the real eigenvalues $\pm 1/\sqrt{\varepsilon\mu}$, showing that the system is hyperbolic. Electromagnetic waves behave exactly as sound waves, except for the propagation speed. In a vacuum

$$c_0 = 1/\sqrt{\varepsilon_0\mu_0} = 299792458 \text{ m/sec}$$

is the speed of light. This means that, if we have a good algorithm for acoustic problems, then we can solve electromagnetic problems as well, by simply changing the parameters.

In 3D, the electric and magnetic fields are vectors

$$\mathbf{E} = \begin{bmatrix} E_x \\ E_y \\ E_z \end{bmatrix}, \qquad \mathbf{H} = \begin{bmatrix} H_x \\ H_y \\ H_z \end{bmatrix},$$

where each element represents the component in the corresponding coordinate direction. The equations are

$$\frac{\partial}{\partial t}\begin{bmatrix} \mathbf{E} \\ \mathbf{H} \end{bmatrix} + A_x\frac{\partial}{\partial x}\begin{bmatrix} \mathbf{E} \\ \mathbf{H} \end{bmatrix} + A_y\frac{\partial}{\partial y}\begin{bmatrix} \mathbf{E} \\ \mathbf{H} \end{bmatrix} + A_z\frac{\partial}{\partial z}\begin{bmatrix} \mathbf{E} \\ \mathbf{H} \end{bmatrix} = \mathbf{0},$$

where

$$A_x = \begin{bmatrix} 0 & 0 & 0 & 0 & 0 & 0 \\ 0 & 0 & 0 & 0 & 0 & 1/\varepsilon \\ 0 & 0 & 0 & 0 & -1/\varepsilon & 0 \\ 0 & 0 & 0 & 0 & 0 & 0 \\ 0 & 0 & -1/\mu & 0 & 0 & 0 \\ 0 & 1/\mu & 0 & 0 & 0 & 0 \end{bmatrix},$$

$$A_y = \begin{bmatrix} 0 & 0 & 0 & 0 & 0 & -1/\varepsilon \\ 0 & 0 & 0 & 0 & 0 & 0 \\ 0 & 0 & 0 & 1/\varepsilon & 0 & 0 \\ 0 & 0 & 1/\mu & 0 & 0 & 0 \\ 0 & 0 & 0 & 0 & 0 & 0 \\ -1/\mu & 0 & 0 & 0 & 0 & 0 \end{bmatrix},$$

$$A_z = \begin{bmatrix} 0 & 0 & 0 & 0 & 1/\varepsilon & 0 \\ 0 & 0 & 0 & -1/\varepsilon & 0 & 0 \\ 0 & 0 & 0 & 0 & 0 & 0 \\ 0 & -1/\mu & 0 & 0 & 0 & 0 \\ 1/\mu & 0 & 0 & 0 & 0 & 0 \\ 0 & 0 & 0 & 0 & 0 & 0 \end{bmatrix}.$$

For a plane wave moving along the x-axis, the y- and z-derivatives are zero, and we have

$$\frac{\partial}{\partial t}\begin{bmatrix} \mathbf{E} \\ \mathbf{H} \end{bmatrix} + A_x\frac{\partial}{\partial x}\begin{bmatrix} \mathbf{E} \\ \mathbf{H} \end{bmatrix} = \mathbf{0}.$$

The six eigenvalues of A_x are

$$\lambda_{1,2} = 1/\sqrt{\varepsilon\mu}, \qquad \lambda_{3,4} = -1/\sqrt{\varepsilon\mu}, \qquad \lambda_{5,6} = 0,$$

i.e., there is only one type of wave moving with the speed of light either forwards or backwards.

Just as for acoustics, we can derive second order forms of the equations. By differentiating the equations, $\mathbf{H}$ can be eliminated, and with $c = 1/\sqrt{\varepsilon\mu}$ we get

$$\frac{\partial^2 \mathbf{E}}{\partial t^2} = c^2 \left(\frac{\partial^2 \mathbf{E}}{\partial x^2} + \frac{\partial^2 \mathbf{E}}{\partial y^2} + \frac{\partial^2 \mathbf{E}}{\partial z^2} \right).$$

The same procedure can be carried out for elimination of $\mathbf{E}$, resulting in an identical system for $\mathbf{H}$. This means that we have six independent scalar equations, one for each component of the electric field and of the magnetic field. However, the normal situation is that there is a coupling between the different variables through the boundary conditions.

There are several effects that can be incorporated in the equations. The immediate one is the introduction of the current density $\mathbf{J}$ which is coupled to the electric field by $\mathbf{J} = \sigma\mathbf{E}$, where σ is the electric conductivity. This introduces a new term:

$$\frac{\partial}{\partial t} \begin{bmatrix} \mathbf{E} \\ \mathbf{H} \end{bmatrix} + A_x \frac{\partial}{\partial x} \begin{bmatrix} \mathbf{E} \\ \mathbf{H} \end{bmatrix} + A_y \frac{\partial}{\partial y} \begin{bmatrix} \mathbf{E} \\ \mathbf{H} \end{bmatrix} + A_z \frac{\partial}{\partial z} \begin{bmatrix} \mathbf{E} \\ \mathbf{H} \end{bmatrix} + \sigma\mathbf{E} = \mathbf{0}.$$

For the Fourier analysis we go back to the 1D-equations

$$\frac{\partial}{\partial t} \begin{bmatrix} E \\ H \end{bmatrix} + \begin{bmatrix} 0 & 1/\varepsilon \\ 1/\mu & 0 \end{bmatrix} \frac{\partial}{\partial x} \begin{bmatrix} E \\ H \end{bmatrix} + \begin{bmatrix} \sigma & 0 \\ 0 & 0 \end{bmatrix} \begin{bmatrix} E \\ H \end{bmatrix} = \mathbf{0}.$$

The Fourier transform is formally obtained by substituting the number ik for the differential operator $\partial/\partial x$:

$$\frac{\partial}{\partial t} \begin{bmatrix} \hat{E} \\ \hat{H} \end{bmatrix} + \begin{bmatrix} \sigma & ik/\varepsilon \\ ik/\mu & 0 \end{bmatrix} \begin{bmatrix} \hat{E} \\ \hat{H} \end{bmatrix} = \mathbf{0}.$$

The eigenvalues of the coefficient matrix are given by

$$\lambda_{1,2} = \frac{\sigma}{2} \pm \sqrt{\frac{\sigma^2}{4} - c^2 k^2} = \frac{\sigma}{2} \left(1 \pm \sqrt{1 - \frac{4c^2 k^2}{\sigma^2}} \right), \quad c^2 = \frac{1}{\varepsilon\mu}.$$

The question is whether the real part of them can be negative, resulting in growing solutions. But this can happen only if the real part of the square-root exceeds one, and it does not. The conclusion is that the Fourier components do not grow with time. On the contrary, there is a certain damping caused by the electric conductivity.

Let us now take a look at the boundary conditions. The zero order term has no influence on the type of boundary condition. According to our recipe for hyperbolic systems, we put the y- and z-derivatives equal to zero and consider the 1D-system (but with all six variables remaining)

$$\frac{\partial}{\partial t} \begin{bmatrix} E_x \\ E_y \\ E_z \\ H_x \\ H_y \\ H_z \end{bmatrix} + \begin{bmatrix} 0 & 0 & 0 & 0 & 0 & 0 \\ 0 & 0 & 0 & 0 & 0 & 1/\varepsilon \\ 0 & 0 & 0 & 0 & -1/\varepsilon & 0 \\ 0 & 0 & 0 & 0 & 0 & 0 \\ 0 & 0 & -1/\mu & 0 & 0 & 0 \\ 0 & 1/\mu & 0 & 0 & 0 & 0 \end{bmatrix} \frac{\partial}{\partial x} \begin{bmatrix} E_x \\ E_y \\ E_z \\ H_x \\ H_y \\ H_z \end{bmatrix} = \mathbf{0}.$$

The diagonal form is here obtained by taking the proper combinations of the variables pairwise, giving

$$
\frac{\partial}{\partial t}
\begin{bmatrix}
\sqrt{\varepsilon} E_y + \sqrt{\mu} H_z \\
\sqrt{\varepsilon} E_z - \sqrt{\mu} H_y \\
\sqrt{\varepsilon} E_y - \sqrt{\mu} H_z \\
\sqrt{\varepsilon} E_z + \sqrt{\mu} H_y \\
E_x \\
H_x
\end{bmatrix}
+
\begin{bmatrix}
c & 0 & 0 & 0 & 0 & 0 \\
0 & c & 0 & 0 & 0 & 0 \\
0 & 0 & -c & 0 & 0 & 0 \\
0 & 0 & 0 & -c & 0 & 0 \\
0 & 0 & 0 & 0 & 0 & 0 \\
0 & 0 & 0 & 0 & 0 & 0
\end{bmatrix}
\frac{\partial}{\partial x}
\begin{bmatrix}
\sqrt{\varepsilon} E_y + \sqrt{\mu} H_z \\
\sqrt{\varepsilon} E_z - \sqrt{\mu} H_y \\
\sqrt{\varepsilon} E_y - \sqrt{\mu} H_z \\
\sqrt{\varepsilon} E_z + \sqrt{\mu} H_y \\
E_x \\
H_x
\end{bmatrix}
= 0.
$$

With the notation used in Sect. 15.1, with $\mathbf{v}^I$ containing the ingoing characteristic variables, $\mathbf{v}^{II}$ containing the outgoing characteristic variables and $\mathbf{v}^{III}$ containing the zero eigenvalue variables, we have

$$
\mathbf{v}^I = \begin{bmatrix} \sqrt{\varepsilon} E_y + \sqrt{\mu} H_z \\ \sqrt{\varepsilon} E_z - \sqrt{\mu} H_y \end{bmatrix}, \qquad
\mathbf{v}^{II} = \begin{bmatrix} \sqrt{\varepsilon} E_y - \sqrt{\mu} H_z \\ \sqrt{\varepsilon} E_z + \sqrt{\mu} H_y \end{bmatrix}, \qquad
\mathbf{v}^{III} = \begin{bmatrix} E_x \\ H_x \end{bmatrix}.
$$

The boundary conditions then have the form (15.3).

Since there are always two positive eigenvalues, two characteristics are pointing into the domain, and there should be two boundary conditions. There are many ways of choosing these variables, but a natural one from a physical point of view is to specify the tangential magnetic field. In the original coordinates with the boundary $x = 0$, this means that H_y and H_z are specified:

$$
H_y = g_y(t),
$$
$$
H_z = g_z(t).
$$

In terms of the characteristic variables $\mathbf{v}$ we have the relations

$$
E_y = \frac{v_1 + v_3}{2\sqrt{\varepsilon}}, \qquad
H_z = \frac{v_1 - v_3}{2\sqrt{\mu}}, \qquad
E_z = \frac{v_2 + v_4}{2\sqrt{\varepsilon}}, \qquad
H_y = \frac{v_4 - v_2}{2\sqrt{\mu}},
$$

so the boundary conditions are in this case

$$
\mathbf{v}^I = \begin{bmatrix} 1 & 0 \\ 0 & 1 \end{bmatrix} \mathbf{v}^{II} + \mathbf{g}^I(t), \qquad
\mathbf{g}^I(t) = 2 \begin{bmatrix} \sqrt{\mu} g_z(t) \\ -\sqrt{\mu} g_y(t) \end{bmatrix},
$$

which has the right form.

For numerical solution of the Maxwell equations, there is also here a choice between the first order Maxwell equations and the second order PDE derived above. Traditionally the first order system has been dominating, and there is a well known difference scheme for solving it, called the Yee scheme. It is based on so-called *staggered grids* which are well suited for these equations, For illustration we use the 1D-equations

$$
\frac{\partial E}{\partial t} + \frac{1}{\varepsilon} \frac{\partial H}{\partial x} = 0,
$$
$$
\frac{\partial H}{\partial t} + \frac{1}{\mu} \frac{\partial E}{\partial x} = 0.
$$

Fig. 15.9 The staggered grid

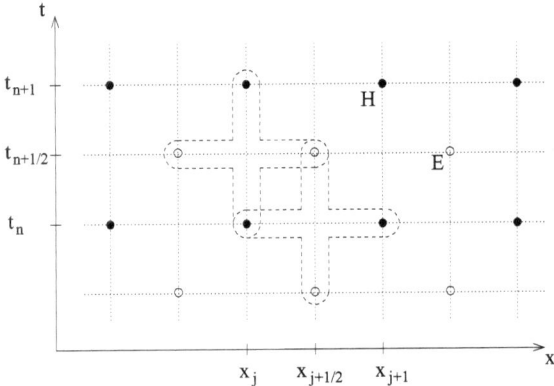

A staggered grid means that the variables E and H are stored at the different points. In addition to the usual grid points (x_j, t_n), where H is stored, another set of grid points $(x_{j+1/2}, t_{n+1/2})$ shifted half a step in both directions is defined for storage of E. The structure of the differential equations is such that a more compact difference stencil is obtained this way, and the scheme is

$$E_{j+1/2}^{n+1/2} = E_{j+1/2}^{n-1/2} - \frac{\Delta t}{\varepsilon \Delta x}(H_{j+1}^n - H_j^n),$$

$$H_j^{n+1} = H_j^n - \frac{\Delta t}{\mu \Delta x}(E_{j+1/2}^{n+1/2} - E_{j-1/2}^{n+1/2}),$$

which is illustrated in Fig. 15.9.

For the standard centered approximation on a standard grid we have for smooth functions $u(x)$

$$\frac{u(x_{j+1}) - u(x_{j-1})}{2\Delta x} = \frac{du}{dx}(x_j) + \frac{\Delta x^2}{6}\frac{d^3 u}{dx^3}(x_j) + \mathcal{O}(\Delta x^4). \tag{15.5}$$

On a staggered grid the approximation is centered at $x = x_{j+1/2}$, and we have

$$\frac{u(x_{j+1}) - u(x_j)}{\Delta x} = \frac{du}{dx}(x_{j+1/2}) + \frac{\Delta x^2}{24}\frac{d^3 u}{dx^3}(x_{j+1/2}) + \mathcal{O}(\Delta x^4), \tag{15.6}$$

showing that the truncation error is four times smaller. This is due to the more compact computing stencil, effectively halfing the step size. The same improvement is obtained in the t-direction, since we have the same type of compact structure also there.

The staggered grid can be generalized to several space dimensions. In 2D the equations are

$$\frac{\partial H_x}{\partial t} + \frac{1}{\mu}\frac{\partial E_z}{\partial y} = 0,$$

$$\frac{\partial H_y}{\partial t} - \frac{1}{\mu}\frac{\partial E_z}{\partial x} = 0,$$

$$\frac{\partial E_z}{\partial t} - \frac{1}{\varepsilon}\frac{\partial H_y}{\partial x} + \frac{1}{\varepsilon}\frac{\partial H_x}{\partial y} = 0,$$

$$\frac{\partial E_x}{\partial t} - \frac{1}{\varepsilon}\frac{\partial H_z}{\partial y} = 0,$$

$$\frac{\partial E_y}{\partial t} + \frac{1}{\varepsilon}\frac{\partial H_z}{\partial x} = 0,$$

$$\frac{\partial H_z}{\partial t} + \frac{1}{\mu}\frac{\partial E_y}{\partial x} - \frac{1}{\mu}\frac{\partial E_x}{\partial y} = 0.$$

The equations have been ordered in two groups independent of each other. There are three variables in each group, and they are represented at different grid points. Figure 15.10 shows the grid structure for the first group of variables H_x, H_y, E_z. It also shows how the variables H_x, H_y are first advanced in time, and then the variable E_z.

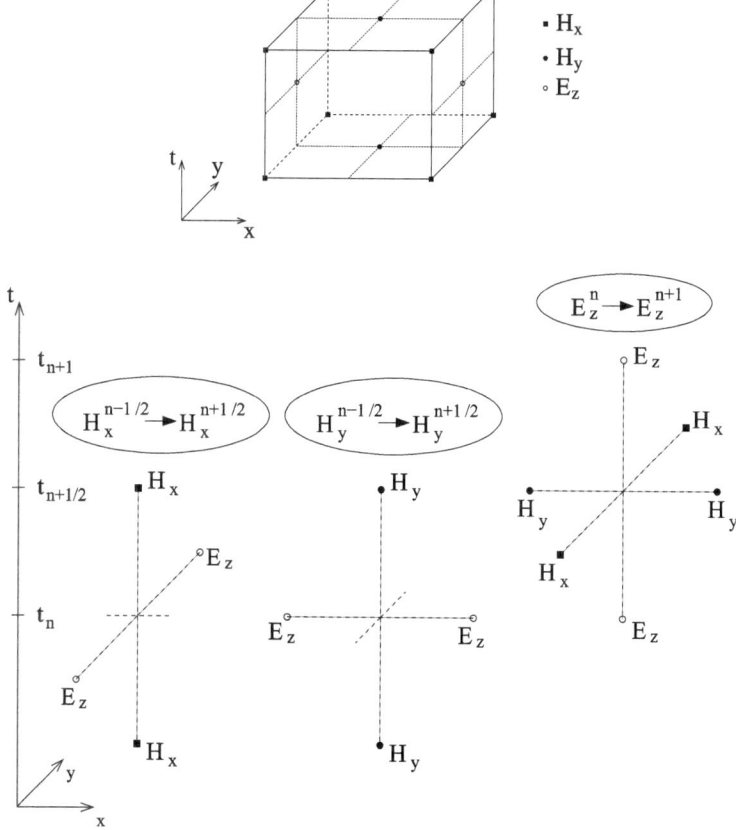

Fig. 15.10 Staggered grid in 2D

15.4 Other Types of Wave Propagation

In this section we shall present the wave equations for a few other types of wave
propagation.

15.4.1 Elasticity

When dealing with elastic materials, the equations are stated in terms of the *displacement* **u**, which is a vector that contains the three components u, v, w representing the displacements in the x-, y-, z-directions. The general formulation of the
equations is

$$\rho\frac{\partial^2 \mathbf{u}}{\partial t^2} = \mu\Delta\mathbf{u} + (\lambda + \mu)\nabla(\nabla \cdot \mathbf{u}),$$

where λ and μ are the Lamé parameters expressed in the unit $\text{Pa} = \text{N/m}^2$. In Cartesian coordinates the equations are

$$\rho\frac{\partial^2 u}{\partial t^2} = \mu\left(\frac{\partial^2 u}{\partial x^2} + \frac{\partial^2 u}{\partial y^2} + \frac{\partial^2 u}{\partial z^2}\right) + (\lambda + \mu)\left(\frac{\partial^2 u}{\partial x^2} + \frac{\partial^2 v}{\partial x\partial y} + \frac{\partial^2 w}{\partial x\partial z}\right),$$

$$\rho\frac{\partial^2 v}{\partial t^2} = \mu\left(\frac{\partial^2 v}{\partial x^2} + \frac{\partial^2 v}{\partial y^2} + \frac{\partial^2 v}{\partial z^2}\right) + (\lambda + \mu)\left(\frac{\partial^2 u}{\partial x\partial y} + \frac{\partial^2 v}{\partial y^2} + \frac{\partial^2 w}{\partial y\partial z}\right),$$

$$\rho\frac{\partial^2 w}{\partial t^2} = \mu\left(\frac{\partial^2 w}{\partial x^2} + \frac{\partial^2 w}{\partial y^2} + \frac{\partial^2 w}{\partial z^2}\right) + (\lambda + \mu)\left(\frac{\partial^2 u}{\partial x\partial z} + \frac{\partial^2 v}{\partial y\partial z} + \frac{\partial^2 w}{\partial z^2}\right).$$

In order to find the typical wave speeds, we make the usual simplification assuming
that all y- and z-derivatives are zero, i.e., plane waves are moving along the x-axis.
We get the reduced system

$$\frac{\partial^2}{\partial t^2}\begin{bmatrix} u \\ v \\ w \end{bmatrix} = \frac{1}{\rho}\begin{bmatrix} \lambda + 2\mu & 0 & 0 \\ 0 & \mu & 0 \\ 0 & 0 & \mu \end{bmatrix}\frac{\partial^2}{\partial x^2}\begin{bmatrix} u \\ v \\ w \end{bmatrix}.$$

Here we have a new feature when comparing to acoustics and electromagnetics,
since there are two different wave speeds:

$$c_1 = \sqrt{(\lambda + 2\mu)/\rho}, \qquad c_2 = \sqrt{\mu/\rho}.$$

The first one corresponds to *longitudinal waves*, where the propagation direction is
the same as the displacement direction. These waves are in direct analogy to sound
waves discussed in Sect. 15.2. Sometimes they are called *compression waves* or
P-*waves*. The second wave speed is lower and corresponds to *transverse waves*,
where the propagation direction is perpendicular to the displacement direction.
Other names for them are *shear waves* or S-*waves*. (P and S stands for Primary
and Secondary.) This type of wave is analogous to surface waves resulting from a

pebble dropped in water with a glazed surface. The water particles move vertically, while the waves propagate horizontally (in a circular pattern in this case).

In 3D, all the variables in the system are fully coupled. However, there is a way to uncouple them. By applying the divergence and curl operator (see Appendix A.2) to the system, we get

$$\frac{\partial^2}{\partial t^2}(\nabla \cdot \mathbf{u}) = \frac{\lambda + 2\mu}{\rho}\Delta(\nabla \cdot \mathbf{u}),$$

$$\frac{\partial^2}{\partial t^2}(\nabla \times \mathbf{u}) = \frac{\mu}{\rho}\Delta(\nabla \times \mathbf{u}).$$

The first one is a scalar equation, the second one is a vector equation with 3 components. They are all independent of each other, but usually coupled through the boundary and initial conditions.

These equations are of fundamental importance in seismology, in particular when it comes to tracing waves from earthquakes. The wave speeds are not constant in the upper part of the earth, but typical values are 6–8 km/s for the P-wave and 3–5 km/s for the S-wave near the surface.

15.4.2 The Schrödinger Equation

The most fundamental differential equation in theoretical physics is probably the *Schrödinger equation*, presented 1926 by the Austrian theoretical physicist Erwin Schrödinger (1887–1961). It is a very general description of how physical systems evolve with time, and is a generalization of Newton's classical mechanics theory to quantum mechanics. It applies to particles on the atomic level, where the position of a certain particle cannot be exactly determined, but rather the probability that it is at a given position.

At center of the theory is the complex wave function $\Psi = \Psi(x, y, z, t)$ which represents this probability in the form of its amplitude. The general form of the PDE is

$$i\hbar\frac{\partial \Psi}{\partial t} = \hat{H}\Psi,$$

where $\hat{H}$ is the Hamiltonian operator and $h = 1.05457162853 \cdot 10^{-34}$ Js is the Planck constant. As usual, a compact mathematical form doesn't mean that the equation is simple. On the contrary, the Hamiltonian is a differential operator that can be given many different forms. Considering a single particle with potential energy $V = V(x, y, z)$, the equation is

$$i\hbar\frac{\partial \Psi}{\partial t} = -\frac{\hbar^2}{2m}\left(\frac{\partial^2}{\partial x^2} + \frac{\partial^2}{\partial y^2} + \frac{\partial^2}{\partial z^2}\right)\Psi + V\Psi,$$

where m is the mass of the particle.

At a first glance the Schrödinger equation has certain similarities with the heat conduction equation, which has no wave propagation properties at all. But the whole difference is made by the imaginary constant on the left hand side. We use the Fourier transform in space to find the essential properties. With V as a constant, we get

$$\frac{d\hat{\Psi}}{dt} = i\alpha\hat{\Psi}, \quad \alpha = -\frac{\hbar}{2m}(k_1^2 + k_2^2 + k_3^2) - \frac{1}{\hbar}V.$$

Since Planck's constant is real, α is real, and we have the solution as the two waves

$$\hat{\Psi}(t) = c_1 e^{i\alpha t} + c_2 e^{-i\alpha t}.$$

When constructing numerical methods, we have to choose between those which are well suited for energy conserving wave propagation problems.

Chapter 16
Heat Conduction

The prediction of heat and temperature distribution in a material is of great importance in many applications. The conduction of heat has certain similarities with diffusion. The mathematical model is in both cases a *parabolic* PDE with the typical property that variations in various quantities are smoothed out with time. We have treated the simplest version of such equations earlier in this book, but here we shall look into its properties a little further. We shall also take a look at the *steady state problem* which leads to *elliptic* equations.

16.1 Parabolic PDE

In Sect. 9.1 the classification of general second order PDE was discussed. It was mentioned that the first order derivatives play an important role in the parabolic case. We have used a simple parabolic PDE earlier in this book for illustration, and we use it here again:

$$\frac{\partial u}{\partial t} = a \frac{\partial^2 u}{\partial x^2}.$$

Here a is a constant, and it was shown in Chap. 2 that it must be positive if the problem is to be solved forward in time. As usual we use the Fourier transform to study the properties of the equation, and we get for the Fourier coefficients

$$\frac{\partial \hat{u}}{\partial t} = -ak^2 \hat{u}, \quad k = 0, \pm 1, \pm 2, \dots.$$

The solution is

$$\hat{u}(k, t) = e^{-ak^2 t} \hat{u}(k, 0),$$

which shows that the magnitude decreases for each wave number k as time increases. Higher wave numbers mean faster oscillations in the x-direction, and obviously there is stronger damping of the amplitude for faster oscillation in space. This has significant implications from a computational point of view. Oscillatory solutions are considered as problematic in computing, since they require fine resolution leading to large scale computational problems. Furthermore, perturbations may lead

B. Gustafsson, *Fundamentals of Scientific Computing*,
Texts in Computational Science and Engineering 8,
DOI 10.1007/978-3-642-19495-5_16, © Springer-Verlag Berlin Heidelberg 2011

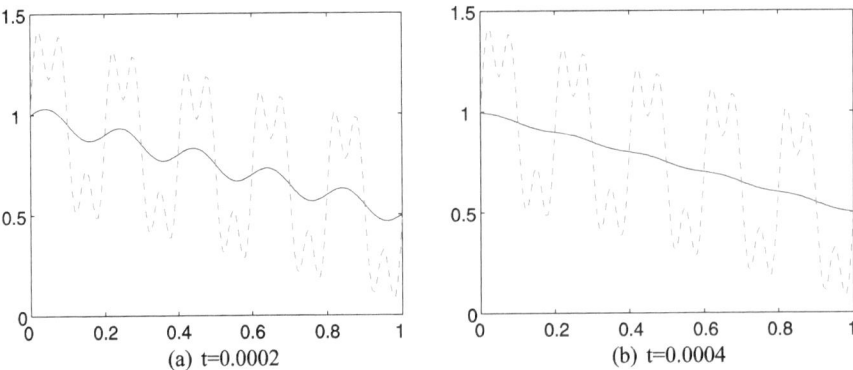

Fig. 16.1 Solution of heat equation (—), initial function (--)

to nonphysical oscillations. However, this equation is forgiving in the sense that the solution becomes smoother. The conclusion is that PDE of this type are "easier" to solve than for example hyperbolic PDE that were discussed in the previous chapter.

For illustrating the smoothing effect, we choose the initial solution

$$u(x, 0) = 1 - 0.5x + 0.4\sin 10\pi x + 0.2\sin 30\pi x,$$

which has two oscillating components. The boundary conditions are

$$u(0, t) = 1,$$
$$u(1, t) = 0.5,$$

i.e., the temperature is held fixed at the boundaries. Figure 16.1 shows clearly that the faster oscillating component (corresponding to the wave number 30) is damped out faster than the other one.

For 2D-problems, the differential equation is

$$\frac{\partial u}{\partial t} = a\frac{\partial^2 u}{\partial x^2} + b\frac{\partial^2 u}{\partial x \partial y} + c\frac{\partial^2 u}{\partial y^2},$$

with the Fourier transform

$$\frac{\partial \hat{u}}{\partial t} = -q(k_x, k_y)\hat{u}, \quad q(k_x, k_y) = ak_x^2 + bk_xk_y + ck_y^2,$$

where k_x and k_y are the wave numbers corresponding to the x- and y-directions. We want to make sure that there are no growing solutions and that there is damping for nonzero wave numbers. The expression multiplying $\hat{u}$ is called a *quadratic form*, and the *parabolicity condition* is

$$q(k_x, k_y) \geq K(k_x^2 + k_y^2), \tag{16.1}$$

where K is a constant independent of k_x and k_y. For 3D-problems, the differential operator in space has the form

$$\frac{\partial u}{\partial t} = a\frac{\partial^2 u}{\partial x^2} + b\frac{\partial^2 u}{\partial y^2} + c\frac{\partial^2 u}{\partial z^2} + d\frac{\partial^2 u}{\partial x \partial y} + e\frac{\partial^2 u}{\partial x \partial z} + f\frac{\partial^2 u}{\partial y \partial z},$$

and the parabolicity condition is generalized in an obvious way.

The classification of the PDE is independent of initial and boundary conditions. It determines the basic behavior of the solutions for any initial functions, but it is restricted to periodic solutions in space, or to infinite space with no boundaries. However, as noted several times earlier in this book, it still tells a good deal about the behavior also for the case where boundary conditions are involved.

For nonperiodic parabolic PDE on a bounded domain, boundary conditions are required and, unlike hyperbolic equations, they are required at all parts of the boundary. For any type of domain, we consider the differential equation locally at each point at the boundary, and define the inward pointing normal $\mathbf{n}$. The general form of the boundary condition is

$$\alpha u + \beta \frac{\partial u}{\partial n} = g(t),$$

where $\partial/\partial n$ denotes differentiation along the normal $\mathbf{n}$. If $\beta = 0$, so that the function u itself is specified, we have a *Dirichlet boundary condition* named after the German mathematician Peter Gustav Lejeune Dirichlet (1805–1859). If $\beta \neq 0$, we have a *Neumann boundary condition*. The name refers to Dirichlet's countryman Carl Gottfried Neumann (1832–1925), and should not be confused with John von Neumann (1903–1957), who is associated with difference methods (among many other things) and the von Neumann condition, see Sect. 10.3.

So far we have discussed PDE with derivatives of at most second order in space. But we can have parabolic equations with derivatives of any even high order, for example

$$\frac{\partial u}{\partial t} = a \frac{\partial^4 u}{\partial x^4}.$$

For the Fourier transform, the differential operator $\partial^4/\partial x^4$ corresponds to $(ik)^4 = k^4$. Consequently, the constant a must be negative to avoid fast growing solutions. A general PDE

$$\frac{\partial u}{\partial t} = a \frac{\partial^{2p} u}{\partial x^{2p}}$$

with a real constant a, is parabolic if

$$a > 0, \quad \text{if } p \text{ is odd,}$$
$$a < 0, \quad \text{if } p \text{ is even.}$$

The generalization to higher space dimensions should be obvious.

When going to systems of PDE, the role of the scalar Fourier transform is taken over by the eigenvalues of the corresponding Fourier matrix. For matrices we must keep in mind that the eigenvalues may be complex even if the elements of the matrix are real. But this is not much of a complication. The condition of nongrowing solutions to a scalar equation $d\hat{u}/dt = \alpha \hat{u}$ is $\text{Re}\,\alpha \leq 0$, and this is applied to the eigenvalues. For example, the system

$$\frac{\partial \mathbf{u}}{\partial t} = A \frac{\partial^2 \mathbf{u}}{\partial x^2} + B \frac{\partial^2 \mathbf{u}}{\partial y^2}$$

is parabolic if the eigenvalues $\lambda(k_x, k_y)$ of the matrix $k_x^2 A + k_y^2 B$ satisfy the condition

$$\mathrm{Re}\,\lambda(k_x, k_y) \geq K(k_x^2 + k_y^2),$$

where K is a constant independent of k_x and k_y.

16.2 Elliptic PDE

Elliptic equations are closely connected to parabolic equations by simply setting $\partial u / \partial t = 0$. For example, the equation

$$a\frac{\partial^2 u}{\partial x^2} + b\frac{\partial^2 u}{\partial x \partial y} + c\frac{\partial^2 u}{\partial y^2} = F(x, y),$$

is elliptic if the condition (16.1) is satisfied. We have here included a forcing function $F(x, y)$ as a right hand side. However, since the classification should be independent of the right hand side, we can of course replace $F(x, y)$ by $-F(x, y)$, and we expect the differential equation to stay in the same elliptic class. The quadratic form $q(k_x, k_y)$ may be either positive definite or negative definite. In the real coefficient case, the condition

$$ac - \left(\frac{b}{2}\right)^2 > 0$$

guarantees that the quadratic form has one and the same sign for all k_x, k_y, be it positive or negative. For example, if $b = 0$, the simple elliptic condition is that a and c have the same sign. This is sufficient for ellipticity.

The definition of elliptic systems of PDE is formulated exactly as for parabolic systems by transferring the conditions to the eigenvalues of the Fourier transform.

Elliptic equations on bounded domains always lead to boundary value problems. There is no way to solve them as initial- or initial–boundary value problems without introducing some extra damping mechanism. This was demonstrated in Sect. 9.1 by using the Fourier transform which can be applied to problems with periodic solutions. Another very similar example is the initial–boundary value problem

$$\frac{\partial^2 u}{\partial y^2} = -\frac{\partial^2 u}{\partial x^2},$$
$$u(0, y) = 0, \qquad u(1, y) = 0,$$
$$u(x, 0) = f(x).$$

In analogy with the example in Sect. 12.1, the solution can be expanded in a sine series $u(x, y) = \sum_k \hat{u}(y) \sin(k\pi x)$, and we get

$$\frac{\partial^2 \hat{u}}{\partial y^2} = (k\pi)^2 \hat{u},$$
$$\hat{u}(k, 0) = \hat{f}(k),$$

which has the solution

$$\hat{u}(k, y) = \alpha e^{k\pi y} + \beta e^{-k\pi y}$$

for some constants α and β. Apparently, there is always one growing component in the positive y-direction and, since the wave number k is arbitrarily large, the problem is ill posed. The conclusion is that we need boundary conditions all the way around the computational domain, not only at three sides as above. The solution has to be tied down by using a boundary condition for some value of $y > 0$. In this way the possibility of an unbounded solution for growing y is eliminated.

16.3 The Heat Equation

Heat conduction in a certain material depends on the thermal conductivity coefficient a, and it may depend on the specific direction we are considering. However, it is very common that the material is *isotropic* with regard to heat conductivity, i.e., a is independent of the direction. In that case the heat equation is

$$\frac{\partial u}{\partial t} = \frac{1}{\rho C}\left(\frac{\partial}{\partial x}\left(a\frac{\partial u}{\partial x}\right) + \frac{\partial}{\partial y}\left(a\frac{\partial u}{\partial y}\right) + \frac{\partial}{\partial z}\left(a\frac{\partial u}{\partial z}\right)\right) + Q,$$

where the various variables and parameters are

u temperature
ρ density
C heat capacity
a thermal conductivity
Q heat source.

In the most general case, all these quantities depend on the temperature, which means that the equation is nonlinear. (In the anisotropic case, a depends also on $\partial u/\partial x$, $\partial u/\partial y$, $\partial u/\partial z$.) However, since they may be independent of u for a large range of temperatures, the linear case is of significant interest. Furthermore, for the analysis of the equation we make ρ, C, a constant, and put $Q = 0$. The equation becomes

$$\frac{\partial u}{\partial t} = \frac{a}{\rho C}\left(\frac{\partial^2 u}{\partial x^2} + \frac{\partial^2 u}{\partial y^2} + \frac{\partial^2 u}{\partial z^2}\right)$$

and, since $a/(\rho C)$ is positive, the equation is parabolic.

There are many different types of boundary conditions. A common one is that the temperature $u = g(t)$ is prescribed. Another one is perfect insulation, which is the Neumann condition

$$\frac{\partial u}{\partial n} = 0,$$

where $\partial/\partial n$ is differentiation in the direction perpendicular to the boundary.

The following example illustrates the heat conduction problem in 2D for a square with 160 cm side. The initial temperature is 20°C everywhere, and the experiment

Fig. 16.2 Triangular grid

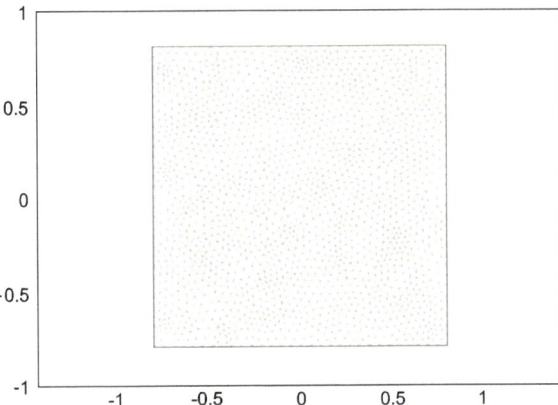

Fig. 16.3 Temperature
distribution after 10 minutes

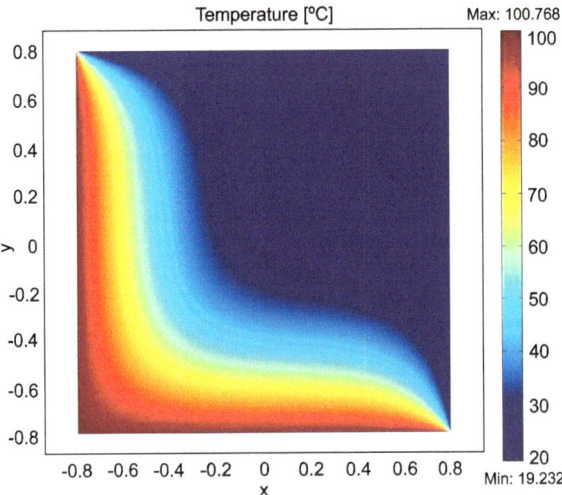

simulates the situation where two edges are suddenly given the temperature 100°C, while the other two edges are held at the initial temperature 20°C. The following parameters are used:

$$\rho = 8700 \text{ kg/m}^3,$$
$$C = 385 \text{ J/(kg·K)},$$
$$a = 400 \text{ W/(m·K)}.$$

The higher temperature is spread into the inner part of the domain with time and there is a strong discontinuity in the temperature at two of the corners. In such a case the solutions of more general PDE are difficult to compute since oscillations are generated near the corners. However, the parabolic character keeps the solution smooth. The finite element method within the COMSOL system is used for solution on a triangular grid. Figure 16.2 shows the grid containing 3784 triangles and 7689

Fig. 16.4 Temperature
profile 1 cm from the bottom
edge after 10 minutes

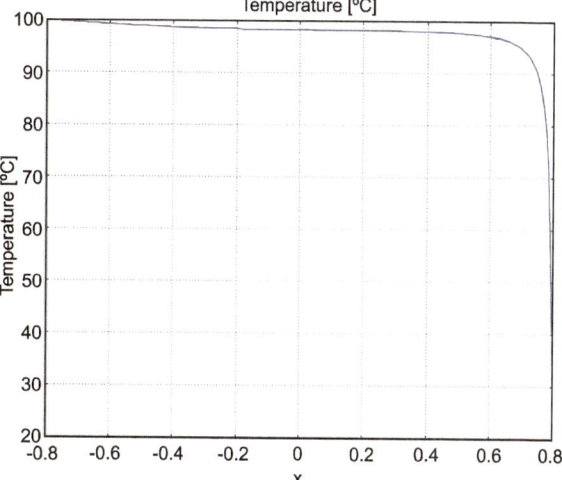

Fig. 16.5 Temperature
distribution at steady state

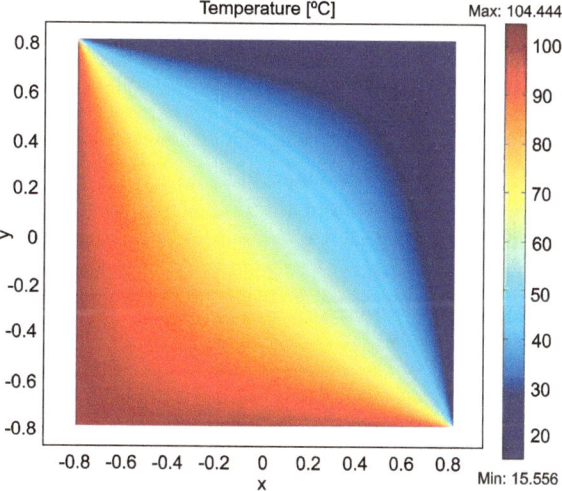

parameters to be determined, i.e., there are 7689 degrees of freedom. Figure 16.3
shows a surface plot of the temperature after 10 minutes, and Fig. 16.4 shows a cut
through the domain 1 cm from the bottom edge at the same point in time. Obviously,
there are no oscillations at all near the corner. The sharp gradient, that is necessarily
there with these boundary conditions, is well represented.

In order to illustrate the solution of an elliptic problem, we choose the steady
state problem for the same heat conduction example. After sufficiently long time,
the temperature distribution settles down to a steady state, and we want to compute
it. The equation is

$$\frac{a}{\rho C}\left(\frac{\partial^2 u}{\partial x^2} + \frac{\partial^2 u}{\partial y^2} + \frac{\partial^2 u}{\partial z^2}\right) = 0,$$

Fig. 16.6 Temperature
profile 1 cm from the bottom
edge at steady state

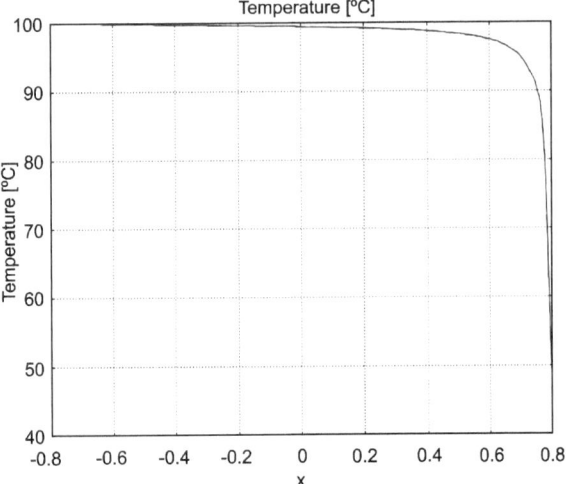

Fig. 16.6 Temperature profile 1 cm from the bottom edge at steady state

with u specified at the boundaries as above. We use again the COMSOL system to solve it, and the result is shown in Fig. 16.5.

One can see that the solution is smoothed out in the central part of the domain. At the south-east and north-west corners there are discontinuities in the boundary data, but the solution still becomes smooth as soon as we move inside the domain. Figure 16.6 shows the same cut as for the time dependent case above.

Another way to compute this solution would be to run the previous program, which solves the time dependent problem. By letting it go for a long time, the steady state solution should be obtained. However, that would require a very long computer time, and it is not a good procedure for this kind of problem, particularly not for problems in 3D.

Chapter 17
Fluid Dynamics

Fluid dynamics is an area which has driven the development of numerical methods probably more than any other area. Computer simulations were done already in the 1940s, but more than 60 years later there are still many remaining challenges that are not yet solved. The computation of turbulent flow around a complete aircraft is still out of reach, even with the gigantic computers that are available today. The mathematical models have been known for more than 150 years, but sufficiently efficient computational methods are still lacking. In this chapter we shall give a brief presentation of the basic equations and methods, and then follow up with a few applications.

17.1 Basic Differential Equations

The most complete model describing the dynamics of a fluid (liquid or gas) is the set of Navier–Stokes equations after the French physicist and engineer Claude–Louis Navier (1785–1836) and the British mathematician and physicist George Gabriel Stokes (1819–1903). These equations are quite complicated, and they are nonlinear. In the effort to derive solutions and to analyze the properties, many different simplifications can be made, and there are a number of less complicated equations that describe different kinds of fluids with special behavior. Any fluid has a certain *viscosity*, i.e., there is some degree of internal resistance to flow forces. Oil has high viscosity and water has low viscosity. For air, the viscosity is even lower, and in the limit one talks about *inviscid flow*. It is always an idealization of a real fluid, but often it is such an accurate assumption that no error is committed from a practical point of view. And in a certain sense it simplifies the mathematical/numerical analysis considerably. On the other hand some extra unwanted complications occur as we shall see.

A fluid is always compressible, which means that the density of the fluid changes as a result of pressure variations. For gases the compressibility effects may be considerable. On the other hand, certain fluids may be almost incompressible, like for example water under normal conditions. If the effects on the density tend to zero,

B. Gustafsson, *Fundamentals of Scientific Computing*,
Texts in Computational Science and Engineering 8,
DOI 10.1007/978-3-642-19495-5_17, © Springer-Verlag Berlin Heidelberg 2011

we talk about *incompressible flow*. It is an idealization of any real fluid, but in analogy with the assumption of inviscid flow, we get a considerable simplification of the differential equations.

In this section we shall discuss the equations for inviscid (but compressible) flow, and for incompressible (but viscous) flow.

17.1.1 The Continuity Equation

In this book we have not yet given the derivation of any models for any nontrivial problem. In order to give a taste of such derivations, we shall take a look at the continuity equation, which is one of the basic ingredients in fluid dynamics. Consider a cube with side length Δx as shown in Fig. 17.1. For convenience we assume that the fluid is moving in the x-direction with no variation in the y- or z-directions.

The density is ρ_1 to the left, and ρ_2 to the right (in kg/m^3), while the particle velocity is u_1 and u_2 m/sec respectively. The average density in the cube is $\overline{\rho}^x$ so that the total mass in the cube is $\overline{\rho}^x \Delta x^3$ kg. We now consider a time interval $[t, t + \Delta t]$, and for any variable v we define the time average $\overline{v}^t$ over this interval. During the time Δt the mass $\overline{u_1 \rho_1}^t \Delta t \Delta x^2$ enters from the left, and the mass $\overline{u_2 \rho_2}^t \Delta t \Delta x^2$ is leaving to the right. Therefore, the change of the mass is $(\overline{u_1 \rho_1}^t - \overline{u_2 \rho_2}^t) \Delta t \Delta x^2$, and we get the equation

$$\left(\overline{\rho}^x(t + \Delta t) - \overline{\rho}^x(t)\right) \Delta x^3 = \left(\overline{u_1 \rho_1}^t - \overline{u_2 \rho_2}^t\right) \Delta t \Delta x^2, \tag{17.1}$$

or equivalently

$$\frac{\overline{\rho}^x(t + \Delta t) - \overline{\rho}^x(t)}{\Delta t} + \frac{\overline{u\rho}^t(x + \Delta x) - \overline{u\rho}^t(x)}{\Delta x} = 0. \tag{17.2}$$

By taking the limit as Δx and Δt tend to zero, the averages turn into point values $\rho(x, t)$ and $u(x, t)$, and we get the final equation

$$\frac{\partial \rho}{\partial t} + \frac{\partial (u\rho)}{\partial x} = 0. \tag{17.3}$$

Here we assumed that there is no variation in the y- and z-directions. When removing this restriction, the partial derivatives in these directions enter as well. With v and w denoting the velocity components in the y- and z-directions, we get

$$\frac{\partial \rho}{\partial t} + \frac{\partial (u\rho)}{\partial x} + \frac{\partial (v\rho)}{\partial y} + \frac{\partial (w\rho)}{\partial z} = 0. \tag{17.4}$$

This is the *continuity equation*, which simply describes the fact that mass is neither created or destroyed. The change in time and the change in space must cancel each other. This is true whether or not the fluid is viscous, and whether or not it is compressible.

Fig. 17.1 Flow through a small cube

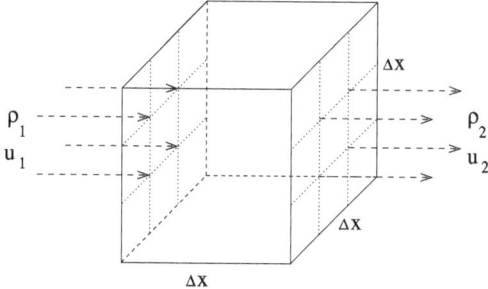

17.1.2 Euler equations

Obviously the continuity equation is not enough to describe a fluid motion. The quantities density and velocity are involved, and we need more differential equations.

Inviscid flow is governed by the Euler system of equations. The 2D-version was presented in the introductory chapter, but here we shall analyze it a little further. The full 3D-equations are

$$\frac{\partial \rho}{\partial t} + \frac{\partial (\rho u)}{\partial x} + \frac{\partial (\rho v)}{\partial y} + \frac{\partial (\rho w)}{\partial z} = 0,$$

$$\frac{\partial (\rho u)}{\partial t} + \frac{\partial (\rho u^2 + p)}{\partial x} + \frac{\partial (\rho u v)}{\partial y} + \frac{\partial (\rho u w)}{\partial z} = 0,$$

$$\frac{\partial (\rho v)}{\partial t} + \frac{\partial (\rho u v)}{\partial x} + \frac{\partial (\rho v^2 + p)}{\partial y} + \frac{\partial (\rho v w)}{\partial z} = 0,$$

$$\frac{\partial (\rho w)}{\partial t} + \frac{\partial (\rho u w)}{\partial x} + \frac{\partial (\rho v w)}{\partial y} + \frac{\partial (\rho w^2 + p)}{\partial z} = 0,$$

$$\frac{\partial E}{\partial t} + \frac{\partial (u(E + p))}{\partial x} + \frac{\partial (v(E + p))}{\partial y} + \frac{\partial (w(E + p))}{\partial z} = 0.$$

The state variables are

ρ density,
u velocity component in x-direction,
v velocity component in y-direction,
w velocity component in z-direction,
E total energy per unit volume,
p pressure.

There are six state variables but only five differential equations. Therefore we need an *equation of state*

$$p = p(\rho, u, v, w, E),$$

which defines the pressure as an explicitly given function of the other variables. Under certain conditions there is a simple relation $p = p(\rho)$ between density and pressure. This is *isentropic flow*, and in that case the last differential equation (called the energy equation) can be left out.

For analysis of the equations, we take the simple case of one-dimensional isentropic flow, where all the y- and z-derivatives are zero, as well as the velocity components v and w:

$$\frac{\partial \rho}{\partial t} + u\frac{\partial \rho}{\partial x} + \rho\frac{\partial u}{\partial x} = 0,$$

$$u\frac{\partial \rho}{\partial t} + \rho\frac{\partial u}{\partial t} + 2\rho u\frac{\partial u}{\partial x} + u^2\frac{\partial \rho}{\partial x} + \frac{\partial p}{\partial x} = 0.$$

In order to get a closed system, we eliminate the pressure variable. It occurs as $\partial p/\partial x$ in the differential equation for u, and according to the chain rule for differentiation we have

$$\frac{\partial p}{\partial x} = \frac{dp}{d\rho}\frac{\partial \rho}{\partial x}.$$

After elimination of $\partial \rho/\partial t$ from the second equation, we get the system

$$\frac{\partial}{\partial t}\begin{bmatrix} \rho \\ u \end{bmatrix} + \begin{bmatrix} u & \rho \\ c^2/\rho & u \end{bmatrix}\frac{\partial}{\partial x}\begin{bmatrix} \rho \\ u \end{bmatrix} = 0,$$

where $c^2 = dp/d\rho$. The system is nonlinear, but it looks very similar to the acoustics equations. We know from Sect. 15.1 that in the linear case wellposedness requires that the eigenvalues λ_j of the matrix are real, and it is natural to require that here as well. The eigenvalues are

$$\lambda_{1,2} = u \pm c,$$

showing that also here there are two waves. The PDE is just the nonlinear version of the acoustics equations, and c denotes the speed of sound here as well. The only difference is that $c = c(\rho)$ depends on the solution itself, and that the speed of the waves is shifted by the particle velocity u. The sound waves are given an extra push by the fluid movement. We also note that, if $u > c$, both waves are propagating in the same direction. We call this *supersonic flow*, and it has great significance for the properties of the solution as we shall show in the next section.

The system is classified as hyperbolic, even if it is nonlinear. The number of boundary conditions is determined by the sign of the eigenvalues. It was demonstrated in Sect. 15.1 how to construct admissible boundary conditions. At any boundary point, the equations are written in terms of the local coordinates, where one direction is along the inward normal $\mathbf{n}$ perpendicular to the boundary surface. The corresponding coefficient matrix then determines the form of the boundary conditions. We can without restriction use the coefficient matrix for $\partial/\partial x$ for the analysis, assuming that the boundary is the plane $x = 0$. It can be shown that the eigenvalues are

$$u, u, u, u + c, u - c.$$

There are four different situations, and the number of boundary conditions are determined by Table 17.1.

Table 17.1 Number of boundary conditions for various flow conditions

Flow condition	Number of boundary conditions
supersonic outflow $u \leq -c$	0
subsonic outflow $\quad -c < u \leq 0$	1
subsonic inflow $\quad 0 < u \leq c$	4
supersonic inflow $\quad c < u$	5

Fig. 17.2 Type of free stream boundary conditions for an ellipse

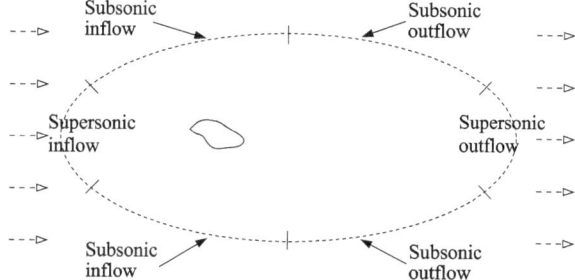

Note that it is always the velocity component perpendicular to the boundary that determines the number of boundary conditions. The flow may well be supersonic in the sense that the particles are moving faster than the sound, i.e., $\sqrt{u^2 + v^2 + w^2} > c$ but, if the normal component is smaller than the speed of sound, there must be less than 5 boundary conditions. Figure 17.2 shows a 2D-example with an ellipse as an open boundary far away from an object. It is assumed here that the flow outside the ellipse is uniformly supersonic and constant without any influence from the flow close to the object. This may not be exactly true on the downstream side, but is often used for practical computations anyway.

A solid wall is a special type of boundary. On physical grounds, the boundary condition should be $u_n = 0$, where u_n is the normal velocity component. The question is whether the theory supports this one as the only necessary condition. Fortunately it does. It is a borderline case with 3 zero eigenvalues, and only one is positive. We recall from Sect. 15.1 that zero eigenvalues don't require any boundary condition for the corresponding characteristic variables. Accordingly, the single solid wall condition on the velocity is sufficient. Actually, the nonlinearity causes a double effect. The condition $u = 0$ makes three eigenvalues zero, and at the same time it provides the necessary boundary condition.

17.1.3 Shocks

The Euler equations are nonlinear and, in order to analyze them, we made them linear by the standard linearization procedure. However, some of the key features of the equations cannot be caught by studying the linearized version. One of those

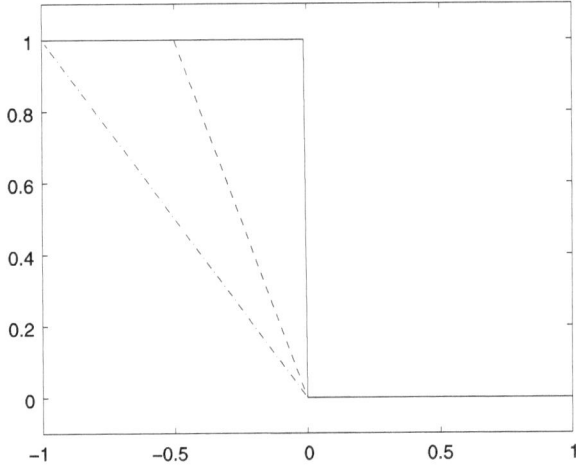

Fig. 17.3 Solution of Burgers' equation, $t = 0$ $(-\cdot)$, $t = 0.5$ $(--)$, $t = 1$ $(—)$

features is the formation of *shocks*, which is the name for a certain type of discontinuity. For illustration we shall use a much simplified but still nonlinear equation, namely the *Burgers' equation*

$$\frac{\partial u}{\partial t} + u \frac{\partial u}{\partial x} = 0,$$

which got its name from the Dutch physicist J.M. Burgers (1895–1981). It is an example of a conservation law that was introduced in Sect. 11.3, which is seen by writing it in the form

$$\frac{\partial u}{\partial t} + \frac{\partial}{\partial x}\left(\frac{u^2}{2}\right) = 0.$$

The first version is very similar to the linear transport equation used many times earlier in this book. Actually it acts much the same as long as the solution is smooth, i.e., the coefficient u multiplying $\partial u / \partial x$ plays the role of wave speed. At any given moment and any given point, a certain feature in the solution is moving with speed $u(x, t)$.

Let us look at an example in the interval $-1 \le x \le 1$ with the boundary condition $u(-1, t) = 1$, and a continuous initial function according to Fig. 17.3. No boundary condition is required at $x = 1$ as long as the solution is non-negative there. The top part of the solution in the left half of the domain is moving at speed one, while it goes down linearly to zero at $x = 0$. In the right part nothing happens. The result at $t = 0.5$ and $t = 1$ is shown in Fig. 17.3.

At $t = 1$, the discontinuity has formed at $x = 0$, and we call it a shock. But this causes big trouble. The upper part wants to continue with speed one, while the lower part stands still. This is an impossible situation; we cannot have a solution that tips over, such that there is no unique value of the function anymore. The question then is: how should the solution be defined for $t > 1$? It seems reasonable that the shock should continue moving to the right, but what should the speed be? For an answer we try to see what nature would do.

Fig. 17.4 Solution of the viscous Burgers' equation, $t = 0$ $(-\cdot)$, $t = 1$ $(--)$, $t = 2$ $(—)$

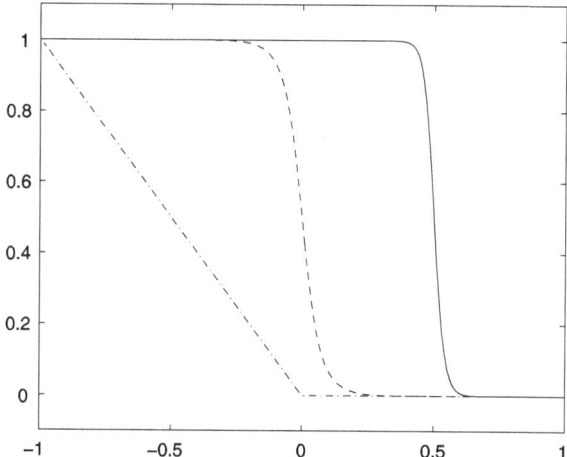

There is no fluid that is truly inviscid, even if the viscosity is extremely small. As we shall see below, viscosity introduces a second derivative in the differential equation. Let us see what happens if we introduce a tiny viscous effect into our simplified model equation:

$$\frac{\partial u}{\partial t} + u\frac{\partial u}{\partial x} = \varepsilon\frac{\partial^2 u}{\partial x^2}.$$

Here the coefficient ε is small. The equation is called the *viscous Burgers' equation*. We choose $\varepsilon = 0.01$ and apply a standard second order difference scheme. The result is shown in Fig. 17.4.

The shock is clearly identified, but now has rounded shoulders. This is typical for viscous flow compared to inviscid flow. The equation has now become parabolic, and the solution is smoother. The really interesting fact is that the differential equation now decided to produce a unique solution, and it decided upon the speed $u = 0.5$ for propagation of the shock. This is the average of the two extreme values when the shock formed.

For decreasing values of the viscosity coefficient ε, the shock will become sharper and sharper, and in the limit we are back to a discontinuous solution. We use this limit process to bring order into the inviscid case. With the notation $u(x, t, \varepsilon)$ for the viscous solution, and $u(x, t)$ for the inviscid solution, we make the definition

$$u(x, t) = \lim_{\varepsilon \to 0} u(x, t, \varepsilon).$$

For smooth solutions of the inviscid equation this definition doesn't introduce anything new and is not necessary. But for a discontinuous solution it does.

For the Euler equations there is the same kind of situation. For supersonic flow, there will be shocks. The viscosity can here be introduced in the physically correct way, and we get the Navier–Stokes equations which have a quite complicated structure. But we can also introduce viscosity in a simpler form. No matter what form is chosen, the physically relevant solutions can also here be defined as the limit of viscous solutions. There are several other ways to define the same unique solution, but in all cases we say that it satisfies the *entropy condition*.

Numerical methods for shock problems can be separated into *shock fitting methods* and *shock capturing methods*. In the first case, the discontinuity is treated as an internal boundary, and the location of it is kept as a special variable that is computed for each time step. On each side of the shock the differential equation is solved by some method that does not use any grid points on the other side. In the second case, the differential equation is solved across the discontinuity, but with some special procedure near the shock. Usually these methods use some kind of artificial viscosity, which has the effect of smoothing the solution in analogy with the example above.

Theory and methods for shock problems is a large and complicated issue. We refer to the final chapter for literature on this topic.

17.1.4 Incompressible Navier–Stokes Equations

It can be shown that for low speed flow the compressibility effects are very small, and they can be neglected without much loss of accuracy. The challenge that is occupying many scientists today is the simulation of turbulence, and the incompressible Navier–Stokes equations is the main model. The problem is to resolve the very small scales for the variation of the variables, compared to the size of the typical object. A common method for getting some insight into a certain large-scale problem, is to reduce the size of the problem by taking out one space dimension. Unfortunately, this doesn't make much sense for turbulence simulation, since the underlying assumption is that there is little or no variation in one of the space directions. Turbulence is a true three-dimensional process.

Here we shall not involve ourselves into this very difficult subject. However, there are certain basic properties of the mathematical and numerical models that still apply to many applications, and we shall now discuss the differential equations for incompressible but viscous flow. The continuity equation (17.4) can be written in the form

$$\frac{\partial \rho}{\partial t} + u\frac{\partial \rho}{\partial x} + v\frac{\partial \rho}{\partial y} + w\frac{\partial \rho}{\partial z} = -\rho\left(\frac{\partial u}{\partial x} + \frac{\partial v}{\partial y} + \frac{\partial w}{\partial z}\right),$$

where the left hand side is called the *total derivative* of ρ, and is denoted by $D\rho/Dt$. It describes a change in a fluid element that is following the flow. The correct definition of incompressibility is

$$\frac{D\rho}{Dt} = 0,$$

i.e., there is no change in the density if one follows the fluid particles. Since $\rho > 0$, it then follows from the continuity equation that an equivalent condition is

$$\frac{\partial u}{\partial x} + \frac{\partial v}{\partial y} + \frac{\partial w}{\partial z} = 0.$$

The left hand side is called the *divergence* of the velocity, and the velocity field is here "divergence free". A special case of incompressible flow is when the density $\rho = \rho_0$ is a constant. It is such a frequent case in applications, that it is sometimes taken as the definition of incompressibility.

In Cartesian coordinates the *incompressible Navier–Stokes equations* are

$$\frac{\partial u}{\partial t} + u\frac{\partial u}{\partial x} + v\frac{\partial u}{\partial y} + w\frac{\partial u}{\partial z} + \frac{1}{\rho}\frac{\partial p}{\partial x} = \nu\left(\frac{\partial^2 u}{\partial x^2} + \frac{\partial^2 u}{\partial y^2} + \frac{\partial^2 u}{\partial z^2}\right),$$

$$\frac{\partial v}{\partial t} + u\frac{\partial v}{\partial x} + v\frac{\partial v}{\partial y} + w\frac{\partial v}{\partial z} + \frac{1}{\rho}\frac{\partial p}{\partial y} = \nu\left(\frac{\partial^2 v}{\partial x^2} + \frac{\partial^2 v}{\partial y^2} + \frac{\partial^2 v}{\partial z^2}\right),$$

$$\frac{\partial w}{\partial t} + u\frac{\partial w}{\partial x} + v\frac{\partial w}{\partial y} + w\frac{\partial w}{\partial z} + \frac{1}{\rho}\frac{\partial p}{\partial z} = \nu\left(\frac{\partial^2 w}{\partial x^2} + \frac{\partial^2 w}{\partial y^2} + \frac{\partial^2 w}{\partial z^2}\right),$$

$$\frac{\partial u}{\partial x} + \frac{\partial u}{\partial y} + \frac{\partial u}{\partial z} = 0.$$

Here ν is the *kinematic viscosity* coefficient and has the dimension m^2/s. Sometimes one is using instead the *dynamic viscosity* coefficient $\mu = \rho\nu$ which has the dimension kg/(m·s) $=$ Pa·s. The variable p is usually called the pressure also here, but it is not the same physical quantity as in the compressible case.

It is often convenient to use vector notation

$$\nabla = \begin{bmatrix} \partial/\partial x \\ \partial/\partial y \\ \partial/\partial z \end{bmatrix}, \qquad \mathbf{u} = \begin{bmatrix} u \\ v \\ w \end{bmatrix},$$

where ∇ is called the *gradient operator*. With $(\cdot, \cdot)$ denoting the vector scalar product, the divergence condition can be written as

$$(\nabla, \mathbf{u}) = 0.$$

By using the vectors defined above, the system takes the compact form

$$\frac{\partial \mathbf{u}}{\partial t} + (\mathbf{u}, \nabla\mathbf{u}) + \frac{1}{\rho}\nabla p = \nu\Delta\mathbf{u},$$

$$(\nabla, \mathbf{u}) = 0.$$

The notation $\nabla\mathbf{u}$ indicates that a vector is acting on a vector, and that is a new type of operation in our presentation so far. With the interpretation

$$\nabla\mathbf{u} = \begin{bmatrix} \nabla u \\ \nabla v \\ \nabla w \end{bmatrix},$$

the formal rule

$$(\mathbf{u}, \nabla\mathbf{u}) = \begin{bmatrix} (\mathbf{u}, \nabla u) \\ (\mathbf{u}, \nabla v) \\ (\mathbf{u}, \nabla w) \end{bmatrix}$$

makes sense, and gives the correct equations.

The operator Δ is the *Laplacian*. It is often written in the form $\Delta = \nabla^2$, where ∇^2 should be interpreted as the scalar product (∇, ∇).

The advantage with the compact notation is not only that it saves space in writing, but also that it is easy to rewrite the equations in other coordinate systems, see Appendix A.2. The expanded form of the operators ∇ and Δ is well known for all standard coordinates.

The incompressible Navier–Stokes equations don't have any of the standard forms that we have seen so far. There are four unknown variables u, v, w, p and four differential equations, but the time derivative of the pressure p doesn't occur anywhere. The pressure is implicitly determined by the continuity equation.

In order to get an idea about the relative size of different variables and parameters, the system is written in dimensionless form. If t_0, L, u_0 are typical values of time, length and velocity respectively, the new variables are obtained by the scaling relations

$$t \to t/t_0, \qquad \mathbf{x} \to \mathbf{x}/L, \qquad \mathbf{u} \to \mathbf{u}/u_0, \qquad p \to p/(\rho u_0^2).$$

With the dimensionless *Reynolds number* defined by

$$Re = \frac{\rho L u_0}{\nu},$$

the dimensionless Navier–Stokes equations take the simple form

$$\frac{\partial \mathbf{u}}{\partial t} + (\mathbf{u}, \nabla \mathbf{u}) + \nabla p = \frac{1}{Re} \Delta \mathbf{u},$$
$$(\nabla, \mathbf{u}) = 0.$$

The Reynolds number was introduced by Stokes, but is named after Osborne Reynolds (1842–1912). For fluids with low viscosity, for example air, and normal speed and size of the object, the Reynolds number is large. A flying aircraft corresponds to Reynolds numbers of the order 10^7. On the other hand, even if air has low viscosity, a flying mosquito corresponds to a much smaller Reynolds number, because L is small. Due to its small size and relatively low speed, the mosquito may experience the air as quite viscous.

High Reynolds numbers means that the right hand side of the PDE system has the character of a *singular perturbation*. By this we mean that second order derivatives are added to a first order system, but with a small coefficient. Second order differential equations have very different properties when compared to first order ones, and this is the reason why the perturbation is called singular.

As an illustration of this, we take the extremely simple problem

$$\frac{du}{dx} = 0, \quad 0 \le x \le 1,$$
$$u(0) = 1,$$

which has the solution $u(x) = 1$. If we now consider the perturbed equation

$$\frac{du}{dx} = \varepsilon \frac{d^2 u}{dx^2}, \quad 0 \le x \le 1,$$

Fig. 17.5 Solution with
boundary layer, $\varepsilon = 0.1$
(——), $\varepsilon = 0.01$ (— —)

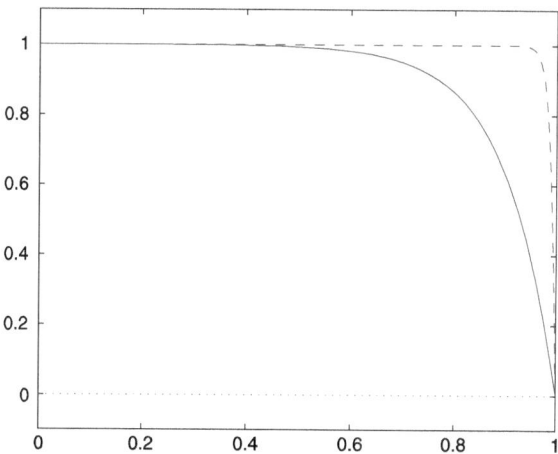

we need two boundary conditions. With

$$u(0) = 1, \qquad u(1) = 0,$$

the solution is

$$u(x, \varepsilon) = \frac{e^{x/\varepsilon} - e^{1/\varepsilon}}{1 - e^{1/\varepsilon}}.$$

For small ε we have

$$u(x, \varepsilon) \approx 1 - e^{-(1-x)/\varepsilon}.$$

This shows that the solution is practically identical to the original unperturbed solution for almost all x, but there is a sharp *boundary layer* near $x = 1$. Figure 17.5 shows the solution for $\varepsilon = 0.1$ and $\varepsilon = 0.01$.

We have here a problem where the limit $u(x, 0)$ of the solutions $u(x, \varepsilon)$ do not converge to the original solution $u(x)$ despite the fact that the differential equation formally converges to the original differential equation. The source of the difficulty is the fact that the perturbed equation requires an extra boundary condition, and it does so no matter how small ε is. Then in the very limit $\varepsilon = 0$ the boundary condition becomes superfluous.

It is important to point out that this is not an academic mathematical exercise with no practical value. On the contrary, if the problem is more realistic such that the solutions must be computed numerically, the analysis above has significant implications on the solution methods. Even if the grid is very fine, there is always an ε small enough such that the numerical solutions don't behave well. We applied the standard second order difference scheme to the problem with $\varepsilon = 0.002$. Figure 17.6 shows the solution for 50 grid points and for 100 grid points. Severe oscillations occur near the boundary layer in both cases.

In the past we have dealt with hyperbolic equations with constant norm solution over time, and with parabolic equations with solutions that are damped with time. Let us now take a look at the incompressible Navier–Stokes equations (which are

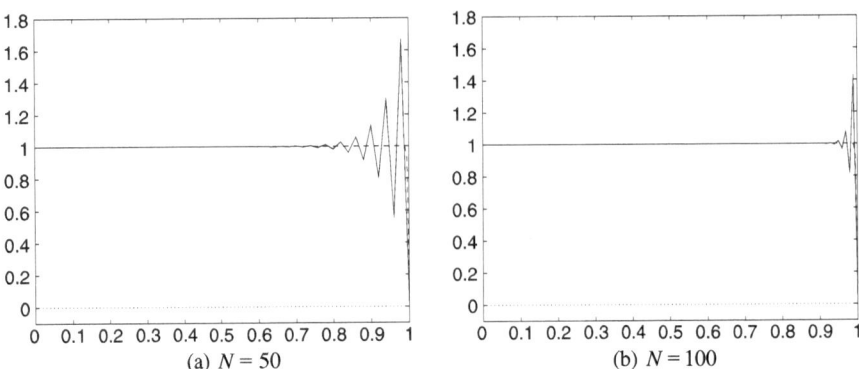

Fig. 17.6 Numerical solution (—) of a boundary layer problem, $\varepsilon = 0.002$. True solution (– –)

neither hyperbolic or parabolic), and again use the Fourier technique to determine their properties.

We take the 2D-equations with constant coefficients $u = a$, $v = b$ and $\varepsilon = 1/Re$:

$$\frac{\partial u}{\partial t} + a\frac{\partial u}{\partial x} + b\frac{\partial u}{\partial y} + \frac{\partial p}{\partial x} = \varepsilon\left(\frac{\partial^2 u}{\partial x^2} + \frac{\partial^2 u}{\partial y^2}\right),$$

$$\frac{\partial v}{\partial t} + a\frac{\partial v}{\partial x} + b\frac{\partial v}{\partial y} + \frac{\partial p}{\partial y} = \varepsilon\left(\frac{\partial^2 v}{\partial x^2} + \frac{\partial^2 v}{\partial y^2}\right),$$

$$\frac{\partial u}{\partial x} + \frac{\partial v}{\partial y} = 0.$$

By differentiating the first equation with respect to x and the second equation with respect to y and adding the two, almost all terms disappear due to the divergence condition. The remaining part is

$$\frac{\partial^2 p}{\partial x^2} + \frac{\partial^2 p}{\partial y^2} = 0,$$

and we replace the divergence condition with this equation. The resulting Fourier transformed system is

$$\frac{d\hat{u}}{dt} + aik_x\hat{u} + bik_y\hat{u} + ik_x\hat{p} = -\varepsilon(k_x^2 + k_y^2)\hat{u},$$

$$\frac{d\hat{v}}{dt} + aik_x\hat{v} + bik_y\hat{v} + ik_y\hat{p} = -\varepsilon(k_x^2 + k_y^2)\hat{v},$$

$$(k_x^2 + k_y^2)\hat{p} = 0.$$

For nonzero k_x or k_y we get $\hat{p} = 0$, and the first two equations become

$$\frac{d\hat{u}}{dt} = -\left(aik_x + bik_y + \varepsilon(k_x^2 + k_y^2)\right)\hat{u},$$

$$\frac{d\hat{v}}{dt} = -\left(aik_x + bik_y + \varepsilon(k_x^2 + k_y^2)\right)\hat{v}.$$

These two scalar equations both correspond to parabolic PDE, with the characteristic damping properties caused by the viscosity.

The original equations are nonlinear, but it turns out that the solutions behave as expected with a smoothing property as long as ε is not too small. This is a striking example, where a much simplified model and a Fourier analysis serve to understand some of the basic properties of a complicated PDE.

When it comes to boundary conditions, the most common type is specification of the velocity:

$$u = u_0(t), \qquad v = v_0(t).$$

In the special case of a solid wall, the viscosity makes the fluid stick to the wall, and we have $u = v = 0$. This is in contrast to the inviscid case, where the normal component vanishes, but the tangential component does not.

The case with outflow open boundaries is the most difficult one. Prescription of the velocity gives a well posed problem with a unique solution, but the problem is that it is very difficult to find the correct data, since it is part of the solution. With a rectangular domain with an outflow boundary $x = x_0$ at one end, it is reasonable to assume that there is no change in the x-direction, and we can use the conditions

$$\frac{\partial u}{\partial x} = 0, \qquad \frac{\partial v}{\partial x} = 0.$$

However, for a more realistic 3D-problem, the computational domain must be kept as small as possible, and these conditions may not be accurate enough. There are many different ways of designing "far field boundary conditions", or general "open boundary conditions". It is still a problem area where much research is going on, and we shall not go further into this here.

For the steady state problem, it can be shown that a unique solution requires that the pressure is prescribed at one point or, alternatively, that the average pressure is prescribed.

The form of the equations is such that an explicit method for time discretization cannot be used. There is no time derivative for the pressure, so it cannot be advanced in a standard way. With an implicit method advancing the solution from t_n to t_{n+1}, terms containing p^{n+1} will be involved. Together with the divergence condition, we get a fully coupled system of equations for all three variables at all grid points. But we need of course an effective method for solution of the system.

For finite element methods used for discretization in space, there is an elegant way of eliminating the difficulty with the missing time derivative. The divergence condition is eliminated before discretization in time by defining the subspace S_N (see Chap. 11) such that all functions in it are divergence free. Such elements have been constructed also for high order accuracy. The remaining differential equations for the velocity components are then solved in the usual Galerkin formulation.

There is another way of eliminating the divergence condition, and it was used for the simplified equations above. The first equation is differentiated with respect to x and the second one with respect to y, and the two are then added. Most of the terms cancel each other by using the divergence condition in its differentiated form, but not all. The final equation is an elliptic equation for the pressure, and the complete system is

$$\frac{\partial u}{\partial t} + u\frac{\partial u}{\partial x} + v\frac{\partial u}{\partial y} + \frac{\partial p}{\partial x} = \varepsilon\left(\frac{\partial^2 u}{\partial x^2} + \frac{\partial^2 u}{\partial y^2}\right),$$

$$\frac{\partial v}{\partial t} + u\frac{\partial v}{\partial x} + v\frac{\partial v}{\partial y} + \frac{\partial p}{\partial y} = \varepsilon\left(\frac{\partial^2 v}{\partial x^2} + \frac{\partial^2 v}{\partial y^2}\right),$$

$$\frac{\partial^2 p}{\partial x^2} + \frac{\partial^2 p}{\partial y^2} + \left(\frac{\partial u}{\partial x}\right)^2 + 2\frac{\partial v}{\partial x}\frac{\partial u}{\partial y} + \left(\frac{\partial v}{\partial y}\right)^2 = 0.$$

The idea is to use an iterative procedure, which is now more natural with the pressure occurring in the third equation. Assuming that all three variables are known at $t = t_n$, the first two equations are advanced one step, providing preliminary values $\tilde{u}^{n+1}$, $\tilde{v}^{n+1}$. These are now used in the third equation to compute new values p^{n+1}. For better accuracy, these iterations are repeated. Note that this type of iteration does not work with the original continuity equation for u and v. With the preliminary values $\tilde{u}^{n+1}$ and $\tilde{v}^{n+1}$, how should we get new values p^{n+1} from the continuity equation?

Since we have raised the order of the differential equation one step, another boundary condition must be specified, to be used for the pressure equation. This condition can be stated in terms of the normal derivative $\partial p/\partial n$, which can be derived from the first two equations since $\partial p/\partial x$ and $\partial p/\partial y$ both are present there.

In applications, one is often interested only in the steady state solution, i.e., with a given initial solution $\mathbf{u}(x, y, 0)$ we want to compute

$$\lim_{t\to\infty} \mathbf{u}(x, y, t) = \mathbf{u}_\infty(x, y).$$

One way of doing this is to use a consistent method for the time dependent equations, and then advance the solution in time until it settles down to a state where there is almost no change at all. There may be a solver available for the time dependent problem, and it is convenient to apply it. This may well be a possible way that works fine, but in general it is not very efficient. First of all, the time integration method may be designed for high accuracy, but we need good accuracy only in the (x, y)-space for the converged solution. This suggest that we can go down in accuracy, and optimize the time integration with respect to convergence rate as time increases. We can also go one step further and disregard the consistency with the correct time behaviour completely. After all, we care only about the final result, not about the way it is arrived at. The situation is illustrated symbolically in Fig. 17.7. It is assumed there that the discretization in space is one and the same, such that all methods produce the same solution. (In Sect. 16.3 we solved the steady state heat conduction problem without going via the time dependent equations.)

No matter what method we are using, the discretization in space leads to some kind of system of equations. In Chaps. 13 and 14 we described some of the most common solution methods for fast convergence.

In the next section we shall apply the Navier–Stokes equations to a few real, but still simple problems.

Fig. 17.7 Solution of steady state problems

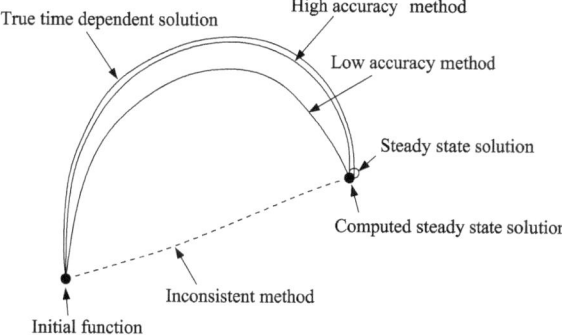

True time dependent solution

High accuracy method

Low accuracy method

Steady state solution

Computed steady state solution

Inconsistent method

Initial function

Fig. 17.8 A driven cavity

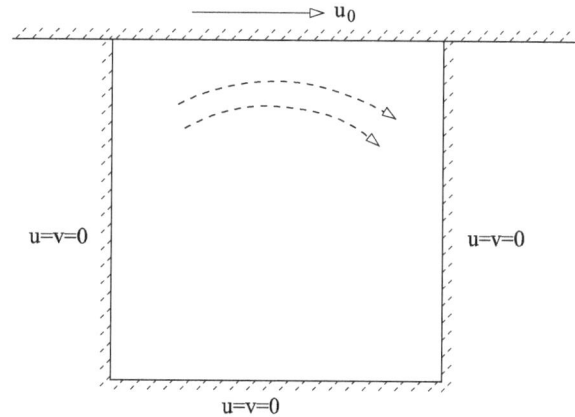

u_0

$u=v=0$

$u=v=0$

$u=v=0$

17.2 Low Speed Flow

As noted above, when it comes to liquids or gases with low particle speed, one can disregard the compressibility effects and still keep quite good accuracy. In this section we shall present two applications where the incompressible Navier–Stokes equations is a good model. The first one is the classic test problem called the *driven cavity* problem, see Fig. 17.8. In this problem we have a 1×1 m square box with four solid walls containing a fluid, with the upper wall moving with a constant velocity u_0.

The fluid is sticking to the wall, such that a swirling motion is taking place in the fluid. We are again using the COMSOL Multiphysics system for computing the steady state solution, and the equations are written in dimensional form:

$$\rho\left(u\frac{\partial u}{\partial x} + v\frac{\partial u}{\partial y}\right) + \frac{\partial p}{\partial x} = \mu\left(\frac{\partial^2 u}{\partial x^2} + \frac{\partial^2 u}{\partial y^2}\right),$$

$$\rho\left(u\frac{\partial v}{\partial x} + v\frac{\partial v}{\partial y}\right) + \frac{\partial p}{\partial y} = \mu\left(\frac{\partial^2 v}{\partial x^2} + \frac{\partial^2 v}{\partial y^2}\right).$$

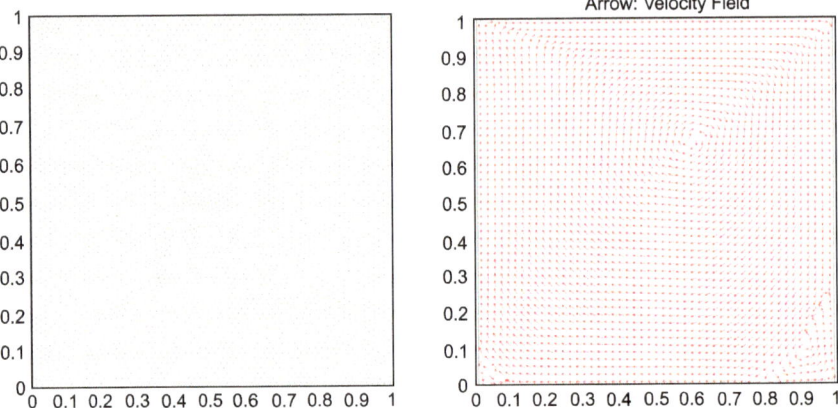

Fig. 17.9 Driven cavity, grid and velocity field

$$\frac{\partial u}{\partial x} + \frac{\partial u}{\partial y} = 0,$$

We are using the parameters $\rho = 1$ kg/m^3 and $\mu = 0.05$ Pa·s. The upper wall moves with $u_0 = 10$ m/s and, based on this normalization speed, the Reynolds number is $Re = 200$. Due to the moving upper wall, there is a discontinuity in the upper corners. However, due to the smoothing property of the equations, the solution becomes continuous away from the corner. In the previous section, it was mentioned that a given point value of the pressure is required for a unique solution, and we prescribe $p = 0$ at one corner.

The geometry of the computational domain suggests a rectangular grid, but we have local refinements in mind, and then a triangular grid is easier to handle. The result of the first computation and its grid is shown in Fig. 17.9. The arrows show the direction of the fluid particles, but not the magnitude of the velocity. The fluid is moving much faster in the upper part of the cavity.

The plot indicates that the velocity field has some special features in the lower corners. We refine the mesh there, and run the program again. The result is shown in Fig. 17.10, with a magnification of the velocity field in the lower left corner.

In order to see in even more detail what is going on very close to the lower corners, we make two more local refinements. The first one is done in an approximate square with sides 0.1, and the final one with sides 0.015. Figure 17.11 shows the result near the lower left corner.

The velocity field figure shows a magnification of a very small part in the left corner. We can see that there is actually a second recirculation area, with the same rotation direction as in the main part of the cavity. This feature would have been practically impossible to reveal by using a uniform grid in the whole domain.

We shall now turn to a second application concerning cross-country skiing. It is well known that a skier close behind the leading skier has an advantage, particu-

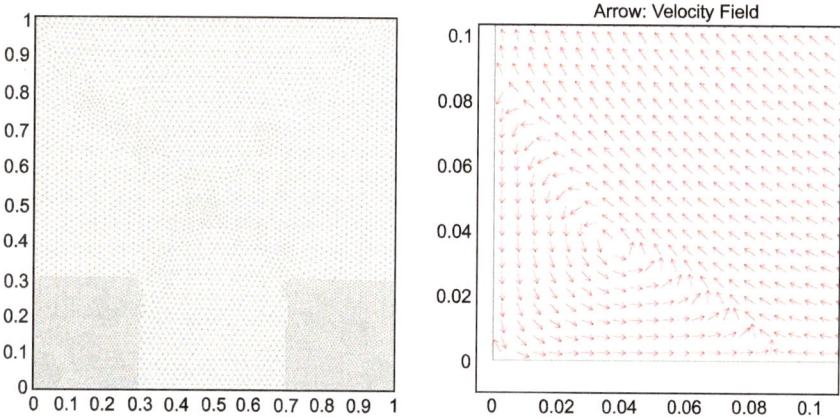

Fig. 17.10 Grid and velocity field with local grid refinement

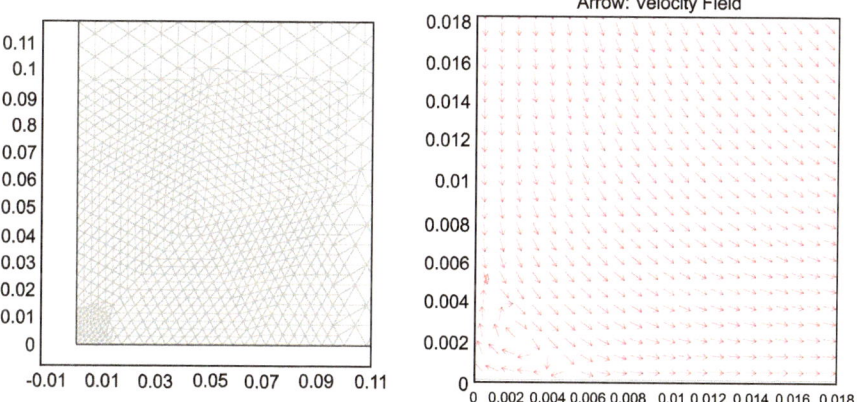

Fig. 17.11 Grid and velocity field in lower left corner with three local grid refinements

larly when the speed is high. The second skier is partially shielded from the wind resistance by the first skier, and we shall compute how large this effect is. We use a simple two-dimensional model with no vertical effects. The first skier is modeled by an ellipse (seen from above) with axis 60 cm along the direction of motion and 40 cm across. The whole computational domain is a 10×4 meter rectangle. We are using the parameters for air at $0°C$: $\rho = 1.292$ kg/m^3 and $\mu = 0.0178$ Pa·s. The constant speed $u_0 = 10$ m/s is prescribed at the forward inflow boundary as well as at the two side boundaries. At the backward outflow boundary we prescribe the pressure $p_0 = 0$. Figure 17.12 shows the grid and the result.

As expected there is a low speed area behind the skier. To see more exactly the strength of the effect, we plot the wind speed from the back of the skier and backwards. The result is shown in Fig. 17.13 for three different positions: right

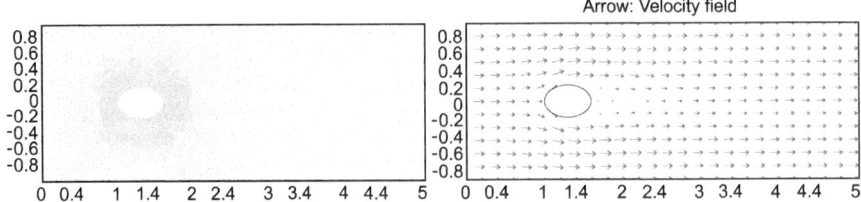

Fig. 17.12 Grid and velocity field for $u_0 = 10$ m/s

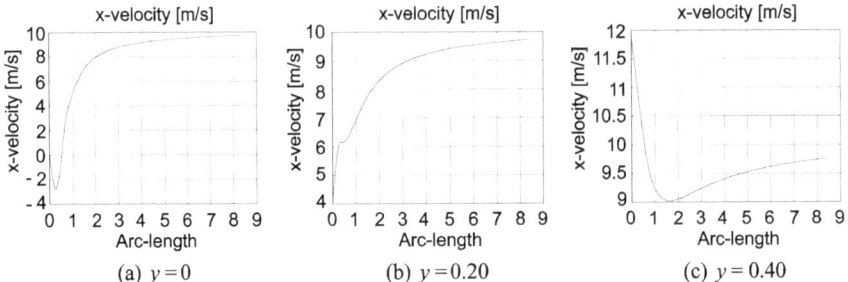

Fig. 17.13 Velocity behind the skier as a function of distance

behind the skier, 20 cm to the side, and 40 cm to the side ($x = 0$ corresponds to the back of the skier).

If the second skier stays in exactly the same track as the first one, there is actually not only a lee area, but he also gets an extra push from behind (negative velocity) if he stays very close! However, if he moves slightly to the side, the effect is much less as shown in the next two subfigures.

The model used here is very coarse, but it can easily be refined with a more detailed geometry and a three-dimensional model. The point we want to make here is that modern mathematical/numerical models can easily be used for estimation of certain effects that may be of great help for practical purposes. (An Olympic gold medal in 50 km cross country skiing may require some other ingredients as well...)

17.3 High Speed Flow and Aircraft Design

In Sect. 1.1 we used aircraft design as an example to illustrate the basics of computational mathematics. Here we shall follow up a little more on this particular problem.

The full Navier–Stokes equations is the correct model for all types of flows, since it takes the nonlinear compressibility effects into account as well as the viscous effects. For high speed flow there is no way of getting any reasonable results with an assumption of incompressible flow. However, for preliminary aircraft design, it makes sense to disregard the viscous effects, and we shall use the Euler equations just as in the introductory chapter.

Fig. 17.14 Transonic airfoil

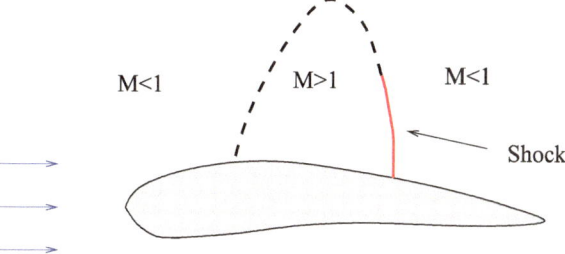

M<1   M>1   M<1

Shock

An aircraft has a complicated geometry, and even with the computers of today it is not possible to do full-scale computations with all details included. As a simplification we reduce the geometry to an airfoil as in the introductory chapter.

From an economic point of view, the flying properties of a passenger aircraft are particularly important at the cruising speed that is held most of the time for long distance flights, since it determines almost completely the fuel consumption. A fundamental quantity is the *speed of sound* c_s as introduced in Sect. 17.1.2. It depends on the various flow quantities, but in air at the cruising altitude of 11,000 meters it is approximately 295 m/sec, or 1062 km/h. If the speed of a moving object is c m/sec, the *Mach number* is defined as $M = c/c_s$. A subsonic passenger jet aircraft has a typical cruising speed around 850 km/h, which corresponds to $M = 0.8$. For our setup with a steady wind blowing from the left towards the fixed airfoil, the corresponding Mach number $M = M_f$ is called the *free stream Mach number*. The speed increases above the airfoil, and it goes *supersonic*, i.e., $M > 1$. Further to the right it has to go subsonic again, and we call this situation *transonic flow*. It can be shown theoretically that there has to be a sudden jump in the pressure where the flow goes from supersonic to subsonic. This discontinuity is a *shock* as described in Sect. 17.1.3. A typical situation is shown in Fig. 17.14. Unfortunately, the presence of the shock causes trouble, since there will be a sharp increase in the drag. This is the reason why manufacturers stay away from cruising speeds very close to the speed of sound, since the shock strength increases with the speed, causing a higher fuel consumption. (For supersonic aircraft, there is an even worse situation, since other even stronger shocks occur.) However, strong efforts are being made to go as high as possible in cruising speed, and this is why the design of the wings is such a crucial problem. Two different geometries may give quite different pressure distributions for the same cruising speed.

Referring to Fig. 1.1 in the introductory chapter, we ask ourselves the same question again: how can the airfoil be optimized? If the shock can be eliminated or be made weak, the flying conditions will improve considerably.

The Euler equations were given for the simplified two-dimensional steady state problem in Sect. 1.1. Here we present them again, but now with the time derivatives included:

$$\frac{\partial \rho}{\partial t} + \frac{\partial (\rho u)}{\partial x} + \frac{\partial (\rho v)}{\partial y} = 0,$$

$$\frac{\partial(\rho u)}{\partial t} + \frac{\partial(\rho u^2 + p)}{\partial x} + \frac{\partial(\rho u v)}{\partial y} = 0,$$

$$\frac{\partial(\rho v)}{\partial t} + \frac{\partial(\rho u v)}{\partial x} + \frac{\partial(\rho v^2 + p)}{\partial y} = 0,$$

$$\frac{\partial E}{\partial t} + \frac{\partial(u(E + p))}{\partial x} + \frac{\partial(v(E + p))}{\partial y} = 0, \quad p = p(\rho, u, v, E).$$

For a long time, the computational procedure was based on trial and error, i.e., the pressure distribution is computed for different geometries and the best one is chosen. The geometries are chosen by skilled engineers with good knowledge about aerodynamics and the influence of certain geometrical features. But the whole procedure is actually an optimization that can be formulated mathematically. For example, for a given required lifting force, the drag is defined as a function of the geometry. But there are infinitely many parameters which define the geometry of the wing and, as usual, it is necessary to discretize in such a way that there are only a finite number of parameters. There are different ways of achieving this discretization. A very direct way is to define a number of coordinates $\mathbf{x}_j$ on the surface, and then interpolate by piecewise polynomials. The optimization is then formulated as

$$\min_{\mathbf{X} \in D} f(\mathbf{X}),$$

where f represents the drag and $\mathbf{X}$ is a vector containing all the subvectors $\mathbf{x}$. But this vector must be limited to some predefined domain D corresponding to certain limitations on the wing shape. For example, the wing must be thick enough for fuel storage etc. There are also other constraints like the requirement on a minimal lift.

This formulation looks simple enough, but it hides a number of severe challenges. The function f is an abstract notation which couples the geometry to the drag. For a given geometry $\mathbf{X}$, the value of $f(\mathbf{X})$ is defined through the solution of the Euler equations. A large optimization problem is always solved by iteration, which for each step requires the computation of $f(\mathbf{X}^{(n)})$. We need an efficient Euler solver.

We have earlier described the three general numerical methods: finite difference, finite element and spectral methods. In fluid dynamics, one often uses still another method called the *finite volume method*, which was mentioned in Sect. 11.3. It is constructed for conservation laws, and we shall describe it briefly here for problems in 2D

$$\frac{\partial u}{\partial t} + \frac{\partial f(u)}{\partial x} + \frac{\partial g(u)}{\partial y} = 0,$$

where $f(u)$ and $g(u)$ are given functions of the unknown solution u. The computational domain is partitioned into quadrilaterals V_j, which are called finite volumes (referring to the 3D case) as shown in Fig. 17.15.

In 1D there is the simple integration formula

$$\int_a^b \frac{df}{dx}\big(u(x)\big)\, dx = f\big(u(b)\big) - f\big(u(a)\big),$$

Fig. 17.15 Finite volume
grid in 2-D

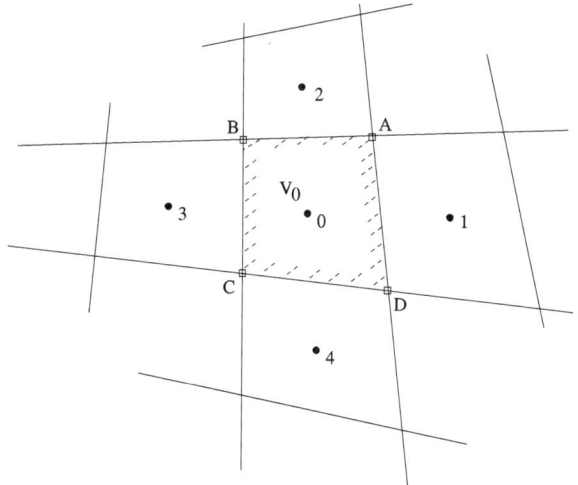

and in 2D it is generalized to the *Green's formula*

$$\int\int_{V_j} \left(\frac{\partial f(u)}{\partial x} + \frac{\partial g(u)}{\partial y} \right) dx\, dy = \int_{\partial V_j} f(u)\, dy - \int_{\partial V_j} g(u)\, dx,$$

where ∂V_j is the boundary of the volume V_j. The conservation law becomes

$$\int\int_{V_j} \frac{\partial u}{\partial t} dx\, dy + \int_{\partial V_j} f(u)\, dy - \int_{\partial V_j} g(u)\, dx = 0,$$

and this is the starting point for the finite volume method. We note that the label conservation law originates from this formula. If V_j is taken as the whole domain for the equation, and the solution is zero at the boundary, then

$$\frac{d}{dt} \int\int u\, dx\, dy = 0,$$

i.e., the integral is independent of time.

Finite difference methods use point values at the grid points, but here we work with the averages of the function over each volume:

$$u_j = \frac{1}{\Delta V_j} \int\int_{V_j} u\, dx dy.$$

The remaining problem is now to approximate the line integrals, which all consist of four parts. For example, referring to Fig. 17.15 and the quadrilateral V_0, we approximate the part between the corner points D and A by

$$\int_D^A f(u)\, dy \approx (y_A - y_D) \frac{f(u_0) + f(u_1)}{2},$$

where y_A denotes the y-coordinate at the point A etc. With the notation $f_0 = f(u_0)$ etc., the method becomes

$$2\Delta V_0 \frac{du_0}{dt} + (y_A - y_D)(f_0 + f_1) + (y_B - y_A)(f_0 + f_2)$$
$$+ (y_C - y_B)(f_0 + f_3) + (y_D - y_C)(f_0 + f_4)$$
$$- (x_A - x_D)(f_0 + f_1) - (x_B - x_A)(f_0 + f_2)$$
$$- (x_C - x_B)(f_0 + f_3) - (x_D - x_C)(f_0 + f_4) = 0.$$

When summing up over the whole computational domain, all terms cancel each other, and we get conservation in the discrete sense

$$\frac{d}{dt} \sum_j u_j \, \Delta V_j = 0$$

if the solution is zero at the outer boundary.

It remains to solve the ODE-system in time, and it is done by applying a difference method as described in Sect. 10.1.

We have described the finite volume method for quadrilaterals here. However, the method can be worked out also for triangular grids, where the line integrals along the sides are approximated by using the function values at neighboring triangles.

This is done in the example we present here. The results have kindly been provided by Olivier Amoignon, see [1], and they were presented already in the introductory chapter. It is hard to see the details of the triangular grid in Fig. 1.3, and a close up near the surface is shown in Fig. 17.16. In order to avoid trouble with the grid and the algorithm, the trailing edge cusp is represented with a small but nonzero cutoff.

The two cases presented in Sect. 1.1 are actually an illustration of the optimization algorithm described above. The dashed curve in Fig. 1.4 shows an initial airfoil shape with the corresponding pressure distribution shown in Fig. 1.5. The latter has a strong destructive shock in the transonic area. The optimization procedure is now applied with a finite volume Euler solver as the key ingredient. The result is the new airfoil shape represented by the solid curve in Fig. 1.4 with the new pressure distribution in Fig. 1.5. The shock is now eliminated, and the pressure distribution provides a weaker drag force.

Fig. 17.16 Close up of the computational grid near the trailing edge

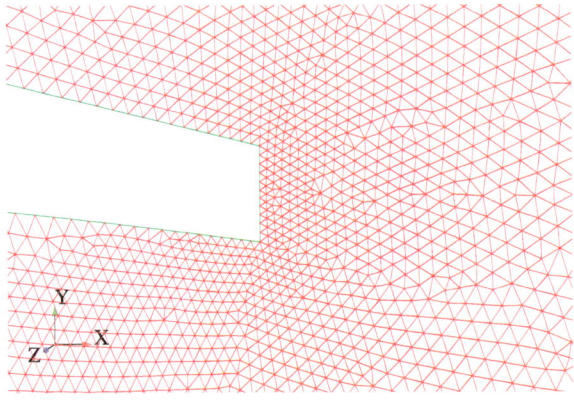

For the steady state problem we are considering here, the time derivatives in the Euler equations are zero, and for each step in the optimization algorithm there is a large system of algebraic equations to be solved as indicated in Sect. 1.1. However, one way of solving this system is to use the time dependent version, and then apply a solver for this type of equation. For each step in the optimization we have a computed solution from the previous computation. This solution is taken as the initial function for the time dependent problem, and a consistent time stepping method is applied. This technique has been applied in the computation presented here.

17.4   Weather Prediction

The use of mathematical models for weather prediction has a long history. After all, the weather is an effect of the atmospheric state, and if the pressure, density, air velocity, temperature and humidity are known, then we know what the weather is. Furthermore, if these state variables are known at a certain time, the basic thermodynamic differential equations determine the state at a later time. These equations were completely known already in the nineteenth century, and some of them have been discussed earlier in this chapter. However, somewhat surprisingly, it took quite a long time before this theoretical approach was accepted in the meteorological community. There was rather a belief that statistical methods should be used, based on weather data from the past. Since there is so much data about old weather patterns, one could search among them way back in time to find a pattern which is similar to the one of the last few days. By assuming that the development for the next few days will be the same as it was in the known case from the past, a prediction could be made.

As a consequence, when the Norwegian scientist Vilhelm Bjerknes suggested in 1904 that weather predictions should be based on the thermodynamical equations, a quite strong scepticism resulted. And even if one accepted the idea, there was the problem that the differential equations are impossible to solve by classic analytical means.

However, the British mathematician and meteorologist Lewis Richardson (1881–1953) believed in Bjerknes' ideas, and he also understood how the differential equations should be solved. He was probably the first one who made a serious effort to solve a set of PDE by using a finite difference method. The computations were enormous, even when limited to a few time steps, and his method of using human beings as the processors organized as a difference scheme, was not realistic. But his work contributed significantly to later fundamental work on PDE-solvers.

With the introduction of electronic computers, the situation changed dramatically, as for many other scientific and engineering areas. The militaries were in the forefront (as is often the case), and the first predictions with at least some degree of reliability were made in the early fifties. But still, after more than fifty years of development, the predictions sometimes fail, and not only in the small details. Why is this?

The thermodynamic physical laws are well known and accurate. The gas-dynamic equations tell it all when it comes to the behavior of the atmosphere. By adding an equation governing the humidity in the air, the physics is complete. The fundamental difficulty is the size of the domain. All parts of the whole atmosphere around the globe interact with each other, and we need initial data everywhere at a given time in order to start the computation. Even with all the weather satellites circling our earth today, the coverage is not fine enough. Furthermore, the advancement in time requires small step sizes in both space and time. We have seen in Sect. 17.3 about aerodynamic computations that, for example, the pressure may be varying on a centimeter scale, and it does so even if high speed aircraft are not involved. This is of course a completely unrealistic degree of resolution. Fortunately, realistic weather computations can be made with a much coarser grid. After all, the often talked about butterfly on the Tienanmen Square in Beijing does not effect the course of a low pressure area traveling across the North Sea from the British Isles to Scandinavia.

But even so, there are certain effects that are important for the dynamic process, and which cannot be represented on the affordable grid resolution. Before discussing the degree of possible resolution, we shall briefly describe the numerical methods.

We take Sweden as an example. A typical forecast is based on a two stage procedure. The first stage is a simulation with the whole globe and its atmosphere as the computational domain. Many institutes use a finite difference method applied to all coordinate directions, others use a combination of a pseudo-spectral and finite difference method. In Chap. 12 we described the principles for the pseudo-spectral method with basis functions that are associated with a grid. The basis functions should be well adapted to the geometry and the solution. In order to describe the atmospheric layer, spherical coordinates are the natural ones. They are the longitude ϕ, the latitude θ and the radius r with the limits

$$0 \le \phi \le 2\pi, \qquad 0 \le \theta \le \pi, \qquad r_0 \le r \le r_1.$$

The interval $[0, \pi]$ for θ is the common one in meteorology, and $\theta = 0$ corresponds to the north pole. The variation in the angular directions is described by the *spherical harmonics* $Y_m^n(\phi, \theta)$, which are closely connected to trigonometric functions and Legendre polynomials. (They are eigensolutions to the Laplace operator in angular coordinates.) For the vertical representation it is inconvenient to use the radial coordinate r, since the surface of the earth plays such an important role. If $r = r_0(\phi, \theta)$ represents the surface, it is better to define this one as $z = 0$, where z is the new coordinate representing the height. Any physical variable, for example the temperature T, is now expressed as

$$T(\phi, \theta, z, t) = \sum_{m=1}^{M} \sum_{n=1}^{N} \hat{T}_{mn}(z, t) Y_{mn}(\phi, \theta).$$

The angular field is now discretized such that $\phi = \phi_j$ and $\theta = \theta_k$. If for certain z and t the temperature is known at the grid points, the derivatives with respect to ϕ and θ are computed by the pseudo-spectral technique. The total number of grid points is MN, but there is a complication here. The longitudal lines (also for

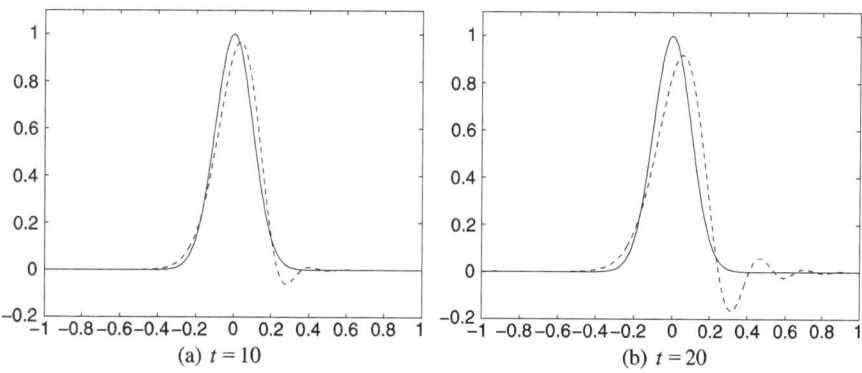

Fig. 17.17 Solution of $\partial u / \partial t = \partial u / \partial x$, exact (—) and numerical (−−)

Fig. 17.18 Global pressure and temperature

difference methods) are converging towards the two poles, which means that the grid points become too close to each other in the northern and southern parts. The remedy is to take out points in those areas, even if it complicates the algorithm a little. The advantage is that for a given number MN one can afford more points in other areas. The final result is that typical grids used today have a step size of the order 25 kilometers.

This is too coarse to catch all features that effect the weather. The character of the Earth's surface is of fundamental importance, and it varies at a smaller scale. For example, there are many lakes that don't show up at all, and there is a significant

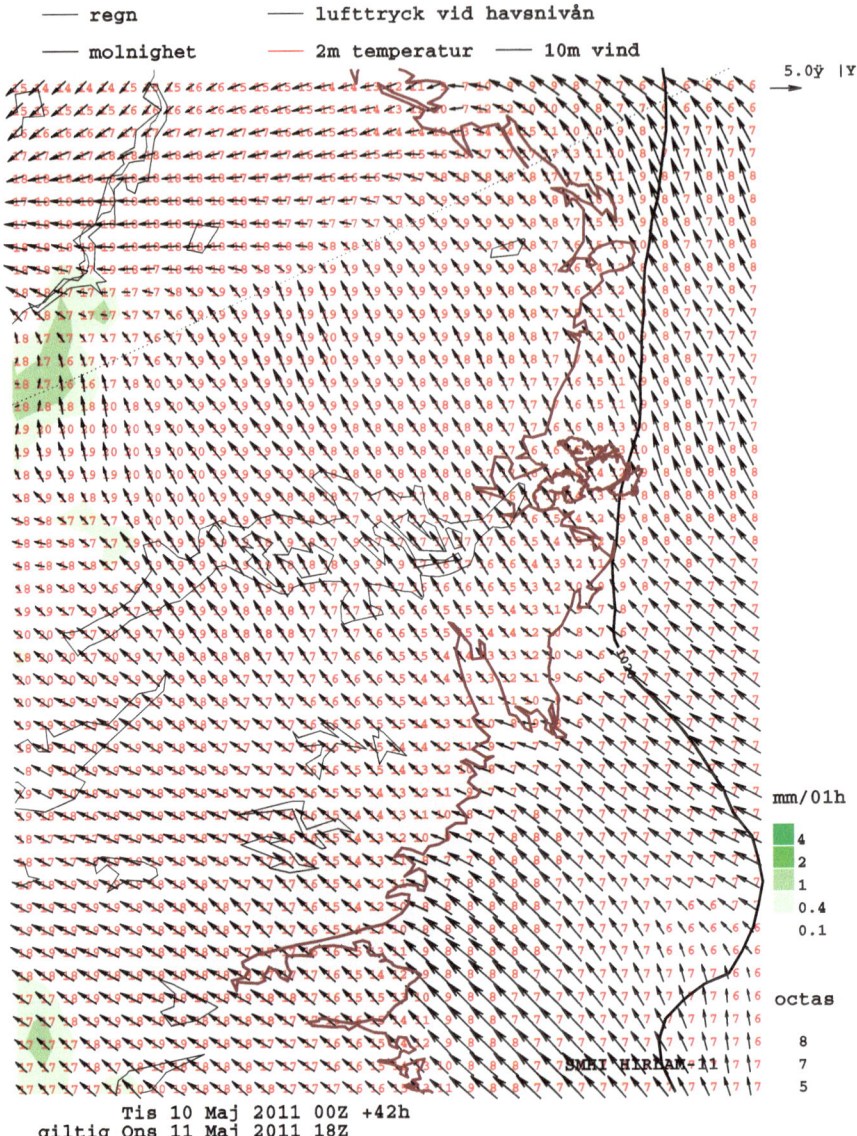

Fig. 17.19 Local pressure, temperature and wind

difference when it comes to heat exchange between the air and a water surface on the one side, and between air and a grass field on the other side. Another feature with a large effect is the presence of clouds, and these may be much smaller than 25 kilometers in diameter.

Fortunately, the lack of resolution doesn't make the situation hopeless. The remedy is to introduce some kind of parameterization that accounts for the desired effects. The effect of the smaller lakes on the heat exchange can be estimated on a larger scale. Assume that the heat exchange coefficient is a_s for a certain type of solid surface, and a_w for a water surface. If the combined water surface area is $w\%$ of a certain (not too large) subdomain, a simple and straightforward method is to adjust the heat exchange coefficient by linear interpolation:

$$a_{sw} = \left(1 - \frac{w}{100}\right)a_s + \frac{w}{100}a_w.$$

In weather prediction, as well as in many other applications, one would like to carry out the simulation over long time intervals. But even with the very best methods, there is always a limit where the result is too inaccurate for any meaningful practical purpose. A high pressure area that moves along a certain path, reaching point A after two weeks, say, will be distorted and dislocated by the numerical model. A two week prediction for point A may well show a low pressure area instead.

As a simplified example, we consider the equation

$$\frac{\partial u}{\partial t} = \frac{\partial u}{\partial x}, \quad -1 \le x \le 1,$$

with 2-periodic solutions and an initial pulse at $x = 0$ that moves to the left with speed one. When it leaves the left boundary, it enters at the right boundary again such that $u(x, t) = u(x, t + 2)$ for any x and t. One complete lap bringing the pulse back to its original position $x = 0$ takes 2 time units, which is called one period. After each one of n periods, where n is any integer, the pulse passes $x = 0$.

Figure 17.17 shows the result of a simulation over 10 periods. The solid curve centered at $x = 0$ represents the original pulse as well as the correct solution after any number of periods n. The dashed curve is the numerical solution. The figure shows how the error increases with time. If we think of positive values of u as the pressure above the normal mean value, we see that the peak is located too far to the right. Furthermore, a low pressure area centered at $x \approx 0.25$ has developed at $t = 10$, and it has become even stronger and shifted further to the right at $t = 20$. Obviously, a prediction based on these data has very poor quality. The longest predictions made today that are considered to have reasonable accuracy, are of the order 5–10 days.

Figures 17.18 and 17.19 have kindly been provided by Per Kållberg at the Swedish Meteorological and Hydrological Institute (SMHI), Norrköping, Sweden. The first one shows the pressure, temperature and percipitation distribution obtained from a 4-day global simulation based on difference methods in all directions. The second one shows a 36-hour local forecast in the Stockholm area where, in addition to pressure and temperature, the wind strength and direction are also shown.

Chapter 18
Computers and Programming

If a certain numerical method has been developed to solve a given problem, a computer is needed to do the work as prescribed by the algorithm. But this requires that the algorithm is transferred into a *program* that can be understood by the computer. In this chapter we shall briefly describe the principles for this programming process.

18.1 Basic Principles for Programming

The central part of a computer is a processor that among other things can do elementary algebraic operations on numbers in binary form, i.e., numbers that are represented as a sequence of zeros and ones. The number 19, which is interpreted as $1 \cdot 10 + 9 \cdot 1$ in the familiar decimal system, is given the representation

$$19 = 1 \cdot 16 + 0 \cdot 8 + 0 \cdot 4 + 1 \cdot 2 + 1 = 1 \cdot 2^4 + 0 \cdot 2^3 + 0 \cdot 2^2 + 1 \cdot 2^1 + 1 \cdot 2^0$$

in the binary system. The coefficients of the powers of 2 result in the number 10011_2, and this sequence is stored in the computer. (The 64-th birthday is really something to celebrate in a binary world, since the number looks like a million: 1000000_2.) The binary form is natural, since the electronic components are made such that they are either *on* for a one, or *off* for a zero.

Noninteger numbers can of course also be represented in binary form. For example,

$$10011.101_2 = 19 + 1 \cdot 2^{-1} + 0 \cdot 2^{-2} + 1 \cdot 2^{-3} = 19.625$$

The precision of computations is limited by the accuracy of the representation of numbers in the computer. How many binary positions (bits) should be used for each number? Most computers and operative systems allow for 32- or 64-bit arithmetic, which is referred to as *single precision* and *double precision*. The numbers are represented in *floating point form* $a2^b$, where the *mantissa* a and the exponent b are stored in binary form with a normalized such that it always is in a certain interval,

B. Gustafsson, *Fundamentals of Scientific Computing*,
Texts in Computational Science and Engineering 8,
DOI 10.1007/978-3-642-19495-5_18, © Springer-Verlag Berlin Heidelberg 2011

usually $1 \leq a < 2$. For single precision the mantissa occupies 24 bits, such that the number above is stored as

$$19.625 = 1.0011101000000000000000000_2 \cdot 2^{00010000_2}.$$

Double precision allows for a 53-bit mantissa.

The representation described here is the common standard, but of course any manufacturer may use its own type of representation. For example, the base 16 is sometimes used in the number representation, such that it takes the form $c16^d$.

It is a long way to go from an algorithm arising from a discretized version of a differential equation to something that can be understood by the computer. It needs a sequence of instructions that tells it step by step how to carry out the arithmetic. Such a set of instructions is called a *computer program* and, since the computer can understand only zeros and ones, these instructions must be given in binary form as well. At the beginning of the electronic computer era, mathematicians were engaged in developing programs. They had to write down the instructions in binary form, which was an extraordinary effort. As we all know, the development of hardware for computers has been extremely fast, but so has the development of software. Programming systems were constructed such that the instructions began looking more like ordinary mathematical and logical symbols. The FORTRAN language was developed already in the fifties, and it was followed by several versions that became more general for each stage. The name is a short form for FORmula TRANslation. For a long period of time it was the dominating programming language for problems in scientific computing, and after 60 years it is still used for many applications. More recent programming languages are C, C++ and Java.

A high level language requires a *compiler*, which is a program by itself, that is able to convert the high level instructions to a binary code. For example, the statement $a = b + c$ is broken down by the compiler to a set of binary instructions that tells the computer to

(1) take the number in a certain memory cell called b into the algebraic unit
(2) take the number in a certain memory cell called c into the algebraic unit
(3) add the two numbers
(4) store the result in a certain cell called a

It is easy to construct electronic circuits for the basic arithmetic operations addition, subtraction, multiplication and division. However, elementary functions such as $\sqrt{x}$ are much harder. Even if there are tables for evaluation, it is impossible to include these in the computer's memory for every possible positive number x. An algorithm is needed to compute it for a given value of x. In Sect. 13.1 the Newton-Raphson method for the solution of algebraic equations was described. This method can be used for finding the square root s. The equation is

$$s^2 - x = 0,$$

and the iterative method becomes

$$s_{j+1} = s_j - \frac{s_j^2 - x}{2s_j} = \frac{1}{2}\left(s_j + \frac{x}{s_j}\right), \quad j = 0, 1, \ldots.$$

(The case where s_j in the denominator approaches zero has to be treated separately.) This algorithm will give the answer very quickly if a good initial guess s_0 is available. This is a problem by itself, and there are several different strategies for determining it. One method is to count the number of digits in the number x represented by its binary form. As an example, consider the computation of $\sqrt{145}$. We have

$$145 = 10010001_2,$$

i.e., there are $n = 8$ digits. The initial value is chosen as $s_0 = 2^{n/2} = 2^4 = 16$. The algorithm above produces the values

$$16.00000000$$
$$12.51250000$$
$$12.05116116$$
$$12.04159838$$
$$12.04159458$$
$$12.04159458$$

Already after three iterations we have five correct decimals. This is what is going on also in the hand calculator when pressing the $\sqrt{}$ key.

This example illustrates that any programming language dealing with mathematics, has to have access to a large number of algorithms stored for the evaluation of elementary functions. Such algorithms are part of the software, and are included in most hand calculators of today.

18.2 MATLAB

Today, there is an abundance of different high level programming languages that are designed for different types of applications. One of the most well known languages for scientific computing is MATLAB. It is different from most other high level languages in the sense that in its basic form no binary program is created by a compiler. Each command is interpreted and carried out by itself in exactly the order in which they are written.

This is not the place to give a tutorial on MATLAB, but we shall present an example for illustration of the structure. The example shows a program for solving the heat equation in its simplest form

$$\frac{\partial u}{\partial t} = \frac{\partial^2 u}{\partial x^2}, \quad 0 \le x \le 1, \ t \ge 0,$$
$$u(0, t) = 0,$$
$$u(1, t) = 0,$$
$$u(x, 0) = 1 - 2|x - 0.5|,$$

by the Euler method:

$$u_j^{n+1} = u_j^n + \lambda(u_{j-1}^n - 2u_j^n + u_{j+1}^n), \quad j = 1, 2, \ldots, N,$$
$$u_0^{n+1} = 0,$$
$$u_{N+1}^{n+1} = 0,$$
$$u_j^0 = 1 - 2|x_j - 0.5|, \quad j = 1, 2, \ldots, N.$$

Here $\Delta x = 1/(N+1)$ and $\lambda = \Delta t/\Delta x^2$. We present first a version with straight-forward programming using so called for-loops. A %-sign indicates that the line is a comment used by the programmer for his own benefit, and it is ignored by the computer. The step size in space is $1/20$, and tf is the final time level. The program is:

```
clear

% Initialization of parameters

tf = 0.1;
N = 19;
lambda = 0.4;
dx = 1/(N + 1);
dt = lambda * dx^2;

% Definition of initial function

for j = 1 : N
    x(j) = j * dx;
    u0(j) = 1 - 2 * abs(x(j) - 0.5);
end
t = 0;
un = u0;

% Advancing the solution from t = 0 to t = tf

while t < tf
    for j = 1 : N
        un1(j) = (1 - 2 * lambda) * un(j);
        if j > 1, un1(j) = un1(j) + lambda * un(j - 1); end;
        if j < N, un1(j) = un1(j) + lambda * un(j + 1); end;
    end
    un = un1;
    t = t + dt;
end

% Adding the end points for the plot

xx = [0 x 1];
u = [0 un 0];
```

$ui = [0 \ u0 \ 0];$

$plot(xx, u, xx, ui,' \ --')$

% *Scaling of the axes*

$axis([0 \ 1 \ 0 \ 1])$

It should not be very difficult to understand the code for somebody who is familiar with some kind of programming language. We make a few comments for those who have never encountered any kind of programming.

A statement $a = expression$ means that the expression on the right hand side is evaluated and the value is then transferred to the variable on the left. This means that it makes sense to write $a = 2 * a$. The value in the variable a is multiplied by two, and the new value is then stored in the same variable a.

The leading *clear*-statement makes all variables and the plot figure empty. If the program has been run before in the same session, the variables are otherwise left with the old values.

The first *for*-loop in the program defines the N vector elements in **x** and the initial vector $\mathbf{u}^0$. The *while* segment is carried out from the beginning to the *end*-statement as long as $t < tf$. It requires that the parameter t is updated inside the segment, otherwise there will be an infinite loop that never stops. In the second *for*-loop it is necessary to separate out the end values of the index j, since otherwise u_{j-1}^n and u_{j+1}^n will be undefined.

The two *if*-statements mean that the instruction before the *end*-statement will be carried out only if the relation in the beginning is true. For the plot at the end of the program we add an extra point at each end of the interval and the corresponding boundary values for u. The parameters in the *axis*-statement specify the end points $[x_b \ x_e \ y_b \ y_e]$ of the x-axis and the y-axis in the plot.

MATLAB was born as an effective system for performing various algebraic operations in matrix algebra. In fact, the name is an acronym for MATrix LABoratory. The current much expanded MATLAB system still has its strength in the fast execution of matrix algebra, and therefore one should try to express as much as possible of the algorithm in terms of vectors and matrices. The central part of our algorithm is the computation of the solution $\mathbf{u}^{n+1}$ in terms of $\mathbf{u}^n$, and this can be formulated as a matrix-vector multiplication. With

$$\mathbf{u^n} = \begin{bmatrix} u_1^n \\ u_2^n \\ \vdots \\ \vdots \\ u_{N-1}^n \end{bmatrix}, \quad A = \begin{bmatrix} 1-2\lambda & \lambda & & & & \\ \lambda & 1-2\lambda & \lambda & & & \\ & \lambda & 1-2\lambda & \lambda & & \\ & & \ddots & \ddots & \ddots & \\ & & & \lambda & 1-2\lambda & \lambda \\ & & & & \lambda & 1-2\lambda \end{bmatrix},$$

the formula is

$$\mathbf{u}^{n+1} = A\mathbf{u}^n.$$

MATLAB has many ways of defining vectors and matrices. The simple direct way is to write them explicitly as in the example $A = [1\ 2\ 4;\ 5\ 3\ 1]$ which means the matrix

$$A = \begin{bmatrix} 1 & 2 & 4 \\ 5 & 3 & 1 \end{bmatrix}.$$

For problems like ours this method doesn't work, since the size of the matrix is governed by a parameter N, which furthermore may be very large. In our case we have a band matrix, and it is convenient to use the MATLAB expression $ones(m, n)$, which means a matrix with m rows and n columns containing only ones. The expression $diag(x, k)$ means a matrix with the vector x in the k-th diagonal, where $k = 0$ for the main diagonal, $k = -1$ for the first subdiagonal, $k = 1$ for the first superdiagonal etc. (If k is omitted, it is assumed to be zero.) All the remaining elements in the matrix are zero. If x has N elements, then A is a square $(N + |k|) \times (N + |k|)$ matrix. Accordingly, the statement

$$A = a * diag(ones(N - 1, 1), -1) + b * diag(ones(N, 1))$$
$$+ c * diag(ones(N - 1, 1), 1)$$

defines the matrix

$$A = \begin{bmatrix} b & c & & & & \\ a & b & c & & & \\ & a & b & c & & \\ & & \ddots & \ddots & \ddots & \\ & & & a & b & c \\ & & & & a & b \end{bmatrix}.$$

We use this method of defining A, and get the following modified version of the program above:

```
clear

% Initialization of parameters

tf = 0.1;
N = 19;
lambda = 0.4;
dx = 1/(N + 1);
dt = lambda * dx^2;
x = dx * (1 : N)';

% Definition of initial function

u0 = ones(N, 1) - 2 * abs(x - 0.5 * ones(N, 1));

% Building the coefficient matrix

A = lambda * (diag(ones(N - 1, 1), -1) + diag(ones(N - 1, 1), 1))
    + (1 - 2 * lambda) * diag(ones(N, 1));
```

```
t = 0;
u = u0;
```

% Advancing the solution from t = 0 to t = tf

```
while t < tf
    u = A * u;
    t = t + dt;
end
```

% Adding the end points for the plot

```
xx = [0 x' 1];
u = [0 u' 0];
ui = [0 u0' 0];
plot(xx, u, xx, ui,' −−')
```

% Scaling of the axes

```
axis([0 1 0 1])
```

––––––––

Here we have also used a simpler way of defining the vector **x**. Note that the vector $(1 : N)$ is a row vector, and we define x as a column vector by writing $(1 : N)'$. Also, in the definition of the initial vector $\mathbf{u}^0$ we use the *abs*-function applied to a vector. In MATLAB this means that the function is defined for each vector element by itself. In the plot statements we work with row vectors, which requires the extra transpose sign $'$ on x and u.

There is still another measure to be taken in order to get a more effective computation. Most of the elements in the matrix A are zero. MATLAB has the ability to store only the nonzero elements of the matrix so that all unnecessary multiplications by zero can be avoided. The command

$$B = sparse(A)$$

stores only the nonzero elements of the matrix in B. The final optimized form of the program is (with the comments taken out):

––––––––

```
clear
tf = 0.1;
N = 19;
lambda = 0.4;
dx = 1/(N + 1);
dt = lambda * dx^2;
x = dx * (1 : N)';
u = ones(N, 1) − 2 * abs(x − 0.5 * ones(N, 1));
plot([0 x' 1], [0 u' 0],' −−')
axis([0 1 0 1])
hold on
```

$B = sparse(lambda * (diag(ones(N - 1, 1), -1) + (diag(ones(N - 1, 1), 1)))$
   $+ (1 - 2 * lambda) * (diag(ones(N, 1), 0)));$
$t = 0;$
$while\ t < tf$
   $u = B * u;$
   $t = t + dt;$
end
$plot([0\ x'\ 1], [0\ u'\ 0])$
$hold\ off$

Here we have saved much storage compared to the previous version. The special initial vector $u0$ is also eliminated, and there is only a single vector u involved in the computation. This requires that the initial function is plotted right after it is defined. The *hold on* statement causes the plot to be frozen to wait for the next plot statement adding the final solution to the figure.

MATLAB has a number of elementary functions. Some of them are listed in Table 18.1.

The argument x in the parentheses can actually be a vector or a matrix. In that case the result is a vector or matrix as well, with the function applied to each element by itself. For example

$$\sin([0\quad pi/2; pi \quad 3*pi/2]) \quad \text{results in the matrix} \begin{bmatrix} 0 & 1 \\ 0 & -1 \end{bmatrix}.$$

Note that pi is a predefined constant with the value π; another one is the imaginary constant i denoted by i.

MATLAB has a large number of special purpose functions, some of them involving advanced numerical algorithms for the type of problems discussed in this book.

Table 18.1 Elementary functions in MATLAB

Function	MATLAB
$\|x\|$	abs(x)
x^a	x^a
e^x	exp(x)
$\ln x$	log(x)
$\log_{10} x$	log10(x)
$\log_2 x$	log2(x)
$\sin x$	sin(x)
$\cos x$	cos(x)
$\tan x$	tan(x)
$\cot x$	cot(x)
$\arcsin x$	asin(x)
$\arccos x$	acos(x)
$\arctan x$	atan(x)
$n!$	factorial(n)

A few examples:

- $y = polyval(a, x)$ evaluates the polynomial at the point x and stores it in y. Here and in the following, a is a vector containing the coefficients of the polynomial $a_n x^n + a_{n-1} x^{n-1} + \cdots + a_0$ in descending order.
- $b = polyder(a)$ computes the coefficients b_{n-1}, b_{n-2}, ..., b_0 of the derivative of the polynomial defined by the coefficients in a.
- $x = roots(a)$ computes all roots of the equation $a_n x^n + a_{n-1} x^{n-1} + \cdots + a_0 = 0$, and stores the result in the vector x (which in general is complex).
- $inv(A)$ computes A^{-1}. Note that the statement $x = inv(A) * b$ is not a good way of solving $A\mathbf{x} = \mathbf{b}$ since the explicit form of A^{-1} stored in $inv(A)$ is not necessary in this case. One should rather use $x = A\backslash b$ that uses Gauss elimination as described in Sect. 14.2.
- $d = eig(A)$ computes all the eigenvalues of A and stores them in the vector d (which in general is complex). The statement $[V, D] = eig(A)$ stores the eigenvalues in the diagonal of the matrix D, and the eigenvectors as column vectors in the matrix V.
- $q = quad(@ f, a, b)$ computes an approximation of the integral $\int_a^b f(x)\, dx$. Here $@f$ is a "handle" which refers to the MATLAB function f defining $f(x)$ for any given x. We shall use the notation $@f$ in this sense also in the following. (f can also be a standard MATLAB function such as sin.)
- $x = fminbnd(@ f, a, b)$ computes the point x giving the minimum value of $f(x)$ in $[a, b]$.
- $b = polyfit(x, y, n)$ computes the coefficients b_j of a least square fit $b_n x^n + b_{n-1} x^{n-1} + \cdots + b_0$ to the data points (x_j, y_j) stored in the vectors x, y.
- $c = fft(f)$ computes the discrete Fourier transform of the vector f. In Sect. 6.2 the transform was described for a discrete function defined on a regular grid in the interval $[0, 2\pi)$. However, the MATLAB function is defined for any sequence $\{f_j\}_1^N$ and any N as

$$c_k = \sum_{j=1}^N f_j e^{-2\pi i (j-1)(k-1)/N}.$$

Note that the normalization is different from the one in (6.9), and that the number of terms is N instead of $N + 1$.
- $f = ifft(c)$ computes the inverse discrete Fourier transform for any sequence stored in c. The definition is

$$f_j = \frac{1}{N} \sum_{k=1}^N c_k e^{2\pi i (j-1)(k-1)/N},$$

such that $ifft(fft(f))$ returns the original vector f.
- There are many solvers for the ODE-system $d\mathbf{u}/dt = \mathbf{f}(t, \mathbf{u})$. The most commonly used one is $ode45(@ f, tspan, u0)$, where $@f$ is the *handle* to a MATLAB function f that evaluates $\mathbf{f}(t, \mathbf{u})$ for any given t and $\mathbf{u}$. The statement $[T, u] = ode45(@ f, [t0\ t1], u0)$ returns the time grid points T (chosen by the solver) in the vector T, and the matrix u with the rows containing the solution

for each time step. Another solver with the same structure is *ode23*, which uses a Runge–Kutta method of order 2 and 3.

- fplot(@f,[a b]) plots the function $f(x)$ for $a \le x \le b$. For example, the statement fplot(@sin,[0 2*pi]) produces the figure

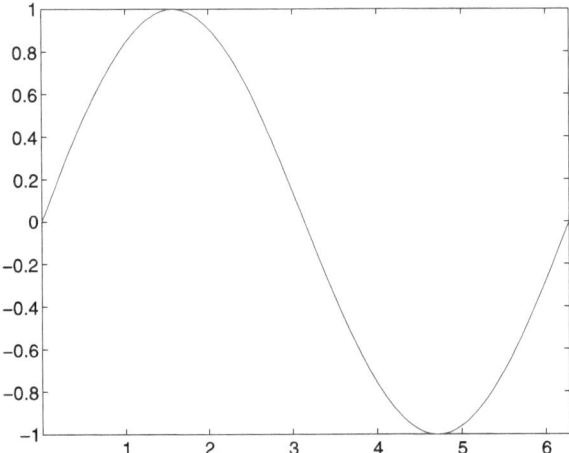

18.3 Parallel Computers

Faster computers are being developed at an enormous speed. A laptop of today has a capacity comparable to the fastest computers available 10 years ago. And the supercomputers, i.e., the fastest ones available at any given time, are breaking new barriers. The computing speed is usually measured in Flops, which means one floating point operation per second. "Floating point" refers to the type of number representation that was mentioned in Sect. 18.1, and "operation" stands for arithmetic operation, like addition. If not specifically mentioned, it is assumed that 64-bit arithmetic is used.

The manufacturers usually announce a theoretical peak performance, which means the number of Flops that can be achieved if the algorithm is such that the program allows for a sufficiently smooth data flow. This theoretical speed is never reached for any practical program. Instead there has been agreement on certain test programs that are used for measuring the speed. (This is analogous to the test cycles for verifying the fuel consumption for cars.) For performance classification, one usually uses a standard program for solving a large system of equations and, based on this test, there is a Top500 list for the 500 fastest computers in the world. In November 2009, the top position was taken by the Cray XT5 Jaguar computer at Oak Ridge National Laboratory, when it was logged at 1.759 PFlops. This means that $1.759 \cdot 10^{15}$ floating point arithmetic operations were performed each second. (An even faster Chinese machine was later installed at the National Supercomputing Center in Tianjin.)

How can such an enormous speed be achieved? Each processor or CPU (Central Processing Unit) is fast, but the main reason for the high speed is that the computer contains a large number of processors. Furthermore, each CPU, which is a physically integrated circuit, may have several *cores*, each one being an independent processor but integrated on the same chip. Each core is fed data by separate threads. The Cray XT5 Jaguar has 224162 cores, and it is what is called a *massively parallel computer*. This type of architecture with a large number of cores has taken over the supercomputer market completely today.

With the introduction of transistors in the fifties, computers became energy efficient and didn't require much power. However, with modern supercomputers the situation has changed once again as a consequence of the high number of processors. The CrayXT5 Jaguar requires 7 MW of power, which is quite high. It is equivalent to almost 10,000 horsepowers, which is what it takes to run a fully loaded 30,000 ton cargo ship at 15 knots! Furthermore, on top of the 7 MW, the computer needs cooling, which requires another 2 MW.

This construction trend towards parallel computers was not clear at the end of the previous century, when there was another competing architecture with few, but very advanced single *vector processors*. In that case the arithmetic is done in a pipeline, much as in an automobile factory. Each operation is broken down into smaller pieces such that, when a certain operation is completed on a certain set of numbers, many more numbers are inside the pipeline, but at earlier stages of the operation. It turned out that later versions of vector processors became very complicated and very expensive to make, and the market for that type of general purpose machine all but disappeared at the beginning of this century.

In order to use a parallel computer effectively, all the processors must be busy working all the time. But this may be impossible depending on the type of algorithm we are using. For example, the simple iterative method

$$u_{j+1} = a_j u_j, \quad j = 0, 1, \ldots$$

with u_0 given, cannot be parallelized. The computation of each new value u_{j+1} requires the previous value u_j in the sequence. Consequently, there is no way to partition the computation between the processors. On the other hand, in practice the algorithm is a large and complicated one, and there is at least some possibility to parallelize parts of it. The important feature of an effective algorithm is that the most time consuming part has a structure such that it can be parallelized. For example, if the central part is a matrix/vector multiplication $A\mathbf{x}$, we are in a good position. If $\mathbf{a}_j$ are the row vectors of the matrix A, we have

$$x_k = \mathbf{a}_k \mathbf{x} = \sum_{j=1}^{N} a_{kj} x_j, \quad k = 1, 2, \ldots, N.$$

If there are p processors, we let them first compute the first p elements x_j, then the next p elements, and so on. In this way we keep all processors busy all the time. The number of processors is often a power of two, and therefore one tries to choose the resolution of the problem such that $N = 2^m$, where m is an integer. In that way no

Fig. 18.1 Parallel computer

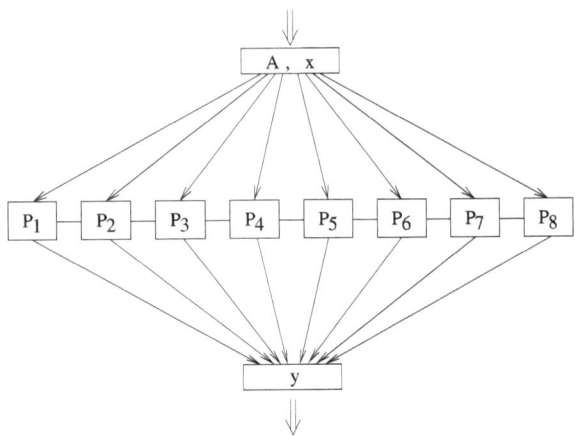

processor gets a chance to rest at any time. Figure 18.1 shows the parallel computer principle for the computation $\mathbf{y} = A\mathbf{x}$.

The development of parallel architectures has had a significant impact on the choice of numerical methods. Many algorithms that were considered as outdated on scalar computers were becoming competitive again. Take the Jacobi iterative method for solving systems of equations as an example. In Sect. 14.3 we described it as a particular version of the general formula

$$\mathbf{x}^{(n+1)} = M\mathbf{x}^{(n)} + \mathbf{f}, \quad n = 0, 1, \ldots,$$

where M is a given matrix. Obviously the matrix/vector multiplication is the core of the algorithm, and we have just shown how to parallelize it. In the scientific community such a problem is called *embarrassingly parallelizable*, since there is no challenge for the researchers.

When using the Gauss–Seidel iteration formula (14.9) we run into trouble when implementing the algorithm on a parallel computer. The computation of each new element $x_j^{(n+1)}$ must wait for all the previous elements to be computed. There are ways to partially overcome the difficulty, but an iteration can never be as fast as the one for the Jacobi method. The improvement of the convergence rate must be big enough to overtake the extra parallelization complication.

Most modern algorithms require extra effort to make them effective on parallel computers. This is where much of the research in numerical linear algebra has been concentrated during the last decades, and it is still going on. The development of new computer architectures poses new challenges all the time, not only for algorithm development, but also for construction of effective operating systems.

Chapter 19
Further Reading

In the text above we have not given many references. Instead of listing a large number of scientific articles, we shall here indicate some books and review articles that we think will give comprehensive material covering the topic we have in mind quite well.

The whole field of numerical analysis is covered almost completely in the book by Dahlquist and Björck [6] which is a follow up of [5]. In particular, the topics discussed in the three chapters of Part II and in Chap. 13 of this book are thoroughly treated. The book is very detailed, and includes also nonstandard material like interval arithmetic.

There are many books on numerical methods for ODE. Finite difference methods are dominating here, and they are usually partitioned into Runge–Kutta type methods and linear multistep methods. For the first class, the world's best specialist is John C. Butcher, who has done impressive development and analysis work. Almost literally everything about these methods up to very high order of accuracy can be found in his book [4].

Another partition of ODE can be made by separating nonstiff problems from stiff problems (see Sect. 10.1). The book [11] covers the first class, while the book [12] covers the second one. These two books together with Butcher's book should cover just about everything that is needed about ODE-solvers.

When it comes to PDE and difference methods, we separate between time-dependent problems and elliptic stationary problems. For the first class, the first complete book was [21] written by one of the most active scholars in this area at that time. The book contains a second part devoted completely to applications. In the second edition [22], the first theoretical part was rewritten with plenty of new material included. In the period that followed, there was a strong development in the theory for initial-boundary value problems, largely due to Heinz-Otto Kreiss and his students. This material, as well as some new concepts in pure initial value problems, are included in [9]. There is an emphasis on the analysis of the PDE themselves and the connection to the difference schemes approximating them. There is also a chapter about nonlinear problems and shocks, which is a large and complicated issue. Peter Lax at the Courant Institute, New York University, is one of the leading researchers in this field (he was awarded the Abel prize in 2005). We recommend his

B. Gustafsson, *Fundamentals of Scientific Computing*,
Texts in Computational Science and Engineering 8,
DOI 10.1007/978-3-642-19495-5_19, © Springer-Verlag Berlin Heidelberg 2011

book [17], and the books [18, 19] by Randall LeVeque. The last one also gives a very good description of finite volume methods that were briefly mentioned in Sect. 17.3.

Difference methods for elliptic problems are hardly used for any large scale problems nowadays, but for special problems with regular geometries they can be effective. A short presentation is found as a chapter Finite Difference Methods for Elliptic Equations in [16]. The area is strongly coupled to the development of fast solvers of systems of algebraic equations, which we shall come back to.

Finite element methods were from the beginning used mostly for problems in structural mechanics, where the design object allows for a partition into different physical elements. In the sixties, FEM got an enormous boost for general PDE, when they were formulated in a unified mathematical manner. A complete presentation of the theory was given in [24] by G. Strang and G. Fix. 35 years later, the second edition [25] appeared, and it contains a second part with more details concerning implementation issues. Many other books have been written on FEM over the years. One of them is [27] written by O.C. Zienkiewics and R.L. Taylor. T. Hughes is a specialist in FEM applications in fluid dynamics, and one of his books is [15].

Discontinuous Galerkin methods are closely related to classic finite element methods as described in Sect. 11.3, and have had a fast expansion period during the last decade. The theory covering this topic is contained in the book [14] by J. Hesthaven and T. Warburton.

Spectral methods came up as an alternative solution method for PDE in the seventies. The first book [8] was written by two of the pioneers in the field, David Gottlieb and Steve Orzsag. A quite recent book that contains almost all new material including polynomial methods is [13].

Numerical methods for linear systems of algebraic equations are improved all the time, partly due to the ever changing architecture of supercomputers. The book [7] by G. Golub and C.F. van Loan is the third edition of the original that was published 1983. It contains rich material on theory and algorithms for all sorts of problems including direct and iterative methods for solution of linear systems. The book [23] by Y. Saad is specialized on iterative methods. Multigrid methods are included in this book as well, but there are many books devoted solely to this particular class of methods. Most of the theoretical foundations are included in the book [10] by W. Hackbusch and in [26] by P. Wesseling. Achi Brandt is one of the most well known representative for multigrid methods, having developed and used it for just about every existing problem class. The article [3] is a good survey of the topic.

Cleve Moler is the father of MATLAB, and for an introduction to this programming language we refer to his book [20], which is a revised version of the original that came out 2004. The major part of the book is actually about numerical methods, but all the time with emphasis on the MATLAB implementation. Since new versions of the system are released frequently (MATLAB 7.10 is the latest when this is written), we refer to the web for detailed descriptions of all features.

Erratum to: Basic Linear Algebra

Erratum to: B. Gustafsson, *Fundamentals of Scientific Computing,*
Texts in Computational Science and Engineering 8, pp. 47–65
DOI 10.1007/978-3-642-19495-5_3,
© Springer-Verlag Berlin Heidelberg 2011

Page 61: The paragraph "The rule can be applied ... $N = 4$" including Fig. 3.5 should be replaced by

"For general $N \times N$ matrices the determinant is defined by

$$\text{Det}(A) = \sum_{\sigma} \text{sgn}(\sigma) \prod_{j=1}^{N} a_{j\sigma_j},$$

where the sum is taken over all possible permutations of the integers 1 to N in the vector σ with the sign determined by the ordering of the elements."

The online version of the original chapter can be found under
doi: 10.1007/978-3-642-19495-5_3.

B. Gustafsson, *Fundamentals of Scientific Computing,*
Texts in Computational Science and Engineering 8,
DOI 10.1007/978-3-642-19495-5_20, © Springer-Verlag Berlin Heidelberg 2011

Appendix
Mathematical Rules

A.1 Rules for Differentiation

f	$\dfrac{df}{dx}$
x^a	ax^{a-1}
e^x	e^x
a^x	$a^x \ln a$
$\ln x$	$\dfrac{1}{x}$
$\log_a x$	$\dfrac{1}{x \ln a}$
$\sin x$	$\cos x$
$\cos x$	$-\sin x$
$\tan x$	$\dfrac{1}{\cos^2 x}$
$\cot x$	$-\dfrac{1}{\sin^2 x}$
$\arcsin x$	$\dfrac{1}{\sqrt{1-x^2}}$
$\arccos x$	$-\dfrac{1}{\sqrt{1-x^2}}$
$\arctan x$	$\dfrac{1}{1+x^2}$
au	$a\dfrac{du}{dx}$
$u+v$	$\dfrac{du}{dx} + \dfrac{dv}{dx}$
uv	$\dfrac{du}{dx}v + u\dfrac{dv}{dx}$
$\dfrac{u}{v}$	$\dfrac{v\,du/dx - u\,dv/dx}{v^2}$
$g(u(x))$	$\dfrac{dg}{du}\dfrac{du}{dx}$ (the chain rule)

B. Gustafsson, *Fundamentals of Scientific Computing*,
Texts in Computational Science and Engineering 8,
DOI 10.1007/978-3-642-19495-5, © Springer-Verlag Berlin Heidelberg 2011

A.2 Differential Operators in Different Coordinate Systems

$$\mathbf{u} = \begin{bmatrix} u \\ v \end{bmatrix} \quad \text{in 2D,} \qquad \mathbf{u} = \begin{bmatrix} u \\ v \\ w \end{bmatrix} \quad \text{in 3D}$$

A.2.1 2D

Cartesian coordinates x, y

$$\nabla = \begin{bmatrix} \partial/\partial x \\ \partial/\partial y \end{bmatrix}$$

$$\operatorname{div} \mathbf{u} = \nabla \cdot \mathbf{u} = \nabla \cdot \begin{bmatrix} u \\ v \end{bmatrix} = \frac{\partial u}{\partial x} + \frac{\partial v}{\partial y}$$

$$\Delta = \frac{\partial}{\partial x^2} + \frac{\partial}{\partial y^2}$$

Polar coordinates $r \in [0, \infty)$, $\theta \in [0, 2\pi)$: $x = r \cos \theta$, $y = r \sin \theta$

$$\nabla = \begin{bmatrix} \partial/\partial r \\ (1/r)\partial/\partial \theta \end{bmatrix}$$

$$\operatorname{div} \mathbf{u} = \frac{1}{r} \frac{\partial (ru)}{\partial r} + \frac{1}{r} \frac{\partial v}{\partial \theta}$$

$$\Delta = \frac{1}{r} \frac{\partial}{\partial r} \left(r \frac{\partial}{\partial r} \right) + \frac{1}{r^2} \frac{\partial^2}{\partial \theta^2} = \frac{\partial^2}{\partial r^2} + \frac{1}{r} \frac{\partial}{\partial r} + \frac{1}{r^2} \frac{\partial^2}{\partial \theta^2}$$

A.2.2 3D

Cartesian coordinates x, y, z

$$\nabla = \begin{bmatrix} \partial/\partial x \\ \partial/\partial y \\ \partial/\partial z \end{bmatrix}$$

$$\operatorname{div} \mathbf{u} = \nabla \cdot \mathbf{u} = \nabla \cdot \begin{bmatrix} u \\ v \\ w \end{bmatrix} = \frac{\partial u}{\partial x} + \frac{\partial v}{\partial y} + \frac{\partial w}{\partial z}$$

$$\operatorname{curl} \mathbf{u} = \nabla \times \mathbf{u} = \nabla \times \begin{bmatrix} u \\ v \\ w \end{bmatrix} = \begin{bmatrix} \partial w/\partial y - \partial v/\partial z \\ \partial u/\partial z - \partial w/\partial x \\ \partial v/\partial x - \partial u/\partial y \end{bmatrix}$$

$$\Delta = \frac{\partial}{\partial x^2} + \frac{\partial}{\partial y^2} + \frac{\partial}{\partial z^2}$$

Cylindrical coordinates $r \in [0, \infty)$, $\theta \in [0, 2\pi)$, $z \in [0, \infty)$: $x = r \cos\theta$, $y = r\sin\theta$, $z = z$

$$\nabla = \begin{bmatrix} \partial/\partial r \\ (1/r)\partial/\partial r \\ \partial/\partial z \end{bmatrix}$$

$$\text{div}\,\mathbf{u} = \frac{1}{r}\frac{\partial(ru)}{\partial r} + \frac{1}{r}\frac{\partial v}{\partial \theta} + \frac{\partial w}{\partial z}$$

$$\text{curl}\,\mathbf{u} = \begin{bmatrix} (1/r)\partial w/\partial\theta - \partial v/\partial z \\ \partial u/\partial z - \partial w/\partial r \\ (1/r)(\partial(rv)/\partial r - \partial u/\partial\theta) \end{bmatrix}$$

$$\Delta = \frac{1}{r}\frac{\partial}{\partial r}\left(r\frac{\partial}{\partial r}\right) + \frac{1}{r^2}\frac{\partial^2}{\partial\theta^2} + \frac{\partial^2}{\partial z^2}$$

Spherical coordinates $r \in [0, \infty)$, $\theta \in [0, 2\pi)$, $\phi \in [0, \pi]$ $x = r \cos\theta \sin\phi$, $y = r\sin\theta\sin\phi$, $z = r\cos\phi$

$$\nabla = \begin{bmatrix} \partial/\partial r \\ (1/r)\partial/\partial\theta \\ (1/(r\sin\theta))\partial/\partial\phi \end{bmatrix}$$

$$\text{div}\,\mathbf{u} = \frac{1}{r^2}\frac{\partial(r^2 u)}{\partial r} + \frac{1}{r\sin\theta}\frac{\partial(v\sin\theta)}{\partial\theta} + \frac{1}{r\sin\theta}\frac{\partial w}{\partial\phi}$$

$$\text{curl}\,\mathbf{u} = \begin{bmatrix} (1/(r\sin\theta))(\partial(w\sin\theta)/\partial\theta - \partial v/\partial\phi) \\ (1/r)((1/\sin\theta)\partial u/\partial\phi - \partial(rw)/\partial r) \\ (1/r)(\partial(rv)/\partial r - \partial u/\partial\theta) \end{bmatrix}$$

$$\Delta = \frac{1}{r^2}\frac{\partial}{\partial r}\left(r^2\frac{\partial}{\partial r}\right) + \frac{1}{r^2\sin\theta}\frac{\partial}{\partial\theta}\left(\sin\theta\frac{\partial}{\partial\theta}\right) + \frac{1}{r^2\sin^2\theta}\frac{\partial^2}{\partial\phi^2}$$

A.3 Trigonometric Formulas

$$\sin(-x) = -\sin x$$

$$\cos(-x) = \cos x$$

$$\sin^2 x + \cos^2 x = 1 \quad \text{(Pythagoras' theorem)}$$

$$\sin(\pi - x) = \sin x$$

$$\cos(\pi - x) = -\cos x$$

$$\sin(x + y) = \sin x \cos y + \cos x \sin y$$

$$\sin(x - y) = \sin x \cos y - \cos x \sin y$$

$$\sin(x + y) = \cos x \cos y - \sin x \sin y$$

$$\cos(x - y) = \cos x \cos y + \sin x \sin y$$

$$\tan(x + y) = \frac{\tan x + \tan y}{1 - \tan x \tan y}$$

$$\tan(x - y) = \frac{\tan x - \tan y}{1 + \tan x \tan y}$$

$$\sin 2x = 2 \sin x \cos x$$

$$\cos 2x = \cos^2 x - \sin^2 x$$

$$\tan 2x = \frac{2 \tan x}{1 - \tan^2 x}$$

$$\sin^2 \frac{x}{2} = \frac{1 - \cos x}{2}$$

$$\cos^2 \frac{x}{2} = \frac{1 + \cos x}{2}$$

$$\sin x + \sin y = 2 \sin \frac{x + y}{2} \cos \frac{x - y}{2}$$

$$\sin x - \sin y = 2 \cos \frac{x + y}{2} \sin \frac{x - y}{2}$$

$$\cos x + \cos y = 2 \cos \frac{x + y}{2} \cos \frac{x - y}{2}$$

$$\cos x - \cos y = -2 \sin \frac{x + y}{2} \sin \frac{x - y}{2}$$

$$2 \sin x \cos y = \sin(x + y) + \sin(x - y)$$

$$2 \cos x \sin y = \sin(x + y) - \sin(x - y)$$

$$2 \cos x \cos y = \cos(x + y) + \cos(x - y)$$

$$2 \sin x \sin y = -\cos(x + y) + \cos(x - y)$$

A.4 Matrix Algebra

In this section we list a number of definitions and rules for matrices of the form $A = (a_{jk})$, $1 \leq j \leq m$, $1 \leq k \leq n$. Most of the rules apply only for square matrices, and this is indicated by the symbol $\square$ to the left. The right column contains one of three types of information: a definition, a statement that is equivalent to the statement to the left, or a statement that follows from the statement to the left. In the latter case there is the leading symbol $\Rightarrow$.

	$C = A + B$	$(c_{jk}) = (a_{jk} + b_{jk}), 1 \le j \le m, \ 1 \le k \le n$
	$C = AB$	$c_{jk} = \sum_{r=1}^{n} a_{jr} b_{rk}, 1 \le j \le m, \ 1 \le k \le n$
☐	A diagonal	$a_{jk} = 0, \ j \ne k$
☐	$\text{diag}(a_{11}, a_{22}, \ldots, a_{nn})$	Diagonal matrix with diagonal elements a_{jj}
☐	Identity matrix I	$I = \text{diag}(1, 1, \ldots, 1)$
☐	A upper triangular	$a_{jk} = 0, \ j > k$
☐	A lower triangular	$a_{jk} = 0, \ j < k$
	A^T	$(a_{kj}), \ 1 \le j \le m, \ 1 \le k \le n$
	A^*	$(\overline{a}_{kj}), \ 1 \le j \le m, \ 1 \le k \le n$
☐	A, B commute	$AB = BA$
☐	$(AB)^T$	$B^T A^T$
☐	$(AB)^*$	$B^* A^*$
☐	Eigenvalue λ	$(A - \lambda I)\mathbf{x} = \mathbf{0}, \ \mathbf{x} \ne \mathbf{0}$
☐	Eigenvector $\mathbf{x}$	$(A - \lambda I)\mathbf{x} = \mathbf{0}, \ \mathbf{x} \ne \mathbf{0}$
☐	Spectral radius	$\rho(A) = \max_{1 \le j \le n} |\lambda_j|$
☐	A^{-1}	$A^{-1}A = I \ \Leftrightarrow \ AA^{-1} = I$
☐	Similarity transformation	$T^{-1}AT$
		$T^{-1}AT$ has the same eigenvalues as A
☐	$(AB)^{-1}$	$(AB)^{-1} = B^{-1} A^{-1}$
☐	$(A^T)^{-1}$	$(A^T)^{-1} = (A^{-1})^T$
☐	$(A^*)^{-1}$	$(A^*)^{-1} = (A^{-1})^*$
☐	A normal	$AA^* = A^*A$
☐	Real part of A, A real[a]	$(A + A^T)/2$
☐	Real part of A, A complex	$(A + A^*)/2$
☐	A orthogonal (unitary)	$AA^* = I$
		$A^{-1} = A^*$
		All eigenvalues of A on the unit circle
☐	A singular	$\text{Det}(A) = 0$
		$A\mathbf{x} = \mathbf{0}$ has a nonzero solution $\mathbf{x}$
		The column vectors are linearly dependent
		The row vectors are linearly dependent
		A has at least one zero eigenvalue
☐	A real, symmetric	$A = A^T$
		$\Rightarrow$ all eigenvalues are real
		$\Rightarrow$ eigenvectors mutually orthogonal
☐	A real, skew-symmetric	$A = -A^T$
		$\Rightarrow$ all eigenvalues are imaginary
		$\Rightarrow$ eigenvectors mutually orthogonal
☐	A complex, Hermitian[b]	$A = A^*$
		$\Rightarrow$ all eigenvalues are real
		$\Rightarrow$ eigenvectors mutually orthogonal
☐	A complex, skew-Hermitian	$A = -A^*$
		$\Rightarrow$ all eigenvalues are imaginary
		$\Rightarrow$ eigenvectors mutually orthogonal
☐	A real, positive definite[c]	A symmetric, $\mathbf{x}^T A \mathbf{x} > 0$, $\mathbf{x}$ real, $\mathbf{x} \ne \mathbf{0}$
		$\Rightarrow$ all eigenvalues positive

☐	A complex, positive definite	A Hermitian, $\mathbf{x}^* A\mathbf{x} > 0$, $\mathbf{x}$ complex, $\mathbf{x} \neq \mathbf{0}$
		$\Rightarrow$ all eigenvalues positive
	Norm of A^{d}	$\|A\| = \max_{\|\mathbf{x}\|=1} \|A\mathbf{x}\|$
	$\|cA\|$, c constant	$\|cA\| = |c| \cdot \|A\|$
	$\|A + B\|$	$\|A + B\| \leq \|A\| + \|B\|$
	$\|AB\|$	$\|AB\| \leq \|A\| \cdot \|B\|$

[a]It may seem strange to define the real part of a real matrix. But it is not the matrix obtained by taking the real part of each element a_{jk}. We take the example

$$A = \begin{bmatrix} 1 & -1 \\ 1 & 1 \end{bmatrix}$$

with eigenvalues $\lambda = 1 \pm i$.
    The real part is

$$\frac{A + A^T}{2} = \frac{1}{2}\begin{bmatrix} 1 & -1 \\ 1 & 1 \end{bmatrix} + \frac{1}{2}\begin{bmatrix} 1 & 1 \\ -1 & 1 \end{bmatrix} = \begin{bmatrix} 1 & 0 \\ 0 & 1 \end{bmatrix}$$

with the double eigenvalue 1. So, the concept real part refers to the fact that the eigenvalues of $(A + A^T)/2$ are the real parts of the eigenvalues of A

[b]An Hermitian complex matrix is often called symmetric, even if it is not symmetric as defined for real matrices

[c]A nonsymmetric real matrix can be defined as positive definite as well. For the example above we have

$$\mathbf{x}^T A\mathbf{x} = [x_1 \; x_2]\begin{bmatrix} 1 & -1 \\ 1 & 1 \end{bmatrix}\begin{bmatrix} x_1 \\ x_2 \end{bmatrix} = x_1^2 + x_2^2 > 0.$$

[d]The matrix norm as defined here is *subordinate* in the sense that it is associated to a certain vector norm, and the rules for the norm holds for any such vector norm. The most common vector norms are

$$\|\mathbf{x}\|_I = |x_1| + |x_2| + \cdots + |x_n|$$

$$\|\mathbf{x}\|_{II} = \sqrt{|x_1|^2 + |x_2|^2 + \cdots + |x_n|^2}$$

$$\|\mathbf{x}\|_\infty = \max_{1 \leq j \leq n} |x_j|$$

The corresponding subordinate matrix norms are

$$\|A\|_I = \max_k \sum_{j=1}^{m} |a_{jk}| \quad \text{(largest column sum)}$$

$$\|A\|_{II} = \rho(A^* A) \quad \text{(spectral radius of } A^* A)$$

$$\|A\|_\infty = \max_j \sum_{k=1}^{n} |a_{jk}| \quad \text{(largest row sum)}$$

References

1. Amoignon, O.: Aesop: a numerical platform for aerodynamic shape optimization. Report, FOI-R-2538-SE (2008)
2. Benzi, M.: Preconditioning techniques for large linear systems: a survey. J. Comput. Phys. **182**, 418–477 (2002)
3. Brandt, A.: Multiscale scientific computation: review 2001. In: Barth, T.J., Chan, T., Haimes, R. (eds.) Multiscale and Multiresolution Methods. Lecture Notes in Computational Science and Engineering, vol. 20, pp. 3–96. Springer, Berlin (2002)
4. Butcher, J.C.: Numerical Methods for Ordinary Differential Equations. Runge-Kutta and General Linear Methods. Wiley, New York (2003)
5. Dahlquist, G., Björck, Å.: Numerical Methods. Prentice-Hall, New York (1974)
6. Dahlquist, G., Björck, Å.: Numerical Methods in Scientific Computing, vol. I. SIAM, Philadelphia (2008)
7. Golub, G., van Loan, C.: Matrix Computations, 3rd edn. John Hopkins University Press, Baltimore (1996)
8. Gottlieb, D., Orzsag, S.: Numerical Analysis of Spectral Methods: Theory and Applications. SIAM, Philadelphia (1977)
9. Gustafsson, B., Kreiss, H.-O., Oliger, J.: Time Dependent Problems and Difference Methods. Wiley, New York (1995)
10. Hackbusch, W.: Multi-grid Methods and Applications. Springer, Berlin (1985)
11. Hairer, E., Nørsett, S.P., Wanner, G.: Solving Ordinary Differential Equations I: Nonstiff Problems, 2nd rev. edn. Springer, Berlin (1993)
12. Hairer, E., Wanner, G.: Solving Ordinary Differential Equations II: Stiff and Differential-Algebraic Problems. 2nd rev. edn. Springer, Berlin (1996)
13. Hesthaven, J.S., Gottlieb, S., Gottlieb, D.: Spectral Methods for Time-Dependent Problems. Cambridge University Press, Cambridge (2006)
14. Hesthaven, J.S., Warburton, T.: Nodal Discontinuous Galerkin Methods. Algorithms, Analysis, and Applications. Texts in Applied Mathematics, vol. 54. Springer, Berlin (2008)
15. Hughes, T.J.R.: The Finite Element Method: Linear Static and Dynamic Finite Element Analysis. Dover, New York (2000)
16. Larsson, S., Thomeé, W.: Partial Differential Equations with Numerical Methods. Texts in Applied Mathematics, vol. 45. Springer, Berlin (2009). 2003
17. Lax, P.D.: Hyperbolic Systems of Conservation Laws and the Mathematical Theory of Shock Waves, II. SIAM Regional Conference Series in Applied Mathematics, vol. 11. SIAM, Philadelphia (1972)
18. LeVeque, R.: Numerical Methods for Conservation Laws. Birkhäuser, Basel (1992)
19. LeVeque, R.: Finite Volume Methods for Hyperbolic Problems. Cambridge University Press, Cambridge (2002)

B. Gustafsson, *Fundamentals of Scientific Computing*,
Texts in Computational Science and Engineering 8,
DOI 10.1007/978-3-642-19495-5, © Springer-Verlag Berlin Heidelberg 2011

20. Moler, C.: Numerical Computing with MATLAB, Revised reprint. SIAM, Philadelphia (2008)
21. Richtmyer, R.: Difference Methods for Initial-Value Problems. Interscience, New York (1957)
22. Richtmyer, R., Morton, K.W.: Difference Methods for Initial–Value Problems, 2nd edn. Interscience, New York (1967)
23. Saad, Y.: Iterative Methods for Sparse Linear Systems. SIAM, Philadelphia (2003)
24. Strang, G., Fix, G.J.: An Analysis of the Finite Element Method. Prentice–Hall, New York (1973)
25. Strang, G., Fix, G.J.: An Analysis of the Finite Element Method, 2nd edn. Wellesley–Cambridge Press, Wellesley (2008)
26. Wesseling, P.: An Introduction to Multigrid Methods. Wiley, New York (1992) (Corrected reprint 2004)
27. Zienkiewics, O.C., Taylor, R.L.: The Finite Element Method, vol. 1: The Basis. Butterworth/Heineman, Stoneham/London (2008)

Index

B. Gustafsson, *Fundamentals of Scientific Computing*,
Texts in Computational Science and Engineering 8,
DOI 10.1007/978-3-642-19495-5, © Springer-Verlag Berlin Heidelberg 2011

Editorial Policy

1. Textbooks on topics in the field of computational science and engineering will be considered. They should be written for courses in CSE education. Both graduate and undergraduate textbooks will be published in TCSE. Multidisciplinary topics and multidisciplinary teams of authors are especially welcome.

2. Format: Only works in English will be considered. For evaluation purposes, manuscripts may be submitted in print or electronic form, in the latter case, preferably as pdf- or zipped ps-files. Authors are requested to use the LaTeX style files available from Springer at: http://www.springer.com/authors/book+authors? SGWID=0-154102-12-417900-0 (for monographs, textbooks and similar) Electronic material can be included if appropriate. Please contact the publisher.

3. Those considering a book which might be suitable for the series are strongly advised to contact the publisher or the series editors at an early stage.

General Remarks

Careful preparation of manuscripts will help keep production time short and ensure a satisfactory appearance of the finished book.

The following terms and conditions hold:

Regarding free copies and royalties, the standard terms for Springer mathematics textbooks hold. Please write to martin.peters@springer.com for details.

Authors are entitled to purchase further copies of their book and other Springer books for their personal use, at a discount of 33.3% directly from Springer-Verlag.

Texts in Computational Science and Engineering

1. H. P. Langtangen, *Computational Partial Differential Equations*. Numerical Methods and Diffpack Programming, 2nd Edition.
2. A. Quarteroni, F. Saleri, P. Gervasio, *Scientific Computing with MATLAB and Octave*, 3rd Edition.
3. H. P. Langtangen, *Python Scripting for Computational Science*, 3rd Edition.
4. H. Gardner, G. Manduchi, *Design Patterns for e-Science*.
5. M. Griebel, S. Knapek, G. Zumbusch, *Numerical Simulation in Molecular Dynamics*.
6. H. P. Langtangen, *A Primer on Scientific Programming with Python*, 2nd Edition.
7. A. Tveito, H. P. Langtangen, B. F. Nielsen, X. Cai, *Elements of Scientific Computing*.
8. B. Gustafsson, *Fundamentals of Scientific Computing*.

For further information on these books please have a look at our mathematics catalogue at the following URL: www.springer.com/series/5151

Monographs in Computational Science and Engineering

1. J. Sundnes, G.T. Lines, X. Cai, B.F. Nielsen, K.-A. Mardal, A. Tveito, *Computing the Electrical Activity in the Heart*.

For further information on this book, please have a look at our mathematics catalogue at the following URL: www.springer.com/series/7417

Lecture Notes in Computational Science and Engineering

1. D. Funaro, *Spectral Elements for Transport-Dominated Equations*.
2. H.P. Langtangen, *Computational Partial Differential Equations*. Numerical Methods and Diffpack Programming.
3. W. Hackbusch, G. Wittum (eds.), *Multigrid Methods V*.
4. P. Deuflhard, J. Hermans, B. Leimkuhler, A.E. Mark, S. Reich, R.D. Skeel (eds.), *Computational Molecular Dynamics: Challenges, Methods, Ideas*.
5. D. Kröner, M. Ohlberger, C. Rohde (eds.), *An Introduction to Recent Developments in Theory and Numerics for Conservation Laws*.
6. S. Turek, *Efficient Solvers for Incompressible Flow Problems*. An Algorithmic and Computational Approach.
7. R. von Schwerin, *Multi Body System SIMulation*. Numerical Methods, Algorithms, and Software.
8. H.-J. Bungartz, F. Durst, C. Zenger (eds.), *High Performance Scientific and Engineering Computing*.

9. T.J. Barth, H. Deconinck (eds.), *High-Order Methods for Computational Physics*.

10. H.P. Langtangen, A.M. Bruaset, E. Quak (eds.), *Advances in Software Tools for Scientific Computing*.

11. B. Cockburn, G.E. Karniadakis, C.-W. Shu (eds.), *Discontinuous Galerkin Methods*. Theory, Computation and Applications.

12. U. van Rienen, *Numerical Methods in Computational Electrodynamics*. Linear Systems in Practical Applications.

13. B. Engquist, L. Johnsson, M. Hammill, F. Short (eds.), *Simulation and Visualization on the Grid*.

14. E. Dick, K. Riemslagh, J. Vierendeels (eds.), *Multigrid Methods VI*.

15. A. Frommer, T. Lippert, B. Medeke, K. Schilling (eds.), *Numerical Challenges in Lattice Quantum Chromodynamics*.

16. J. Lang, *Adaptive Multilevel Solution of Nonlinear Parabolic PDE Systems*. Theory, Algorithm, and Applications.

17. B.I. Wohlmuth, *Discretization Methods and Iterative Solvers Based on Domain Decomposition*.

18. U. van Rienen, M. Günther, D. Hecht (eds.), *Scientific Computing in Electrical Engineering*.

19. I. Babuška, P.G. Ciarlet, T. Miyoshi (eds.), *Mathematical Modeling and Numerical Simulation in Continuum Mechanics*.

20. T.J. Barth, T. Chan, R. Haimes (eds.), *Multiscale and Multiresolution Methods*. Theory and Applications.

21. M. Breuer, F. Durst, C. Zenger (eds.), *High Performance Scientific and Engineering Computing*.

22. K. Urban, *Wavelets in Numerical Simulation*. Problem Adapted Construction and Applications.

23. L.F. Pavarino, A. Toselli (eds.), *Recent Developments in Domain Decomposition Methods*.

24. T. Schlick, H.H. Gan (eds.), *Computational Methods for Macromolecules: Challenges and Applications*.

25. T.J. Barth, H. Deconinck (eds.), *Error Estimation and Adaptive Discretization Methods in Computational Fluid Dynamics*.

26. M. Griebel, M.A. Schweitzer (eds.), *Meshfree Methods for Partial Differential Equations*.

27. S. Müller, *Adaptive Multiscale Schemes for Conservation Laws*.

28. C. Carstensen, S. Funken, W. Hackbusch, R.H.W. Hoppe, P. Monk (eds.), *Computational Electromagnetics*.

29. M.A. Schweitzer, *A Parallel Multilevel Partition of Unity Method for Elliptic Partial Differential Equations*.

30. T. Biegler, O. Ghattas, M. Heinkenschloss, B. van Bloemen Waanders (eds.), *Large-Scale PDE-Constrained Optimization*.

31. M. Ainsworth, P. Davies, D. Duncan, P. Martin, B. Rynne (eds.), *Topics in Computational Wave Propagation*. Direct and Inverse Problems.

32. H. Emmerich, B. Nestler, M. Schreckenberg (eds.), *Interface and Transport Dynamics*. Computational Modelling.

33. H.P. Langtangen, A. Tveito (eds.), *Advanced Topics in Computational Partial Differential Equations*. Numerical Methods and Diffpack Programming.
34. V. John, *Large Eddy Simulation of Turbulent Incompressible Flows*. Analytical and Numerical Results for a Class of LES Models.
35. E. Bänsch (ed.), *Challenges in Scientific Computing – CISC 2002*.
36. B.N. Khoromskij, G. Wittum, *Numerical Solution of Elliptic Differential Equations by Reduction to the Interface*.
37. A. Iske, *Multiresolution Methods in Scattered Data Modelling*.
38. S.-I. Niculescu, K. Gu (eds.), *Advances in Time-Delay Systems*.
39. S. Attinger, P. Koumoutsakos (eds.), *Multiscale Modelling and Simulation*.
40. R. Kornhuber, R. Hoppe, J. Périaux, O. Pironneau, O. Widlund, J. Xu (eds.) *Domain Decomposition Methods in Science and Engineering*.
41. T. Plewa, T. Linde, V.G. Weirs (eds.), *Adaptive Mesh Refinement – Theory and Applications*.
42. A. Schmidt, K.G. Siebert, *Design of Adaptive Finite Element Software*. The Finite Element Toolbox ALBERTA.
43. M. Griebel, M.A. Schweitzer (eds.), *Meshfree Methods for Partial Differential Equations II*.
44. B. Engquist, P. Lötstedt, O. Runborg (eds.), *Multiscale Methods in Science and Engineering*.
45. P. Benner, V. Mehrmann, D.C. Sorensen (eds.), *Dimension Reduction of Large-Scale Systems*.
46. D. Kressner, *Numerical Methods for General and Structured Eigenvalue Problems*.
47. A. Boriçi, A. Frommer, B. Joó, A. Kennedy, B. Pendleton (eds.), *QCD and Numerical Analysis III*.
48. F. Graziani (ed.), *Computational Methods in Transport*.
49. B. Leimkuhler, C. Chipot, R. Elber, A. Laaksonen, A. Mark, T. Schlick, C. Schütte, R. Skeel (eds.), *New Algorithms for Macromolecular Simulation*.
50. M. Bücker, G. Corliss, P. Hovland, U. Naumann, B. Norris (eds.), *Automatic Differentiation: Applications, Theory, and Implementations*.
51. A.M. Bruaset, A. Tveito (eds.), *Numerical Solution of Partial Differential Equations on Parallel Computers*.
52. K.H. Hoffmann, A. Meyer (eds.), *Parallel Algorithms and Cluster Computing*.
53. H.-J. Bungartz, M. Schäfer (eds.), *Fluid-Structure Interaction*.
54. J. Behrens, *Adaptive Atmospheric Modeling*.
55. O. Widlund, D. Keyes (eds.), *Domain Decomposition Methods in Science and Engineering XVI*.
56. S. Kassinos, C. Langer, G. Iaccarino, P. Moin (eds.), *Complex Effects in Large Eddy Simulations*.
57. M. Griebel, M.A Schweitzer (eds.), *Meshfree Methods for Partial Differential Equations III*.
58. A.N. Gorban, B. Kégl, D.C. Wunsch, A. Zinovyev (eds.), *Principal Manifolds for Data Visualization and Dimension Reduction*.
59. H. Ammari (ed.), *Modeling and Computations in Electromagnetics: A Volume Dedicated to Jean-Claude Nédélec*.

60. U. Langer, M. Discacciati, D. Keyes, O. Widlund, W. Zulehner (eds.), *Domain Decomposition Methods in Science and Engineering XVII.*

61. T. Mathew, *Domain Decomposition Methods for the Numerical Solution of Partial Differential Equations.*

62. F. Graziani (ed.), *Computational Methods in Transport: Verification and Validation.*

63. M. Bebendorf, *Hierarchical Matrices.* A Means to Efficiently Solve Elliptic Boundary Value Problems.

64. C.H. Bischof, H.M. Bücker, P. Hovland, U. Naumann, J. Utke (eds.), *Advances in Automatic Differentiation.*

65. M. Griebel, M.A. Schweitzer (eds.), *Meshfree Methods for Partial Differential Equations IV.*

66. B. Engquist, P. Lötstedt, O. Runborg (eds.), *Multiscale Modeling and Simulation in Science.*

67. I.H. Tuncer, Ü. Gülcat, D.R. Emerson, K. Matsuno (eds.), *Parallel Computational Fluid Dynamics 2007.*

68. S. Yip, T. Diaz de la Rubia (eds.), *Scientific Modeling and Simulations.*

69. A. Hegarty, N. Kopteva, E. O'Riordan, M. Stynes (eds.), *BAIL 2008 – Boundary and Interior Layers.*

70. M. Bercovier, M.J. Gander, R. Kornhuber, O. Widlund (eds.), *Domain Decomposition Methods in Science and Engineering XVIII.*

71. B. Koren, C. Vuik (eds.), *Advanced Computational Methods in Science and Engineering.*

72. M. Peters (ed.), *Computational Fluid Dynamics for Sport Simulation.*

73. H.-J. Bungartz, M. Mehl, M. Schäfer (eds.), *Fluid Structure Interaction II – Modelling, Simulation, Optimization.*

74. D. Tromeur-Dervout, G. Brenner, D.R. Emerson, J. Erhel (eds.), *Parallel Computational Fluid Dynamics 2008.*

75. A.N. Gorban, D. Roose (eds.), *Coping with Complexity: Model Reduction and Data Analysis.*

76. J.S. Hesthaven, E.M. Rønquist (eds.), *Spectral and High Order Methods for Partial Differential Equations.*

77. M. Holtz, *Sparse Grid Quadrature in High Dimensions with Applications in Finance and Insurance.*

78. Y. Huang, R. Kornhuber, O. Widlund, J. Xu (eds.), *Domain Decomposition Methods in Science and Engineering XIX.*

79. M. Griebel, M.A. Schweitzer (eds.), *Meshfree Methods for Partial Differential Equations V.*

80. P.H. Lauritzen, C. Jablonowski, M.A. Taylor, R.D. Nair (eds.), *Numerical Techniques for Global Atmospheric Models.*

81. C. Clavero, J.L. Gracia, F. Lisbona (eds.), *BAIL 2010 – Boundary and Interior Layers, Computational and Asymptotic Methods.*

For further information on these books please have a look at our mathematics catalogue at the following URL: www.springer.com/series/3527